# Coherent Radiation Generation and Particle Acceleration

# RESEARCH TRENDS IN PHYSICS

# Coherent Radiation Generation and Particle Acceleration

A. M. Prokhorov
Editor-in-Chief

J. M. Buzzi, P. Sprangle & K. Wille
Editors

American Institute of Physics New York

Papers included in this volume are reproduced from camera-ready copy supplied by the contributors.



L.C. Catalog Card No. 92-82562
ISBN 0-88318-926-7
DOE CONF-910249

Printed in the United States of America.

## CONTENTS

**Part II. PARTICLE ACCELERATION**

**Chairmen: J. M. Buzzi**
**École Polytechnique,**
**Palaiseau, France**
**A. Dyson**
**University of California**
**Los Angeles, California**
**N. A. Ebrahim**
**Chalk River Laboratories,**
**Chalk River, Ontario, Canada**
**A. Ogata**
**KEK National Laboratory for High Energy Physics,**
**Tsukuba, Japan**
**G. J. Peters**
**United States Department of Energy,**
**Washington, DC**

## Part III. ABSTRACTS

# PREFACE

The International Topical Conference on Research Trends in Coherent Radiation Generation and Particle Acceleration, held February 11–13, 1991, in La Jolla, California, provided an up-to-date account of research in the field. It covered experimental technological and theoretical problems addressed by prominent researchers around the world. The conference was hosted by La Jolla International School of Physics, a division of The Institute for Advanced Physics Studies, La Jolla, California.

The International Steering Committee identified trends, suggested invited talks, and selected the final list of invited speakers. It was composed of the following members:

J. M. Buzzi (France)
J. M. Dawson (U.S.A.)
A. E. Dangor (United Kingdom)
L. M. Gorbunov (U.S.S.R.)
C. Joshi (U.S.A.)
N. Kroll (U.S.A.)
A. Litvak (U.S.S.R.)
K. Mima (Japan)
C. Pellegrini (U.S.A.)
A. M. Prokhorov (U.S.S.R.)
A. Renieri (Italy)
R. Rooth (U.S.A.)
N. Rostoker (U.S.A.)
A. L. Schawlow (U.S.A.)
P. Sprangle (U.S.A.)
R. N. Sudan (U.S.A.)
R. J. Temkin (U.S.A.)
K. Wille (Germany)
E. J. N. Wilson (Switzerland)

The Conference Executive Committee coordinated the decisions of the Steering Committee. It consisted of:

T. Antonsen (*University of Maryland, College Park, Maryland*)

P. Chen (*Stanford Linear Accelerator, Palo Alto, California*)

S. Eckhouse (*Maxwell Laboratories, Inc., San Diego, California*)

F. Felber (*Jaycor, Inc., San Diego, California*)

B. Levush (*University of Maryland, College Park, Maryland*)

J. Marx (*Lawrence Berkeley Laboratory, Berkeley, California*)

V. Stefan, Conference Chairman (*The Institute for Advanced Physics Studies and Tesla Laboratories, Inc., La Jolla, California*)

J. S. Wurtele (*Massachusetts Institute of Technology, Cambridge, Massachusetts*)

A number of crucial topics were addressed at this meeting: free electron lasers, microwave generators, advanced light sources, x-rays, and laser accelerators. The subject material was formatted under two headings and ten nonparallel sessions:

The conference and the publication of the proceedings were supported by the La Jolla International School of Physics, La Jolla, California; Naval Research Laboratory, Washington, DC; and Tesla Laboratories, Inc., La Jolla, California. We are indebted to Mr. Tarek Salawy and Ms. Lara Blau for handling the meeting arrangements and manuscripts, and to Mrs. Karen Bierstedt and Mr. Richard L. Knauel for processing the manuscripts of the proceedings.

We would like to express our deepest gratitude to all those who contributed to and/or attended the conference for their work and efforts which made this meeting a success.

*La Jolla, California*
*February 1992*

*Editorial Board*

*J. M. Buzzi*
*A. M. Prokhorov, Editor-in-Chief*
*P. Sprangle*
*K. Wille*

*Research Trends in Physics* Series

This is a forum for the publication of edited topical books within key fields of physics. Volumes in the series are based largely on invited symposia organized by the La Jolla International School of Physics, a division of The Institute for Advanced Physics Studies, La Jolla, California. These small, select meetings bring together prominent researchers from around the world to exchange ideas and approaches, discuss critical issues, critique a field's present status, and point the way toward new avenues in research. Some volumes, however, are based on selection of topics and contributors made solely by appointed editor(s).

Contributors in the series—all of whom appear by invitation—are asked to go beyond the narrow particulars of their work and to reflect on larger questions, problems, and trends. The resulting volumes will document the latest thinking regarding the current and future states of physics research for the international physics community.

V. Stefan
Series Editor
*La Jolla International School of Physics*
*The Institute for Advanced Physics Studies*
*La Jolla, California*

# COHERENT RADIATION GENERATION AND PARTICLE ACCELERATION

## Editorial Board



This volume documents the latest thinking of prominent researchers from around the world in the field of coherent radiation generation and particle accelerators. It covers the following topics: free electron lasers, microwave generators, and particle accelerators.

The applied nature of the presentations and their interdisciplinary character make this book a most valuable resource to research scientists in coherent radiation generation and particle acceleration physics. This book can be used by graduate students in physics research, by faculty and research staffs in applied physics at universities and research institutions, by research staffs in industry and national laboratories, and by the staffs of science-supporting agencies as a document of some of the most advanced research in the field.

### ABOUT THE EDITOR-IN-CHIEF

A. M. Prokhorov has made outstanding contributions to research in physics. He won the 1964 Nobel Prize in Physics with C. Townes and N. G. Basov for his work in quantum electronics, which led to the construction of lasers and masers. He began his research career at the P. N. Lebedev Physics Institute, Moscow, Russia, where he eventually became head of the laboratory. A. M. Prokhorov is the editor-in-chief of the *Encyclopedia of Physics*, published by Bolshaya Rossiiskaya Entsiklopediya. He is a member of the Russian Academy of Sciences and an honorary member of the American Academy of Arts and Sciences. A. M. Prokhorov is a founder and director of the General Physics Institute of the Russian Academy of Sciences in Moscow.

## List of Participants

T. M. Antonsen, Jr.
University of Maryland at College Park
Laboratory for Plasma Research
College Park, MD 20742
Tel: (301) 405-5020
Fax: (301) 314-9437

Lara Blau
The Institute for Advanced Physics Studies
La Jolla International School of Physics
7825 Fay Avenue, Suite 350
La Jolla, CA 92037
Tel: (619) 456-5737
Fax: (619) 456-1679

A. Bourdier
Laboratoire de Physique des Milieux Ionisés
École Polytechnique
91128 Palaiseau, France
Tel: 011-33 1 69 41 82 00
Fax: 011-33 1 69 41 33 92

B. N. Breizman
Institute of Nuclear Physics
630090 Novosibirsk, Russia
Tel: 011-7-38 32 359 701
Fax: 011-7-38 32 355 635

K. A. Brueckner
Department of Physics, 0319
University of California, San Diego
9500 Gilman Drive
La Jolla, CA 92093-0319
Tel: (619) 534-3290
Fax: (619) 534-0173

Jean-Max Buzzi
Laboratoire de Physique des Milieux Ionisés
École Polytechnique
91128 Palaiseau, France
Tel: 011-33 1 69 41 82 00
Fax: 011-33 1 69 41 33 92

David Cline
Astroparticle, Accelerator,
and High Energy Physics Department
University of California, Los Angeles
405 Hilgard Avenue
Los Angeles, CA 90024-1547
Tel: (310) 825-4649
Fax: (310) 206-1091

Bruce G. Danly
Massachusetts Institute of Technology
NW16-172
Cambridge, MA 02139
Tel: (617) 253-8454
Fax: (617) 253-6078

M. C. Downer
Department of Physics
University of Texas at Austin
Austin, TX 78712-1081
Tel: (512) 471-6054
Fax: (512) 471-9637

Anthony Dyson
Laser Accelerator Group
University of California, Los Angeles
405 Hilgard Avenue
Los Angeles, CA 90024
Tel: (310) 825-8453
Fax: (310) 825-4789

Nizar A. Ebrahim
Station III
Chalk River Laboratories
AECL Research
Chalk River, Ontario K0J 1J0, Canada
Tel: (613) 584-3311 Ext. 3962
Fax: (613) 584-2634

S. Eckhouse
Maxwell Laboratories, Inc.
8888 Balboa Avenue
San Diego, CA 92123-1506
Tel: (619) 576-7708
Fax: (619) 279-7794

Eric Esarey
Naval Research Laboratory
Washington, DC 20375-5000
Tel: (202) 767-3493
Fax: (202) 767-0631

F. Felber
Jaycor, Inc.
P.O. Box 85154
San Diego, CA 92138
Tel: (619) 453-6580
Fax: (619) 452-7899

L. M. Gorbunov
P. N. Lebedev Physics Institute
Russian Academy of Sciences
Leninsky Prospect 53
Moscow 117924, Russia
Tel: 011-7-095-135 78 08
Fax: 011-7 095-135 02 70

L. Henlein
Publishing Consultant
La Jolla, CA
Tel: (619) 454-8176

Tom Katsouleas
Department of Physics
University of California, Los Angeles
405 Hilgard Avenue
Los Angeles, CA 90024
Tel: (310) 206-0372
Fax: (310) 206-5668

Kwang-Je Kim
Lawrence Berkeley Laboratory
MS 71-259
1 Cyclotron Road
Berkeley, CA 94720
Tel: (415) 486-7224
Fax: (415) 486-7981

Yoneyoshi Kitagawa
Institute of Laser Engineering
Osaka University
2-6 Yamada-oka
Suita, Osaka 565, Japan
Tel: 011-81 06 877-5111 Ext. 6598
Fax: 011-81 06 877 4799

Stefano Lami
Instituto Nazionale di Fisia Nucleare
Via Livornese 582/A
56010 S. Piero A Grado
Pisa, Italy
Tel: 011-39 50 546 111
Fax: 011-39 50 589 047

Y. Y. Lau
Naval Research Laboratory
Code 4790
Washington, DC 20375-5000
Tel: (202) 767-2765
Fax: (202) 767-0631

Alexander Litvak
Institute of Applied Physics
Russian Academy of Sciences
603600 N-Novgorod, Russia
Tel: 011-7-8312 367 291
Fax: 011-7-8312 362 061

Xavier K. Maruyama
Department of Physics
Naval Postgraduate School
PH/MX
Monterey, CA 93943-5000
Tel: (408) 646-2431
Fax: (408) 646-2834

Jay Marx
Lawrence Berkeley Laboratory
MS 46-161
1 Cyclotron Road
Berkeley, CA 94720
Tel: (415) 486-5244
Fax: (415) 486-4873

Atsushi Ogata
KEK
National Laboratory
for High Energy Physics
1-1 Oho, Tsukuba 305, Japan
Tel: 011-81-298-64-1171
Fax: 011-81-298-64-3182

M. I. Petelin
Institute of Applied Physics
Russian Academy of Sciences
603600 N-Novgorod, Russia
Tel: 011-7-8312 367 291
Fax: 011-7-8312 362 061

Gerald J. Peters
United States Department of Energy
Division of High Energy Physics
ER-224 GTN
Washington, DC 20585
Tel: (301) 353-3233
Fax: (301) 353-5079

N. Rostoker
Department of Physics
University of California, Irvine
Irvine, CA 92717
Tel: (714) 856-6949
Fax: (714) 856-5903

A. S. Sakharov
General Physics Institute
Russian Academy of Sciences
117333 Moscow, Russia
Tel: 011-7-095-135 13 31
Fax: 011-7-095-135 02 70

Tarek Salawy
The Institute for Advanced Physics Studies
La Jolla International School of Physics
7825 Fay Avenue, Suite 350
La Jolla, CA 92037
Tel: (619) 456-5737
Fax: (619) 456-1679

Diana Sarabia
Advanced Studies, Inc.
P.O. Box 2946
La Jolla, CA 92038
Tel: (619) 456-9228
Fax: (619) 456-1679

Richard Savage
Department of Electrical Engineering
University of California, Los Angeles
56-124 Engineering IV
405 Hilgard Avenue
Los Angeles, CA 90024-1594
Tel: (310) 825-2068
Fax: (310) 206-5668

Philip Sprangle
Naval Research Laboratory
Code 4790
Washington, DC 20375-5000
Tel: (202) 767-3493
Fax: (202) 767-0631

V. Stefan
The Institute for Advanced Physics Studies
La Jolla International School of Physics
7825 Fay Avenue, Suite 350
La Jolla, CA 92037
Tel: (619) 456-5737
Fax: (619) 456-1679

L. A. Steinert
McDonnell Douglas Astronautics
P.O. Box 3973
Long Beach, CA 90803
Tel: (714) 896-3874
Fax: (714) 896-3874

E. J. Sternbach
Laboratoire de Physique des Milieux Ionisés
École Polytechnique
91128 Palaiseau Cedex, France
Tel: 011-33-1 69 41 82 00
Fax: 011-33-1 69 41 33 92

J. J. Su
Department of Physics
University of California, Los Angeles
405 Hilgard Avenue
Los Angeles, CA 90024
Tel: (310) 825-7385
Fax: (310) 206-8220

Amalia Torre
ENEA, Dip. Sviluppo Technologie di Punta
P.O. Box 65,
00044 Frascati (Rome), Italy
Tel: 011-39 6 9400 1
Fax: 011-39 6 9400 5607

L. Thode
Los Alamos National Laboratory
P.O. Box 1663, MS P234
Los Alamos, NM 87545
Tel: (505) 667-6554
Fax: (505) 665 4092

W. B. Thompson
Department of Physics, 0319
University of California,
San Diego
9500 Gilman Drive
La Jolla, CA 92093-0319
Tel: (619) 534-4173
Fax: (619) 534-0173

N. L. Tsintsadze
Institute of Physics
Academy of Sciences of Georgia
117333 Georgia
Tel: 011-7 88 32 212 236
Fax: 011-7 88 32 983 853

K. Wille
Institute of Physics
University of Dortmund
P.O. Box 500500
46 00 Dortmund, Germany
Tel: 011-49-231-7551
Fax: 011-49-231-755 4283

# Part I

# COHERENT RADIATION GENERATION

Research Trends in Physics: Coherent Radiation Generation and Particle Acceleration

*La Jolla International School of Physics*, The Institute for Advanced Physics Studies, La Jolla, California

# Mode Competition and Cooperation in Relativistic Electron Microwave Devices

**V.L. Bratman, N.S. Ginzburg, N.F. Kovalev, M.I. Petelin**

Institute of Applied Physics
N-Novgorod, Russia

Methods to provide the single mode operation in high power microwave generators and amplifiers are reviewed, a special attention being paid to devices with irregular ( in particular, sectioned ) interaction space. Limitations on the interaction space cross-section and on the electron current are estimated.

Some fresh experimental results by IAP and related Russian institutes are presented to illustrate the analysis.

Quasi-optical methods to match high-power high-mode microwave devices to consuming electrodynamic systems ( e.g., accelerators ) are discussed.

To satisfy the most of consumers, the microwave radiation delivered by high power generators and amplifiers [1-3] should be space-time coherent: any generator should produce the monochromatic radiation and any amplifier should reproduce the spectrum of the input signal ; in the both cases the output wave should be of a stable space structure, i.e. should represent a) a single mode of the "cold" ( empty of the electron beam ) electrodynamic system or b) a single "hot" mode, i.e. a definite composition of "cold" modes.

Let us emphasize, that if a high-power microwave device operates at a high mode,, it is necessary to use a more or less sophisticated method to match this mode with an outer ( consuming) electrodynamic system.

 

## MODE SELECTION

*Selective excitation of a single mode of "cold " cylindrical ( or periodic ) electrodynamic system.* The simplest method to obtain the single-mode regime is to provide conditions, when the operation mode is a) of the lowest type and b) the only one synchronous ( or resonant ) to electrons. However, even in this case , according to the Brillouin diagram ( Fig. 1, cases 1 ), the interaction space can be broad in the wavelength scale, if the translational velocity of electrons be commensurable with that of light.

One of recent applications of this obvious ( and widely used [1-3] ) method is a relativistic electron 35 GHz travelling wave amplifier ( [4], Fig. 2 ). The amplifier operated at $E_{01}$ mode. To exclude the self-excitation , the wall corrugation was tapered adiabatically to zero at the both ends of interaction space and a distributed absorber was inserted in the middle of the interaction space.

To excite selectively a mode of a relatively high type, the simplest method is to organize its synchronism with electrons near the cut-off ( Fig. 1 , cases 2 ) : the mode with the lowest group velocity couples to the electron beam stronger than others. The method is used, in particular, in relativistic orotrons [2], where, to provide an additional mode selectivity, a) longitudinal slots or b) a helical corrugation are usually introduced ( Fig.3 ).

A rather effective mode selection method used in Cherenkov microwave devices consists in the cyclotron absorption introduces in spurious modes by the electron beam itself [5].

Fig.1. SINGLE MODE REGIMES:

A) IN REGULAR SYSTEMS B) IN PERIODIC SYSTEMS

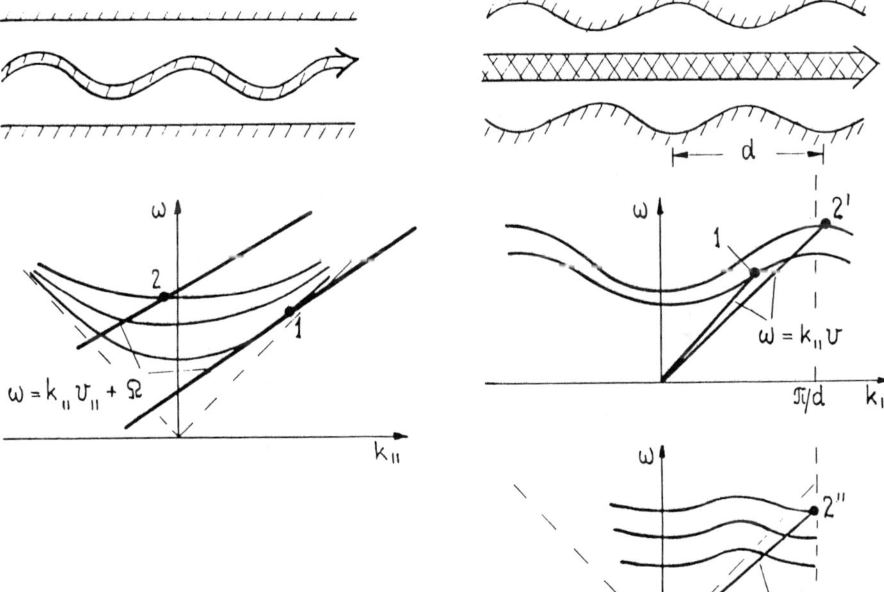


SYNCHRONISMS OF ELECTRONS WITH:

1) A MODE OF THE LOWEST TYPE;

2) A LOW GROUP VELOCITY MODE (2- GYROTRON, 2'- CHERENKOV π-MODE GENERATOR, 2''- CHERENKOV 2π-MODE GENERATOR - OROTRON)

C) CANALIZATION OF EM-WAVE BY E-BEAM - OPTICAL GUIDING

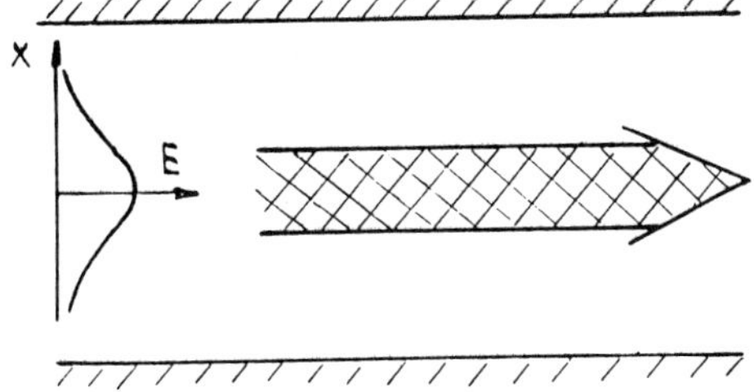

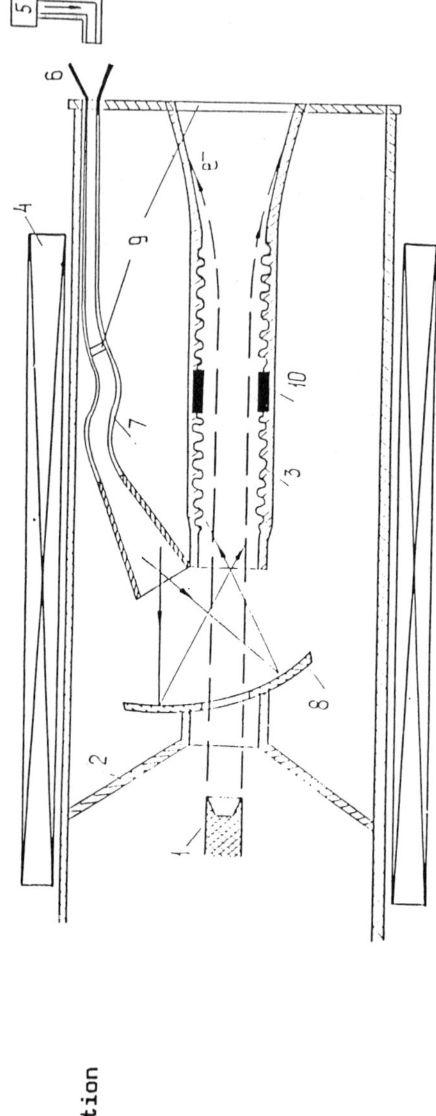


Fig. 2. TW amplifier

8 mm, 100 MW, 5 ns, 44 dB

1. cathode
2. anode
3. slow-wave section
4. solenoid
5. nagnetron
6. horn
7. $H_{11}$-$E_{01}$ node converter
8. mirror
9. windows
10. absorber

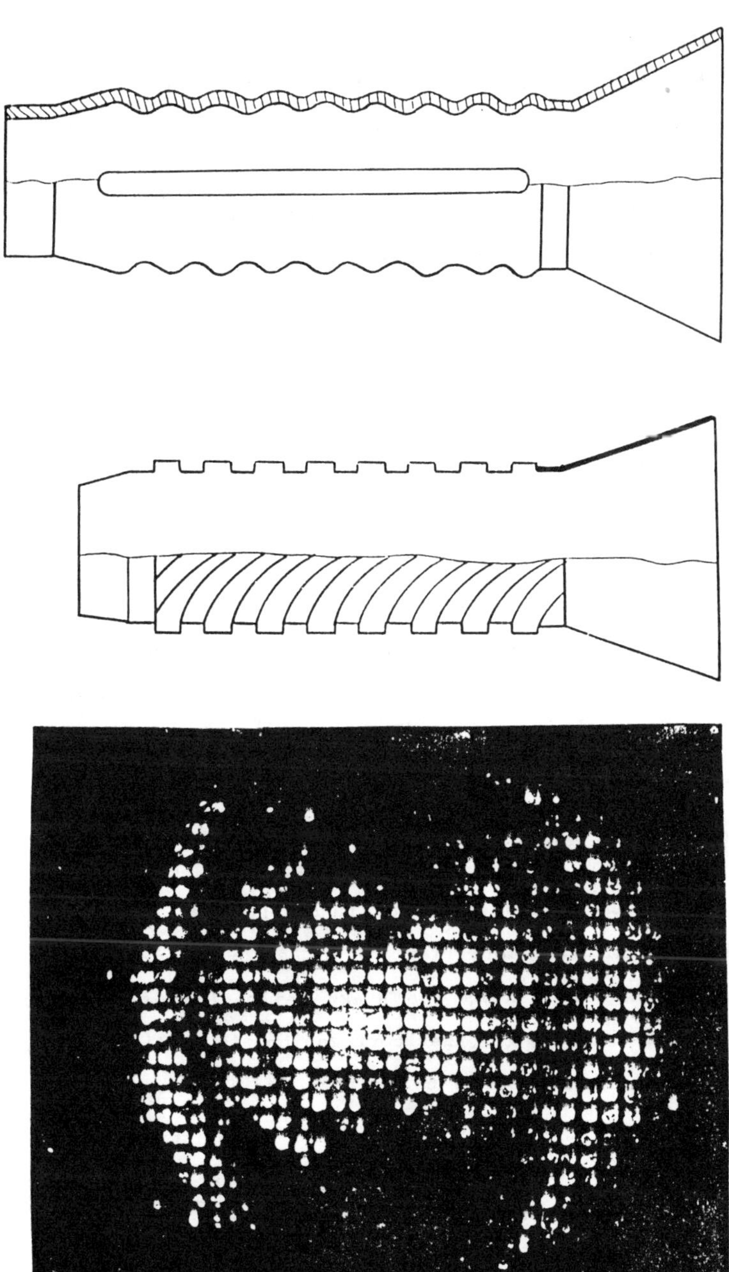

Fig. 3. Relativistic orotron cavities and an output radiation pattern ( $\lambda$ = 8*mm*, P = 100 MW, $\tau$ = 15 *ns* ).

*Selective excitation of "cold" modes in microwave devices with irregular electrodynamic systems.* In a number of cases a very small irregularity introduced in an electrodynamic system proves to be favorable to avoid the self-excitation of spurious modes. For example , in CARMs and ubitrons with Bragg reflection cavities [6] a very weak conicity of the cavity is usually sufficient to suppress modes of gyrotron type. To optimize the main mode performance, an additional ( compensating ) irregularity can be introduced in the static magnetic field. Analogous methods of spurious mode suppression are used in Cherenkov relativistic electron devices ( BWO , TWA, twystrons etc.)[2] as well.

In a number of cases a stronger irregularity is necessary. For example, the gyrotron usually operates at a TE mode. But if the electrons have the translational velocity commensurable with that of light, TM modes are coupled to the electron beam as strongly as TE ones. But a TM mode can be operated, only if rival TE modes are suppressed. In [7] ( Fig. 4 ) it was provided using an echelette cavity .

Another version of the idea is exploiting of sectioned electrodynamic systems with separated functions. For example, it is possible to turn an amplifier to an auto-oscillator by a selective feedback provided by the Bragg reflection of waves in a corrugated waveguide. If the corrugation is helical , an incident wave can be reflected to another one with a change of transverse indexes. This method was used in an auto-oscillator with the amplification section of the Cherenkov type ( [8], Fig. 5 ):the amplified wave, namely, hybrid $EH_{11}$ was coupled with the feedback wave ,namely, rotating $H_{41}$ in a helically

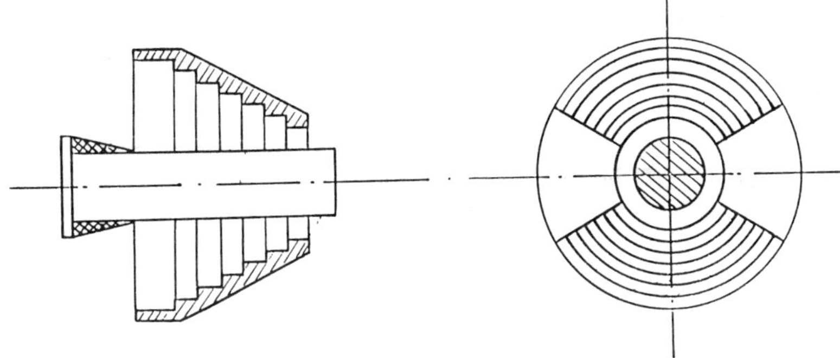

Fig.4. Echelette cavity for relativistic gyrotron with E-mode ( $\lambda$ = 3 *mm*, P = 20 - 30 MW, $\tau$ = 40 *ns* ).

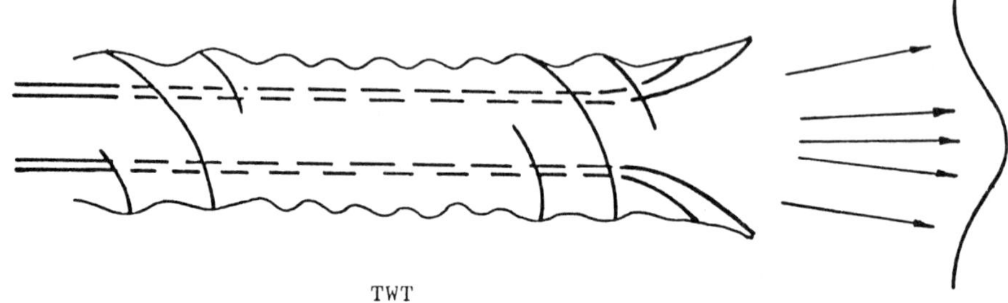


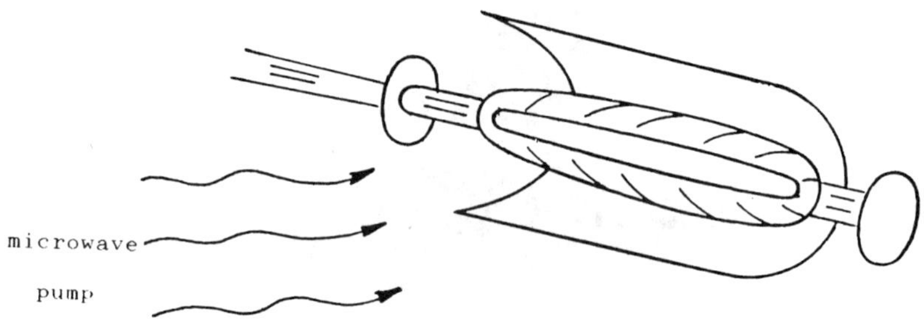


Fig.5. Sectioned microwave generator and microwave-pumped ultraviolet laser

corrugated section of azimuthal index m = 3. A convenient feature of the microwave generator is that the output wave $EH_{11}$ has the structure close to that of the Gaussian beam and, so, can be immediately used in physical experiments. And indeed, the generator has been successfully used to make a gas discharge UV laser.

An additional possibility to produce high-power coherent microwaves consists in a separation of functions among electron sub-beams. This method was used in a device ( [9] , Fig. 6 ) combining a) a BWO excited by the inner electron beam and b) travelling wave amplifier excited by the outer electron beam and driven ( through a periodic system of holes ) by the wave produced in the BWO section.

*Selective excitation of a single " hot " mode.* In microwave generators and amplifiers of broad cross-section the electron stream can interact simultaneously with a large amount of "cold " modes , which can compose independent combinations - "hot" modes. In the limiting case of distant walls, these "hot" modes turn into eigen-modes of the electron stream, i.e. into waves canalized by the stream ( due to a balance of diffraction and refraction ) without participation of the walls [10-12]. This very promising regime called the optical guiding still waits for its experimental realization . Let us note, however, that even in systems of the sort mode selection should vanish, if the electron stream cross-section become too large ( Appendix 1 ).

## NON-LINEAR MODE INTERACTION

To provide the coherent, single-mode regime , the most reliable method seems to exclude self-excitation of parasitic

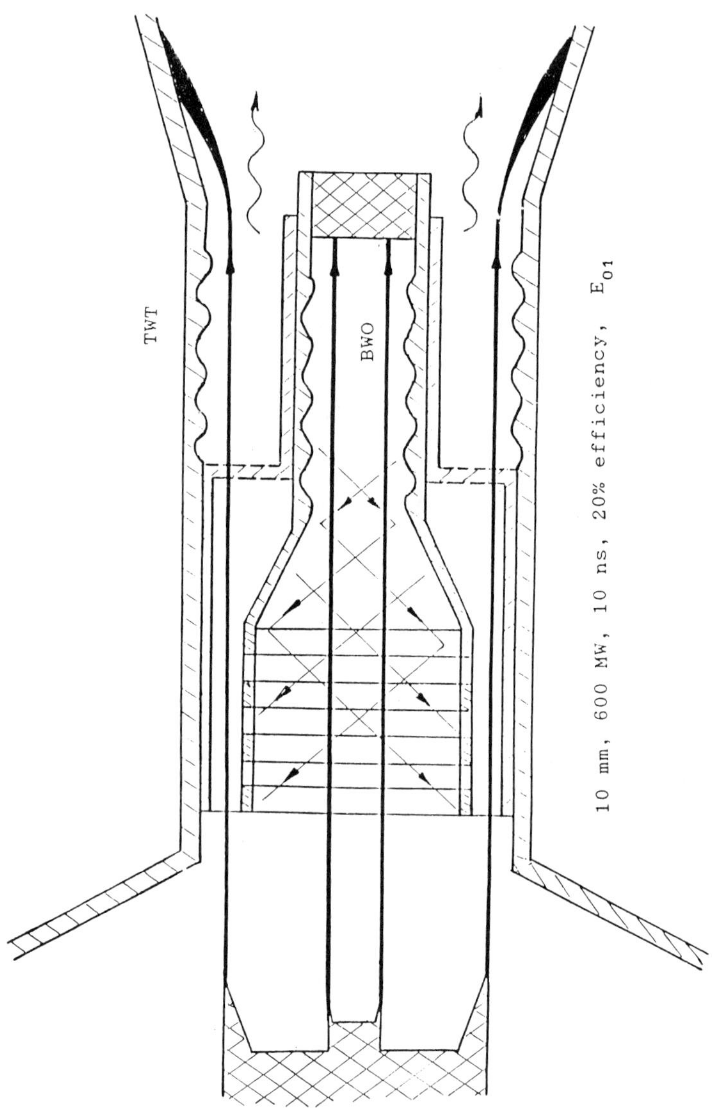


Fig.6. Double-stream Cherenkov microwave generator.

modes at the electron current the main mode operates at. However , in principle, the same goal can be achieved under less rigid conditions: one can admit a number of modes to start simultaneously; however , at the final ( stationary ) stage one of them can suppress its rivals ( the effect discovered by Van der Pol in 1922). Moreover, a rather fresh theory [13] has revealed a ( maybe exotic ) case , when the operating mode does not self-excite itself, but is launched by other ones and finally extinguish its producers.

And, of course, in microwave generators with space-developed interaction space there are a lot of cases when multi-mode multi-frequency ( in particular, stochastic ) regimes establish. These regimes are paid with great attention by theorists [14-16], but usually ( with a small exception [17]) avoided by experimentalists, mostly because of a lack of applications.

## MODE CONVERSION AND MATCHING TO CONSUMING DEVICES

Methods to convert high order modes produced by high-power microwave devices into simpler modes fit for distant transportation [18] can be set in the following sequence:

- relatively low order modes can be converted by a slow ( adiabatic) variation of the waveguide cross-section ,
- higher modes can be converted by scattering inside a corrugated waveguide section ,
- still higher modes can be converted by a ray-tracing curved-mirror technique.

A new method [19] based on a synthesis of the three ones mentioned gives a possibility to convert high order modes ( e.g., in an experiment , $TE_{15.4}$) into the Gaussian beam with efficiency over 95 % .

The Gaussian beam is convenient to feed a number of radio technical and plasma devices. It can be proposed to feed microwave particle accelerators [20] as well ( Appendix 2).

ACKNOWLEDGMENT

The authors are grateful to the Organizing Committee of the Conference on Coherent Radiation Generation and Particle Accelerators for a fine possibility to discuss the problems concerned in the report with leading specialists in the field.

## Appendix 1. DIFFRACTION LIMITATION ON ELECTRON BEAM TRANSVERSE DIMENSION FOR OPTICAL GUIDING SYSTEMS.

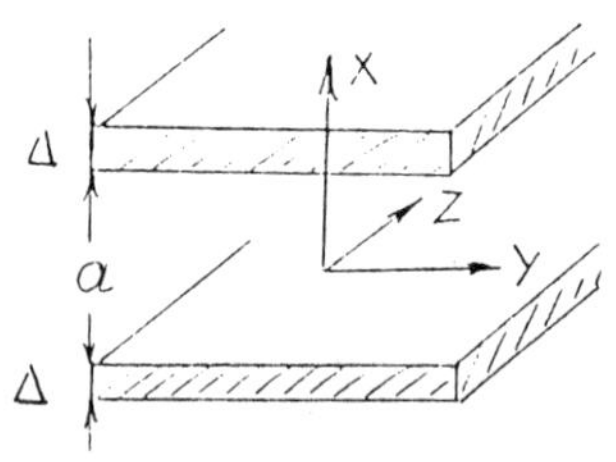


Fig.7. Open waveguide composed of "dielectric" plates.

*Canalization of waves by coupled open dielectric waveguides.* Let us take (Fig.7) two parallel indefinite plates with dielectric permittivities $\varepsilon_m$ (m=1,2). Let the plates be separated by a distance a large compared to their thickness $\Delta \ll a$. Eigen-modes

$$\vec{E} = \mathrm{Re}(\vec{x}_0 E_0 e^{i\omega t - ihz}) \qquad (1)$$

of the system satisfy the wave equation

$$\frac{d^2 E_0}{dx^2} + [k^2 \varepsilon(x) - h^2] E_0 = 0 , \qquad (2)$$

where $k=\omega/c$ . According to Eq.(2), outside the plates ( in the vacuum) the function $E_0(x)$ has the form $E_0 = A_s e^{-px} + B_s e^{px}$, $p^2 = h^2 - k^2$, (s=1,2,3). At opposite sides of a thin plate magnitudes of the RF field can be taken equal

$$E_{+} = E_{-}$$

and the x-derivatives of the field are related by

$$\left.\frac{\partial E}{\partial x}\right|_{+} - \left.\frac{\partial E}{\partial x}\right|_{-} = -\,2q_m E \quad ,$$

here

$$q_m = (\varepsilon_m - 1)k^2\Delta\ /\ 2\ . \tag{3}$$

With these boundary conditions we arrive to a dispersion equation

$$(p-q_1)(p-q_2) = q_1 q_2 e^{-2pa} \tag{4}$$

If the separation of plates is small

$$|p|a \ll 1\ , \tag{5}$$

expansion of the exponent function in the right hand side of Eq.(4) reduces the latter to

$$p = q_1 + q_2 - 2aq_1q_2\ ,$$

approximate solutions of which are

$$p \approx q_1 + q_2 \tag{6}$$

and, for the frequency-resonant permittivities $\varepsilon_m$ ,

$$q_1 + q_2 \approx 2aq_1q_2\ . \tag{7}$$

If the plates are identical $\varepsilon_1 = \varepsilon_2$ , the Eq.(6) corresponds to the syn-phase (x-symmetric) mode and Eq.(7) corresponds to the counter-phase (x-antisymmetric) one.

For distant plates

$$|e^{-2pa}| \ll 1 \tag{8}$$

solutions of Eq.(4) are close to

$$p \approx q_m\ . \tag{9}$$

If the plates have equal or almost equal dielectric permittivities

$$|\varepsilon_1 - \varepsilon_2|/\varepsilon_{1,2} \ll |e^{-2pa}|, \tag{10}$$

the waves canalized by "separate" plates are mutually coupled and compose syn-phase and counter-phase modes. In the opposite case

$$|\varepsilon_1 - \varepsilon_2|/\varepsilon_{1,2} >> |e^{-2pa}| \tag{11}$$

eigen-modes (9) of the different plates can be regarded as independent.

*Modes in coupled waveguides composed of electron-oscillator streams.* For "plates" ( Fig. 1A ) composed of electrons oscillating with frequency $\Omega$ and proceeding with translational velocity $v_{\parallel}$, the dielectric permittivity takes the form

$$\varepsilon_m = 1 - \frac{\varkappa^2 \omega_m^2}{(\omega - h v_{\parallel} - \Omega)^2} , \tag{12}$$

where $\omega_m$ is an effective plasma frequency of the electron layer, $\omega_m^2$ is proportional to the electron beam density, $\varkappa$ is of the order of $v_{\sim}/c$, $v_{\sim}$ is the electron oscillation velocity.

Under the resonance condition

$$\omega - kv_{\parallel} = \Omega \tag{13}$$

formulae (3) and (12) give

$$q_m = - I_m p^{-4} , \tag{14}$$

where $I_m = 2\Delta k^4(\varkappa\omega_m/v_{\parallel})^2$. Correspondingly Eqs.(6),(7) and (9) acquire forms

$$p^5 = -(I_1 + I_2) , \tag{6a}$$

$$p^4 = -2aI_1I_2/(I_1+I_2) , \tag{7a}$$

$$p^5 \approx - I_m . \tag{9a}$$

Let us note, that with account of $p^2 \approx 2k(h-k)$ the limitation (5) being the base for (6a),(7a) can be reduced to the form

$$\left(ka^2/L\right)^{1/2} < 1 , \tag{5a}$$

where $L \sim |\mathrm{Im}\ h|^{-1}$ is the longitudinal amplification scale.

Each of equations (6a), (7a) and (9a) has a complex root corresponding to a wave canalized ( Re p > 0 ) and amplified

(Im h > 0) by the electron stream (the optical guiding).

*Wave filtration and coherence loss in excessively broad stream.* If the electron sub-streams are identical, the larger is the distance *a* between them, the less is the difference of increments for syn-phase and counter-phase modes (Fig.2A), and, so, the weaker is the wave structure filtration.

In addition, according to Eq. (10) and (11), the larger is the distance *a* between the layers, the more sensitive is the wave field structure to redistributions of the current among the electron sub streams .

Thus, if the electron stream cross-section is excessive so, that the diffraction limitation ( 5a ) is violated, the RF field coherence cannot be obtained in practice.

## Appendix 2. QUASI-OPTICS FOR MICROWAVE PULSE COMPRESSION AND INPUT INTO LINAC.

In the electron-positron colliders, accelerating sections must be pre-pumped by pre-compressed microwave pulses during a time limited by ohmic losses in metallic walls. So, the electrodynamic system should be of minimum working surface and, hence, of the quasi-optical type ( Fig. 8 ) :

1)The frequency-modulated pulse can be compressed by a chain of coupled open resonators ( for example, composed of plane grills *1* or mirror resonators with periodically corrugated mirror couplers [21]). The eigen-frequencies and coupling of resonators should be optimized to reduce ohmic and diffraction losses.

2) The microwave-storage and particle-acceleration cavity *3* is expedient to be fed through a periodic set of waveguide

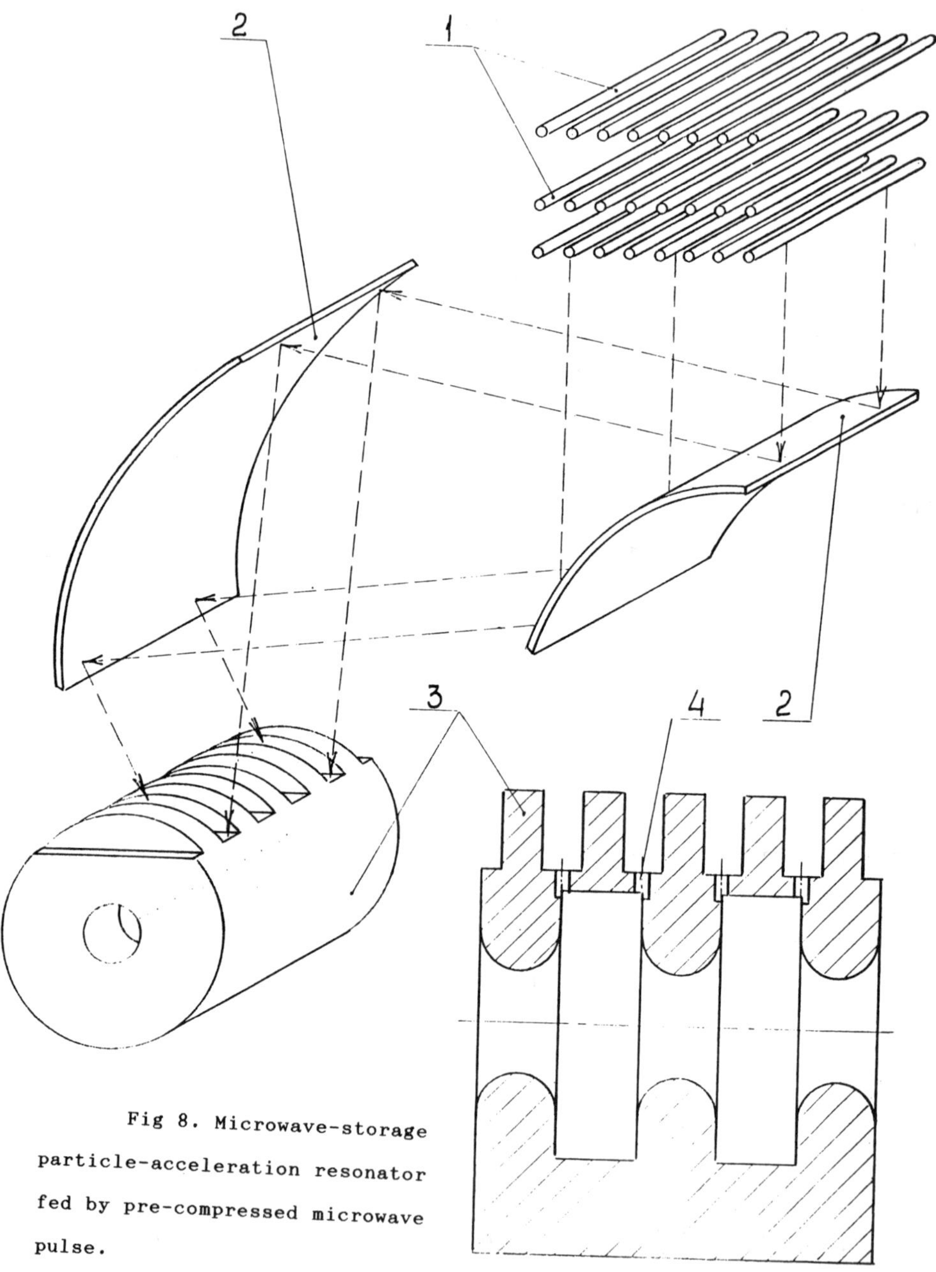


Fig 8. Microwave-storage particle-acceleration resonator fed by pre-compressed microwave pulse.

canals *4.* To suppress the Wood anomaly outside the cavity, the period of canals should be equal to the half-period of the accelerating structure.

3) The pulse compression section can be matched to the storage-acceleration one by a system of mirrors *2* .

REFERENCES

1. " Physics of Quantum Electronics, 1980,1982, Vol 7-9, Free-Electron Generators of Coherent Radiation". Edited by S.F. Jacobs et al. (Reading, Mass: Addison-Wesley).

2. "High-Frequency Relativistic Electronics" (collections of reviews). Edited by A.V. Gaponov ,1979, 1981,1983,1984,1987,1990, Gorky, IAP.

3. IEEE Trans. on Plasma Sci.,1990, Vol. PS-18, No. 3 (special issue on high-power microwave generation, W.W. Destler and B. Levush Eds. ).

4. E.B. Abubakirov et al., "High-Power Millimeter Wave Cherenkov Amplifier with Intense REB", ZhTF( Journ.Tech. Phys.), 1990, Vol. 60, No.11, pp. 186-190.

5. E.B. Abubakirov et al., "Experimental Realization of a Cyclotron Resonant Mode Selection Method in Relativistic HF Cherenkov Oscillators", Pisma v ZhTF ( Lett. to Journ. Tech. Phys.), 1983, Vol. 9, No.9, pp. 533-538.

6. V.L. Bratman et al., " FELs with Bragg Reflection Resonators. Cyclotron Resonance Masers vs Ubitrons", IEEE J. Quant. Electron., Vol. QE-19, No.3, pp.282-296,1983.

7. V.L. Bratman et al., "Millimeter-wave HF Relativistic Electron Oscillators", IEEE Trans. on Plasma Sci., Vol.PS-15, No.1, pp2-15, 1987.

8. E. B. Abubakirov et al., "Cherenkov Relativistic Electron Generators of Coherent Electromagnetic Radiation with Multi-mode Electrodynamic Systems " , Int. Conf. Beams-90, Novosibirsk, Jul 2-5, 1990, Proceed., part 2, pp.1105-1110.

9. V.L. Bratman et al., " Experimental Investigation of a Sectioned Microwave Generator with Relativistic Electron Beam", Pisma v ZhTF ( Lett. to Journ. Tech. Phys.),1988, Vol. 14,No.1, pp. 9-13.

10. A.M. Kondratenko and E.L. Saldin, " Generation of Coherent Radiation by Relativistic Electron Beam in Undulator", Journ. of Tech. Phys. ( Russian ),1981,Vol. 51, pp.1633-1642.

11. A.V. Gaponov et al., " Achievements and Problems of High-Frequency Relativistic Electronics", Proc. of 10th Europ. Conf. on Controlled Fusion and Plasma Physics, Moscow, 1981, Vol.2, pp. 48-53.

12. C. M. Tang and P. Sprangle, "3-D Non-linear Theory of FEL Amplifier", Physics of Quantum Electronics, 1982, Vol.9, Free-Electron Generators of Coherent Radiation, edited by S.F. Jacobs (Reading, Mass: Addison-Wesley) pp.627-650.

13. M. I. Petelin, "High - Current Relativistic - Electron Generators of Millimeter and Longer Waves", Int. J. Electronics, 1989, Vol. 67, No. 1, pp. 137-145.

14. W.B.Colson, "The Trapped-Particle Instability in FEL's Oscillators and Amplifiers", Nucl. Instr. Meth., 1986. Vol. A-250,

pp. 168-175.

15. N.S.Ginzburg, M.I.Petelin, "Multi-frequency Generation in FEL's with Quasi-optical resonators", Int. J. Electr., 1985. Vol. 59, pp. 291-314.

16. T.M.Antonsen, Jr. and B. Levush, "Mode Competition and Suppression in FEL's", Phys. Fluids, 1989. Vol. B.1, pp. 1097-1108.

17. B. P. Bezruchko, S. P. Kuznetsov, D. I. Trubetskov, " Experimental Study of Stochastic Oscillations in Dynamic System: Electron Beam - Electromagnetic Wave", Pisma v ZhETF ( JETP Letters),1979, Vol. 29, No. 3, p. 180-183.

18. M.Thumm, "Electrodynamic Systems for Mode Conversion, Transmission and Diagnostcs of High-power Millimeter-Wave Radiation". Int. Workshop on Strong Microwaves in Plasmas, Suzdal, 1990.

19. D.V.Vinogradov et al., "Adiabaticity Criteria of Irregular Quasi-Optical Waveguide Structures", Izv. VUZov. Radiofizika, 1990, Vol. 33, No. 11.

20. "New Developments in Particle Acceleration Techniques", ed. S.Turner (CERN Report No. CERN-87-11, Geneva,1987).

21. V.I. Belousov et al., " Auxiliary elements for high-power quasi-optical waveguide ducts", Gyrotron (collection of papers, Russian) , V.A. Flyagin ed.,Gorky, 1989, pp.155-160.

Research Trends in Physics: Coherent Radiation Generation and Particle Acceleration

*La Jolla International School of Physics*, The Institute for Advanced Physics Studies, La Jolla, California

# Nonlinear Interaction of Ultra Short Relativistically Strong Laser Pulses with Nonuniform Plasma

**S.V. Bulanov, I.N. Inovenkov*, V.I. Kirsanov, N.M. Naumova, and A.S. Sakharov**

General Physics Institute
Russian Academy of Sciences
117942 Moscow, Russia

*M.V. Lomonosov Moscow State University
Moscow, Russia

The theory and computer simulations of the interaction of an ultrashort relativistically strong laser pulse with uniform and nonuniform plasmas are presented. The distortion and fast depletion of the pulse due to the nonlinear plasma wake excitation are considered. The role of the backward stimulated Raman scattering in the process of the leading front steepening is indicated. Electron acceleration and heating in the course of plasma wave breaking are demonstrated. The generation of relativistically strong locked electromagnetic modes at the final stage of the pulse depletion is shown up.

## I. Introduction

The progress of the laser technology has made available for investigators the subpicasecond laser pulses at the terawatt power [1-3]. Being focused the electric field of such a pulse is strong enough not only to ionize the matter fully transforming it into the plasma but is capable to make the plasma electrons to oscillate at about the speed of light [3].

 

One of the possible application of mentioned above laser pulses is the Laser Wake Field Accelerator (LWFA) in which the field of the plasma wave behind the pulse is supposed to accelerate charged particles [4-7] with the extremely high acceleration gradients more than two orders over those in conventional accelerators [8]. A successive development of LWFA could lead to the construction of rather compact accelerators with the energy of accelerated particles over 1 TeV.

If the excited wave is weakly nonlinear the maximum particle energy gain will be given by the formula [4]:

$$W_{max} \simeq 2\varepsilon mc^2\gamma_{ph}^2 , \tag{1}$$

where $\gamma_{ph} = ( 1 - v_{ph}^2/c^2 )^{-1/2} \simeq \omega_0/\omega_p$ , $v_{ph}$ is the phase velocity of the plasma wave, $\varepsilon = eE_p/m\omega_p c$ is the dimensionless amplitude of the plasma wave electric field ( for a weak nonlinearity $\varepsilon \ll 1$ ), $\omega_0$ is the carrying frequency of the laser pulse, and $\omega_p = (4\pi e^2 n_0/m)^{1/2}$ is the plasma electron frequency ( $\omega_p \ll \omega_0$ ).

The condition $\varepsilon \ll 1$ being combined with formula (1) leads to the limitation on the maximum energy of the accelerated particles for a given radiation wavelength $\lambda_0$:

$$W_{max}[\text{eV}] \ll 10^{13}\left(E_p[\text{GV/m}]\lambda_0[\mu\text{m}]\right)^{-2} . \tag{2}$$

As it follows from nonequality (2), for a weakly nonlinear laser driven plasma wave with sufficiently high electric field $E_p$ = 10 GV/m ( much over that in conventional accelerators ) it is not possible to obtain the final energies over TeVs for the available laser radiation wave band ( $0.26\mu m < \lambda_0 < 10.6\mu m$ ).

The limitation on the final energy of accelerated electrons for the plasma based accelerators to certain extent can be avoided by using a multi-stage system [9] or by the proper plasma density profiling [10,11]. However, in the both variants additional restrictions appear and it seems that concept of laser-plasma based accelerator could not be improved radically on this way..

The more was the understanding of the discussed above problem the more was the interest to the possibility of a strong nonlinear plasma wave excitation ( $\varepsilon > 1$ ) [6,7,12-16,24] as a way to provide the advantages of laser based accelerators.

For the strong nonlinear plasma wave the limitation (2) is no more valid and it is principally possible to obtain at the same time the fairly high accelerating gradients and high final output particle energies. From this point of view LWFA is very attractive concept to provide the generation of strongly nonlinear plasma wakefield with $\varepsilon > 1$. As for this purpose the extremely high power of laser field in ultrashort pulses is required the specific phenomena of the processes could appear in the course of laser-plasma interaction.

The interest to the interaction of relativistically strong electromagnetic pulses with a plasma is not due only to the development of the new acceleration technique. Other aspects of such an interest are the fundamental problems of laser field evolution and of plasma heating. For an ultrashort pulse of ultrarelativistic amplitude some new impressing phenomena in the process of laser plasma interaction occur. Note that even for one dimensional case and for a homogeneous plasma the consistent picture describing the strongly nonlinear ultrashort laser pulse-plasma interaction is far from being completed.

In this paper we consider the propagation of an intense arbitrary polarized electromagnetic pulse both in homogeneous and inhomogeneous collisionless plasmas and discuss various aspects of the interaction.

## II. Basic equations

The interaction of a short intense laser pulse with a plasma can be described by the set of Maxwell equations and the fully relativistic cold electron fluid equations. In the one dimensional case ( for high enough transversal size of the pulse $L_{tr} \gg k_p^{-1} = c/\omega_p$ ) all variables can be considered as functions of the longitudinal coordinate $x$ and time $t$. Since the laser field is transverse one we take the transverse component of the vector $\mathbf{q}=\mathbf{p}/mc$ as a variable characterizing these fields and write it in the form

$$\mathbf{q}_{tr} = \frac{1}{2}\left[ \mathbf{a} \exp(-i\omega_0 t + ik_0 x) + \mathbf{a}^* \exp(i\omega_0 t - ik_0 x)\right], \qquad (3)$$

where $k_0$ is the wave vector of the radiation, which is related to $\omega_0$ by the linear dispersion relation ( $\omega_0^2 = k_0^2 c^2 - \omega_p^2$ ), and $\mathbf{a}$ is a complex amplitude which is convenient to choose as a function of the variables $t$ and $\xi = x - v_g t$ ( $v_g = c^2 k_0/\omega_0$ is the group velocity of a laser radiation in the linear approximation ). As usual, the envelope approximation approach (3) assumes:

$$\partial/\partial\xi \ll k_0, \qquad \partial/\partial t \ll c\,\partial/\partial\xi \ll \omega_0. \qquad (4)$$

The ponderomotive force of HF field produces the longitudinal redistribution of the plasma electrons at the background of heavy immobile ions. The field of charge separation we describe by the dimensionless electrostatic potential $\Phi = e\varphi/mc^2$, considering it in the form $\Phi = \Phi_0 + \Phi_{hf}$ , where $\Phi_0$ and $\Phi_{hf}$ are slowly varying and high frequency parts of the potential, respectively.

Assuming that the plasma density is fairly low [7,15]

$$\omega_0^2/\omega_p^2 \gg (1+q^2)/(1+\Phi)^2 , \qquad (5)$$

we can neglect $\Phi_{hf}$ and obtain the set of equations for $\Phi_0$ and $\mathbf{a}$ [7,14-16]

$$\frac{\partial^2\Phi_0}{\partial\xi^2} - k_p^2 \frac{1+|\mathbf{a}|^2/2 - (1+\Phi_0)^2}{2(1+\Phi_0)^2} = 0 , \qquad (6)$$

$$2i\omega_0 \frac{\partial \mathbf{a}}{\partial t} + \frac{\omega_p^2}{\omega_0^2} c^2 \frac{\partial^2 \mathbf{a}}{\partial \xi^2} + 2v_g \frac{\partial^2 \mathbf{a}}{\partial \xi \partial t} = - \omega_p^2 \frac{\Phi_0}{1+\Phi_0} \mathbf{a} \ . \tag{7}$$

It is worth to mention, that the condition (5) is equivalent to the plasma electron oscillations to be far from breaking.

## III. Plasma Wave Excitation by Given Electromagnetic Wave Packet

For the high intensity ( $|\mathbf{a}| \gg 1$ ) ultra short pulse ( with the duration $\tau \leqslant 2\sqrt{2}/|\mathbf{a}|\omega_p$ ) inside the pulse location region we have from (6)

$$\Phi_0 = \frac{1}{4} k_p^2 \int_{\infty}^{\xi'} d\xi' \int_{\infty}^{\xi} d\xi'' \, |\mathbf{a}(\xi'', t)|^2 \leqslant 1 \ . \tag{8}$$

Joining the solution (8) and the solution for free plasma wave behind the pulse we can find the wakefield amplitude

$$\Delta\Phi = \sqrt{A(A+4)}, \quad A = \left[\int |\mathbf{a}|^2 k_p d\xi/4\right]^2 \ . \tag{9}$$

It can be seen from (9) that though the electric potential inside the pulse is relatively small, the amplitude of the wakefield can be fairly great. For example, for $(\omega_p \tau)|\mathbf{a}| \simeq 2\sqrt{2}$ it follows from (9) that $\Delta\Phi \simeq |\mathbf{a}|^2/2 \gg 1$.

The potential of charge separation field generated by short laser pulse is shown in Fig.1a [15].

Going now to pulses with a relatively larger duration ( $\tau \gg 2\sqrt{2}/|\mathbf{a}|\omega_p$, $|\mathbf{a}| \gg 1$ ) we note that the amplitude of the wake is a maximum ( $\Phi_{max} \simeq |\mathbf{a}_1|^2/2$ ) for the short enough rise time of the pulse ( $\delta\tau_1 \omega_p < |\mathbf{a}_1|^{-1}$ ) [7,15]. The corresponding maximum electric field is about

$$E_p = \frac{mc\omega_p}{e} |\mathbf{a}_1|/\sqrt{2} \ . \tag{10}$$

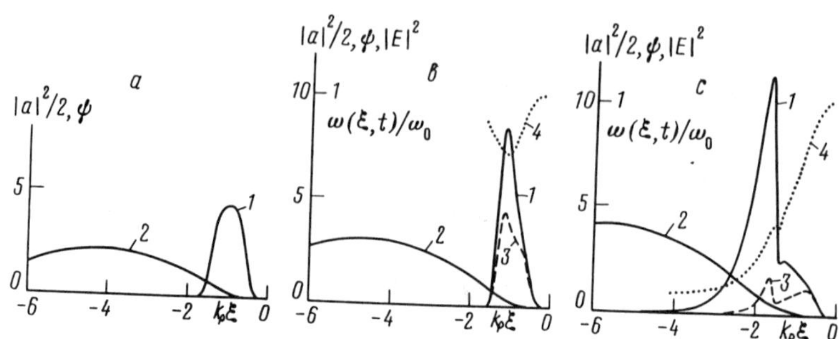


**Fig.1.** Three plots of the nonlinear plasma-wave excitation by short laser pulse for $t = 0$, $t_{nl}$, $2t_{nl}$ ($t_{nl}=8 \cdot 10^3 \omega_p^{-1}$). Curve 1 - the quantity $|\mathbf{a}|^2/2$ ; curve 2 - the scalar potential $\phi$; curve 3 -the electromagnetic field energy density in units of $8\pi(m\omega_o c/e)^2$; curve 4 (dotted line) - the ratio of the value of local frequency of the electromagnetic radiation $\omega(\xi,t)$ to $\omega_o$. Here $a_o=3$, $\omega_o/\omega_p=100$ and $\omega_p\tau \simeq 0.7$.

Here and below the low case 1 corresponds to the quantities on the leading front of the pulse.

Under the condition $\delta\tau_1\omega_p > |\mathbf{a}_1|^{-1}$, an increase of the rise time of the pulse will result in a decrease in the amplitude of the excited plasma wave. Finally, if the pulse is smooth enough at the both edges ( $\delta\tau\ \omega_p \gg |\mathbf{a}|^{1/2}$ ), no wakefield will be excited at all.

Thus, in the ultrarelativistic case the nonlinear wakefield of the maximum amplitude will be excited by short laser pulse ( or pulse with short leading front ) with a duration about $(|\mathbf{a}|\omega_p)^{-1}$, that determines its spectrum width $\Delta\omega \geqslant \omega_p|\mathbf{a}|$.

## IV. Theory of Laser Field Evolution in Course of Plasma Wave Excitation

Eq.7 has the first integral

$$\int d\xi\left\{|\mathbf{a}|^2-\frac{iv_g}{\omega_0}\left[\mathbf{a}^*\frac{\partial \mathbf{a}}{\partial \xi}-\mathbf{a}\frac{\partial \mathbf{a}^*}{\partial \xi}\right]\right\}=\text{const.} \qquad (11)$$

Taking $\mathbf{a}$ in the form $\mathbf{a}=|\mathbf{a}|e^{i\vartheta(\xi,t)}$ and utilizing the usual definition of the local frequency of electromagnetic radiation $\omega(\xi,t)$ as the time derivative of the whole phase in the laboratory frame $\omega=\partial(\omega_0 t-\vartheta(x-v_g t,t))/\partial t\simeq\omega_0+v_g\partial\vartheta/\partial\xi$, one can rewrite (11) in the form

$$\int d\xi\,|\mathbf{a}|^2\omega=\text{const.} \qquad (12)$$

The combination of variables $|\mathbf{a}|^2\omega$ in (12) ( since the modulus of amplitude is defined as $|\mathbf{a}|=e|E|/m\omega c$ ) is proportional to the density of transverse radiation photon number $N=|E|^2/\omega$ . Thus, Eq.(12) ( or Eq.(11) ) may be interpreted as the conservation law of the total photon number.

In addition, from Eq.7 one can get the relationship

$$\frac{\partial}{\partial t}\int d\xi\left(\frac{iv_g}{\omega_0}\frac{\partial \mathbf{a}}{\partial \xi}+\mathbf{a}^*\right)\left(\frac{iv_g}{\omega_0}\frac{\partial \mathbf{a}^*}{\partial \xi}-\mathbf{a}\right)=\frac{1}{2}v_g\frac{\omega_p^2}{\omega_0^2}\int d\xi\,|\mathbf{a}|^2\frac{\partial}{\partial \xi}\frac{1}{1+\Phi_0}, \qquad (13)$$

that can be transformed to the expression for the laser field energy losses

$$\frac{\partial}{\partial t}\int d\xi\left(\frac{|E|^2+|B|^2}{16\pi}\right)=-\frac{1}{\pi}\left(\frac{m\omega_p c}{4e}\right)^2\int d\xi\,|\mathbf{a}|^2\frac{\partial}{\partial \xi}\frac{1}{1+\Phi_0}. \qquad (14)$$

Using Eq.(6) it is easy to show, in agreement with the total energy conservation law, that the right hand side of Eq.(14) is equal to the rate of the energy input into the plasma wakefield ( taken with the opposite sign ).

To study qualitatively the evolution of the fairly short laser pulse ( $\omega_p\tau|\mathbf{a}| \ll 1$ ) one can use a rough model based on the conservation laws mentioned above. Assuming that the laser pulse is characterized by a certain average value of frequency $\langle\omega(t)\rangle$ and by the integral $Q(t) = \int d\xi|\mathbf{a}|^2$, it is possible to obtain ( using also Eq.(8) ) the next formulae

$$Q(t)=(1-t/t_{nl})^{-1/3}\,Q(0), \qquad \langle\omega(t)\rangle=(1-t/t_{nl})^{1/3}\langle\omega(0)\rangle. \tag{15}$$

where the characteristic time of the nonlinear pulse energy depletion is given by

$$t_{nl} = (16/3)(k_pQ(0))^{-1}\omega_0^2/\omega_p^3 \quad . \tag{16}$$

The approximate solution (15) predicts, that the decrease in the laser pulse frequency $\langle\omega(t)\rangle$ ( and in the pulse total energy ) is accompanied by the growth of $Q$ as well as by the increase in the amplitude of the excited wakefield ( according to (9) ). The considered model is valid as far as the inequalities $t < t_{nl}$ and $(\omega_p\tau)|\mathbf{a}(t)| < 1$ are satisfied.

The linear spreading being taken into consideration does not affect the process of wake excitation as in this case as for pulses of a low intensity [17].

The results of the numerical solution of the set of Eqs.(6),(7) are shown in Figs.1a,b,c ( for $t = 0$, $t = t_{nl}$ and $t = 2t_{nl}$, where, for the chosen parameters, $t_{nl} \simeq 8{\cdot}10^3\omega_p^{-1}$ ) and in Fig.2. The dependencies presented in Fig.1 as well as the temporal changes in normalized energy of the laser pulse and in the plasma wakefield normalized amplitude ( Fig.2 ) corroborate the general tendencies previously discussed in the frame of the rough model. In addition, as it can be seen in Fig.2, both the rate and the total value of the laser pulse energy losses are

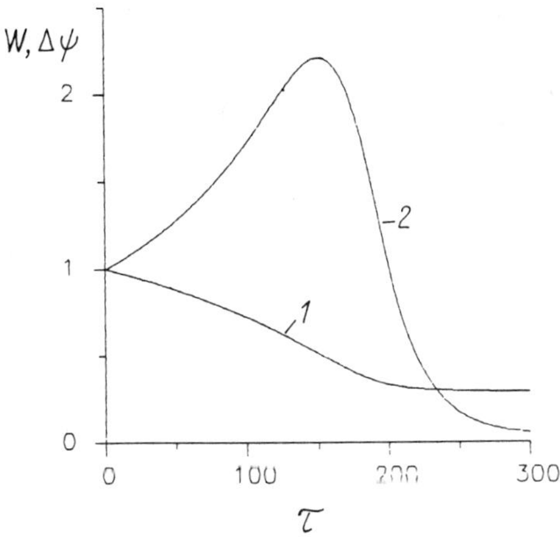


**Fig.2.** The laser pulse total energy $W$ (1) and the plasma wake-wave amplitude $\Delta\phi$ (2) normalized to their initial values at $t = 0$ versus time ($\tau=\omega_p^2 t/\omega_0$). The same parameters as for Fig.1 .

fairly considerable for an ultrarelativistic pulse in contrast with the low intensity case [5,17]. For the chosen parameters the laser pulse loses about one half of its total energy per time $t \simeq 2t_{nl}$.

As the local frequency measurements are hardly possible to carry for picasecond laser pulses it is reasonable to present here the temporal evolution of the pulse spectrum. In Fig.3 one can see the down shift of the laser pulse frequency and the broadening of its spectrum.

The set of Eqs.(6),(7) is valid for the description of the plasma wakefield excitation and the distortion of the pulse shape and spectrum until the minimum value of the pulse local frequency $\omega_{min}(t)$ would become comparable with the plasma electron frequency $\omega_p$.

Let us now consider the plasma wave excitation by a considerably longer laser pulse: $\tau\omega_p \gg 2\sqrt{2}|\mathbf{a}|^{-1}$ with a short

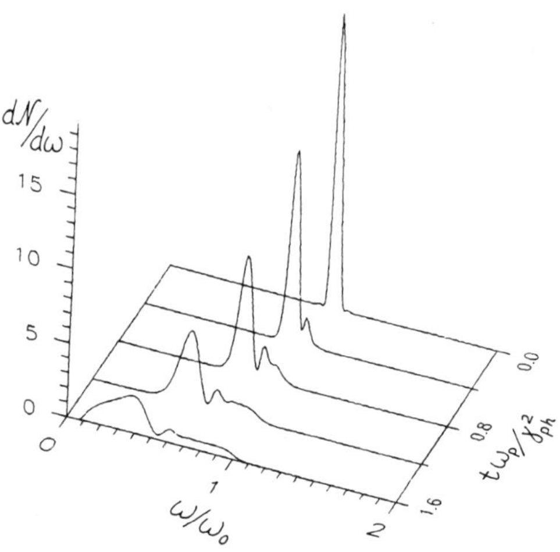


**Fig.3.** The plots of normalized spectral density of photon number of the laser pulse (given in Fig.1) for different times. The figure demonstrates the down shift in the laser pulse frequency and broadening of its spectrum.

leading front $\delta\tau_1\omega_p \leqslant |\mathbf{a}_1|^{-1}$. In this case the considerable nonlinear distortion, similar to those of the short pulse, appear only just behind the leading edge of the pulse in the region where $\phi \leqslant 1$ [15]. The width of this region is about $\delta\xi_{nl} \simeq 2\sqrt{2}(k_p|\mathbf{a}|)^{-1}$, and the value of $t_{nl}$ is determined by expression (16), where the integration in $Q(0)$ is carried over this region. Hence, it is approximately given by the expression

$$t_{nl} \simeq (8/3\sqrt{2})|\mathbf{a}|^{-1}\omega_0^2/\omega_p^3 \quad . \tag{17}$$

It is curious, that this expression has the form similar to that obtained for characteristic time of cascading in a weakly nonlinear beat-wave excitation [18], though the nature of the processes differs and they correspond to quite different intensity limits.

The expression (17) determines also the time of energy depletion in this region. As this process does not result in the decreasing of amplitude gradients at the leading front, the energy depletion does not terminate at $t > t_{nl}$. The region of effective interaction of radiation with plasma shifts backward with a relative velocity $d\xi/dt = -\delta\xi_{nl}/t_{nl} \simeq -c\omega_p^2/\omega_0^2$, which does not depend on the amplitude. Therefore, one can expect the total depletion of the long pulse ( with an initially sharp leading front ) due to the plasma wake excitation to occur at the time $t_{dep} \simeq \tau\omega_0^2/\omega_p^2$. It is remarkable, that the time of resonant depletion of a long two-frequency pulse of a moderate intensity $t_{dep} \simeq \tau\omega_0^2/\omega_p^2|\mathbf{a}|^{2/3}$ ($|\mathbf{a}| < 1$) [19] coincides with our formula at $|\mathbf{a}| \simeq 1$.

## V. Pulse Erosion and Depletion in the Slab of Underdense Plasma.

In the previous sections we have shown that the short laser pulses or the laser pulses with sharp leading edge are capable to excite the plasma wave with maximum amplitude. More realistic ultrashort laser pulse has duration of leading and tailing fronts comparable with the pulse duration and, in addition, can appear not to be short in comparison with a plasma wave length. The amplitude of the plasma wake wave for such a pulse is fairly low but the question is to answer whether the process of the pulse evolution would enhance the plasma wave generation.

The more general approach is to study the interaction of an intense laser pulse with a plasma of a finite size. In the present section we consider the interaction of a laser pulse with a slab of underdense plasma. The slab is assumed to be transparent for the electromagnetic radiation of a chosen wavelength in the low intensity limit. For a strong laser pulse the relativistic electron-mass growth suppresses the interaction. Hence, in this case one can expect the plasma to be even more transparent than for the moderate radiation power. From the other hand, the evolution of the laser pulse considered in previous sections shows that the turn to the strong nonlinear case can bring new aspects to the problem, for example, such as the extremely fast pulse depletion.

To clarify the situation one need to understand the dominating physical processes of the pulse evolution and to answer a number of questions of the practical interest. Among these questions are ones about the amount of the laser field energy to be absorbed, transmitted and reflected in the course of interaction.

Basing on the analysis of the Section IV one can find out that even for a homogeneous plasma the laser-plasma interaction essentially depends on the pulse shape which can change due to the nonlinear effects. As about the collisionless losses of the pulse energy considered above, they will be the strongest in the region of a plasma where the width of the pulse leading front $\delta\xi_1(t)$ can be regarded to be a relatively small:

$$\omega_p \delta\xi_1 \leqslant 2\sqrt{2}\, c/|\mathbf{a}|. \tag{18}$$

For the nonuniform plasma one should need to compare the time of the pulse propagation through the slab $t_{pr}$ with the time of the pulse distortion $t_d$. If $t_d \gg t_{pr}$ the pulse can be considered as a given one. Then the regions of the most effective pulse energy absorption due to the plasma wave generation will be determined by the inequality (18) which in this case can be considered as a restriction on a plasma density.

If $t_d$ becomes comparable with $t_{pr}$, then even for the long and initially smooth pulse, the distortion can result in the formation of jumps of laser field amplitude and hence lead to the strong plasma wave generation.

Let us discuss in brief the main physical mechanisms which could determine the pulse evolution for a case under consideration.

For the adiabatically slow rise of the laser field amplitude, as it is predicted by the previous quasi-stationary analysis, the excitation of the plasma waves does not appear. Still, the laser pulse change its form. The rise of the laser field amplitude at the leading edge of the pulse produces the relativistic electron mass variation along the pulse. The far from the leading edge the more is the local group velocity. As a result one can expect the compression or the steepening of the leading front of the pulse. In contrast with the moderate intensity case the pulse of ultrarelativistic amplitude can not be prevented from steepening by the dispersion and its leading front even can contract up to the size $1/(k_p|\mathbf{a}|)$. The steepening is the most fast for the region with the maximum gradient of the group velocity. Naturally, it occurs in the vicinity of the point with $|\mathbf{a}| \simeq 1$. The characteristic time of the pulse leading front steepening in this region can be expressed as:

$$t_{st} \simeq \omega_0^2/\omega_p^2 \; (c\; \partial|\mathbf{a}|/\partial\xi)^{-1}\Big|_{|\mathbf{a}|=1} \; . \tag{19}$$

The other physical mechanisms which can affect the laser field evolution are the forward and backward stimulated Raman scattering (SRS) [20] which can appear for the pulse with sufficiently long rise and fall time. In homogeneous plasma the growth rate of the forward SRS in the small amplitude limit $|\mathbf{a}| \ll 1$ is given by the following expression:

$$\gamma = \omega_p(\omega_p/\omega_0)^{1/2}|\mathbf{a}|/4 \; . \tag{20}$$

It achieves a maximum at $|\mathbf{a}| \simeq 1$ and decreases with a further enlarging of $|\mathbf{a}|$. This instability can be initiated by a seed plasma wave of a small amplitude which sill is excited even for practically adiabatic profile of the pulse intensity.

The backward SRS can be even more important than the forward one because of its larger growth rate. The extrapolation of the theory [20] up to strongly relativistic intensities predicts that its maximum growth rate is achieved at $|\mathbf{a}| \simeq 1$ for which

$$\gamma \simeq \omega_0(\omega_p/\omega_0)^{2/3} \quad . \tag{21}$$

Considering laser field evolution one should be aware of the possible role of the large scale modulation instability. One can obtain from Eqs.(6),(7) that for a given perturbation wave number $æ \ll k_p$ the growth rate of this instability is maximum for $|\mathbf{a}| \simeq 1$ and is given by expression

$$\gamma \simeq æc\ (\omega_p/\omega_0)^2. \tag{22}$$

All mentioned above mechanisms can be strongly modified in the plasma with heated and accelerated electrons.

Now we turn to discussion of the results of the numerical simulations.

At first we take the electromagnetic wave packet short enough not to be affected by the previously discussed instabilities. At the same time we choose the pulse long enough ( $\tau > \omega_p^{-1}|\mathbf{a}|^{1/2}$ ), so that for the fixed profile of the pulse the energy losses due to the excitation of plasma waves would be negligibly small. The initial polarization of the pulse is taken to be circular.

The different stages of interaction of such a laser pulse with the slab of plasma are demonstrated in the series of Figs.4-7 for $\omega_0 t$ = 200, 400, 800 and 1400, respectively. The coordinate $x$ is normalized to $k_0^{-1}$. The initial half-width of the pulse is equal to $8\pi k_0^{-1}$. In the top plots (a) of each figure the value of transverse electric field $E_\perp=(E_y^2+E_z^2)^{1/2}$ is given by solid line; the dashed lines show the profile of the immobile ion density normalized to its maximum value. The half-width of the plasma slab equals $400k_0^{-1}$. At $t=0$ the pulse is located near the left boundary of the slab ($x = 800k_0^{-1}$). For all considered here and below cases the maximum plasma density is equal to one ninth of the critical density $n_c= m\omega_0^2/4\pi e^2$ and the maximum value of $|\mathbf{a}|$ equals to 3.

In each figure under the top plots there are three other plots for longitudinal electric field (b), transverse (c) and longitudinal (d) electron momenta normalized to $mc$. The electric fields are given in units of $mc\omega_0/e$.

In the first figure of the series (Fig.4, $t=200\omega_0^{-1}$) the pulse just has penetrated into plasma. No serious distortion can be seen though it already excites the plasma wake. This is due to the fact that in the rare plasma region the pulse can be regarded as fairly short ( $\tau \leqslant \omega_p^{-1}|\mathbf{a}|^{1/2}$ ).

The next figure (Fig.5, $t=400\omega_0^{-1}$) demonstrates the strong erosion of the pulse, but up to this moment the pulse still exist as the whole. The considerable growth of the transverse electron momentum in the region occupied by the pulse is accompanied by the pulse frequency downshift. In the bottom phase-space plot for the longitudinal electron momentum there can be seen a group of accelerated electrons just behind the pulse.

Fig.6 ($t=800\omega_0^{-1}$) displays the moment when the total destruction of the pulse has occurred. This moment just corresponds to the time which is necessary for small amplitude wave packet to pass through the slab. Up to this moment the plasma electron thermalization is in progress (bottom plot), but the regular structure of the transverse momentum demonstrates the presence of the strong localized transverse field. This trapped transverse mode has been formed previously on the stage of the most fast pulse distortion and can be traced even in the previous figure.

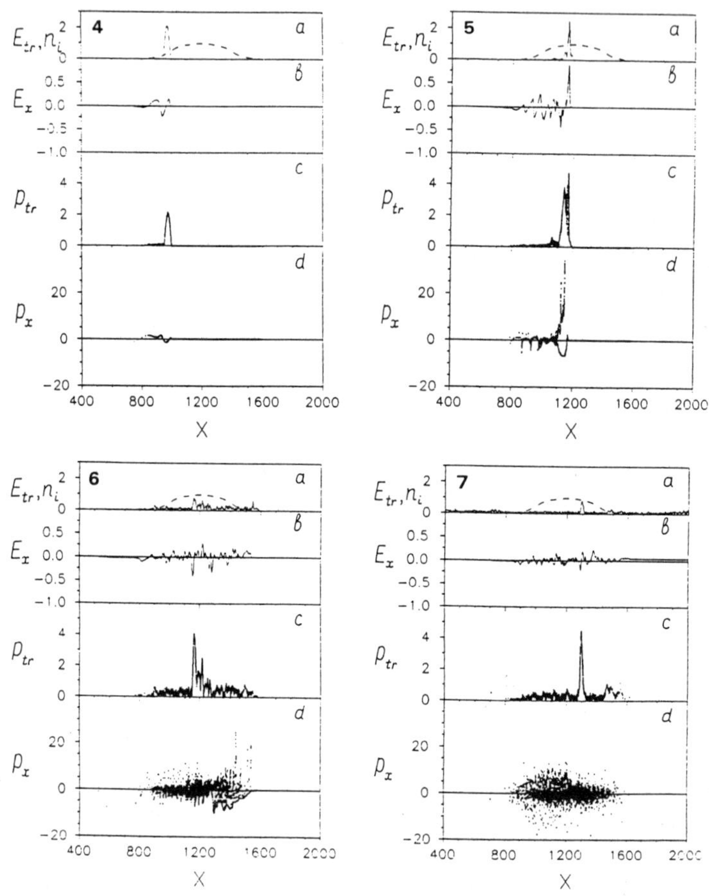


**Fig.4,5,6,7** Computer simulations, showing the the different stages of the laser pulse-plasma interaction. Series of Figs.9,10,11,12 corresponds to $\omega_0 t$ = 200, 400, 800 and 1400, respectively. The coordinate $x$ is in units of $k_0^{-1}$. The plots (from the top to the bottom) show: (a)-transverse electric field $E_\perp = e(E_y^2+E_z^2)^{1/2}/(m\omega_0 c)$ (solid) and profile of the immobile ion density $n(x)/n_{max}$ (dashed), $n_{max}=1/9(m\omega_0^2/4\pi e^2)$; (b)-longitudinal electric field $E_\parallel = eE_x/(m\omega_0 c)$; (c)-transverse and (d)-longitudinal electron momenta normalized to $mc$.

Fig.7 ($t$=1400$\omega_0^{-1}$) shows that the locked transverse mode does not disappear. Up to this moment the electrons are practically thermalized thought the strong anisotropy of the longitudinal and transverse electron momentum distribution remains.

Fig.8 is for the same run as the previous ones. It shows the temporal behavior of portions of energy stored in transverse (curve 1) and in longitudinal (curve 2) field and in the plasma electrons motion (curve 3). Note, that the time dependence of the pulse energy losses ( see the decrease of energy stored in transverse field ) is much alike that for the homogeneous case for short pulse (Fig.2). The interaction in the low density part of the plasma slab has appeared to be fairly strong to distort the laser pulse and thus to give start to the excitation of intense Langmuir wave. As a result the main part of the initial pulse energy is transferred to the plasma electrons. We emphasize that if one supposes that while propagating the pulse does not change its shape, the energy losses would be about only one percent.

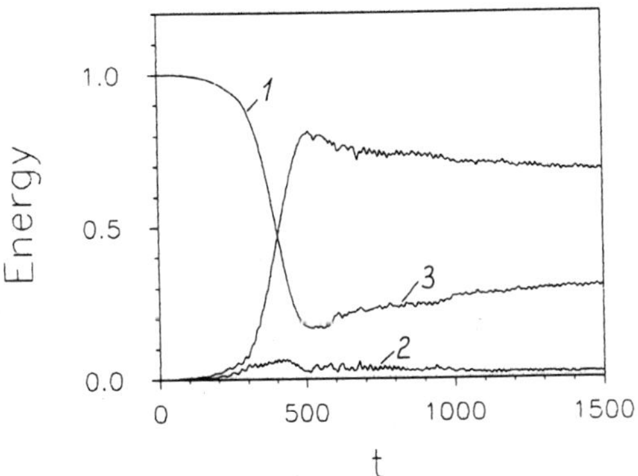


**Fig.8** Evolution of energy stored in transverse (1) and longitudinal (2) field and in the plasma electrons motion (3).

To complete the picture of the pulse destruction we give in Fig.9 the spectrum of the reflected radiation ( passed through the left boundary ) shown in the upper part of the plot and the spectrum of the transmitted radiation at the right boundary shown in the bottom part. The spectra are given for the radiation passed through the boundaries up to $t=1400\omega_0^{-1}$. For comparison, the initial spectrum of the laser pulse is given by dashed line in arbitrary units.

The Fig.10 gives additional details of behavior of the mentioned above locked mode. In this figure there are given the history plots of the maximum and minimum values of the $y$-component of electron transverse momentum along the plasma. These values correspond to the electron momenta in the transverse field of the locked mode. The polarization of the transverse field in the locked mode is the same as the initial one and the frequency is below the local plasma frequency ( even with the relativistic mass growth being taken into account ).

The standard approach to consider the relativistically strong envelope solitary wave [21,22] gives the relationship between the frequency $\Omega$ ( normalized to the plasma electron frequency ), the amplitude $a_{max}$ and the width $\Delta x$ of the standing solitary wave envelope:

$$a_{max}^2 = 4(1-\Omega^2)/\Omega^4, \quad (\kappa_p \Delta x)^2 = 4\,\Omega^2\left((2-\Omega^2)(1-\Omega^2)\right)^{-1}. \qquad (23)$$

This expression corresponds to the limit of the cold plasma electrons. Though in the numerical simulations the plasma is heated anisotropically up to the relativistic temperature the parameters of the demonstrated locked mode is in a qualitative accordance with the relationship (23). The Figs.6,7 show in addition, that the locked mode drifts slowly down the plasma density gradient without sufficient distortions and, thus, demonstrates solitone-like behavior.

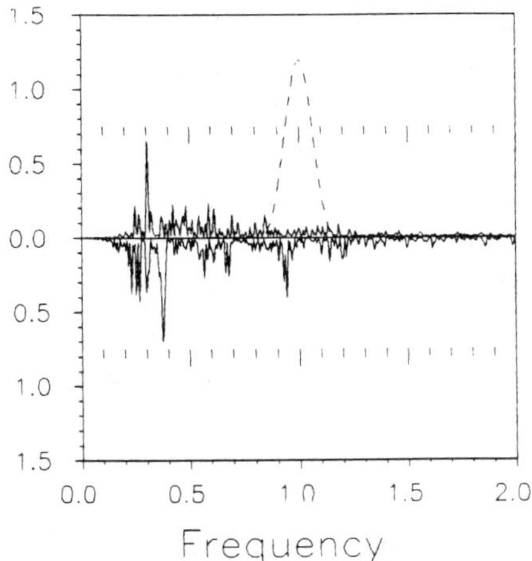


**Fig.9** The spectra of the reflected (top) and transmitted (bottom) radiation trough the boundaries up to $t = 1400\omega_0^{-1}$. Dashed line is initial pulse spectrum.

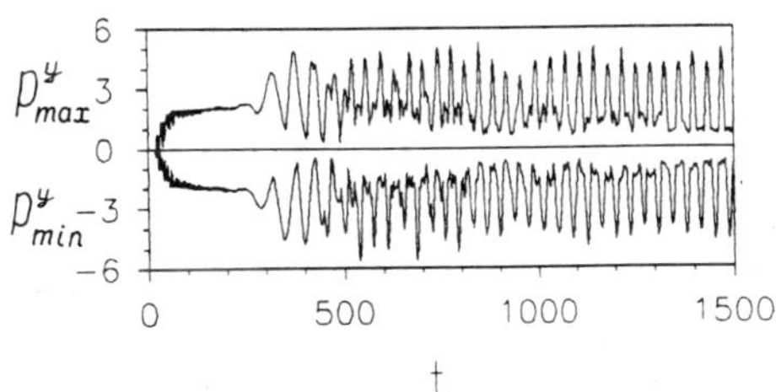


**Fig.10** Maximum (top curve) and minimum (bottom) values of electron transverse momentum component $p_y$ along the plasma slab are given.

Figs.11-13 demonstrate the interaction of the same pulse with the plasma slab two time tinner than the previous one. In Fig.11 the electric field of the pulse is shown for the moments before interaction and after it. In Figs.12,13 there are shown the same dependencies for energy and spectrum as previously in Fig.8,9. One can see that in this case the pulse looses a significant part of its energy ( about sixty percents ) in the course of interaction with plasma but it still exists as the whole after leaving the plasma. In contrast to the previous case there is practically no reflection from the plasma slab.

In both considered cases the strong interaction at the left edge of a plasma appears to be capable to switch on the process of plasma wave excitation. From one hand, the propagation of the pulse up the density gradient enhances the interaction because of the growing electron density. From the other hand, the propagation into more dense plasma could depress the plasma wave generation as the size of the pulse ( or its leading front width ) can become much greater than the local plasma wavelength. In the both considered examples the first of the possibilities dominates triggering extremely fast depletion of the pulse. This results in the collisionless transformation of the main part of the pulse energy to plasma electrons. The time of the total pulse depletion in the first case agrees with an estimation given above in the end of Sec.IV. In the second case the plasma is too narrow to provide the total depletion.

As one more example of the strong laser-plasma interaction we consider the interaction of the fairly long pulse with the half-width equal to $120\pi k_0^{-1}$, which is fifteen time over the previously considered one. The half-width of the plasma slab was equal to 800 $k_0^{-1}$. In this case nor the variation of the group velocity along the laser pulse neither the forward SRS which should be suppressed by the plasma inhomogeneity can not lead to serious erosion of the laser pulse. In addition, the generation of

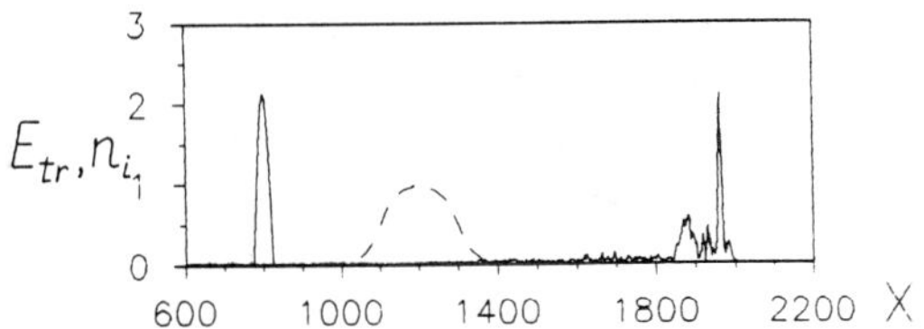


**Fig.11** Same as in top plots of Figs.4-7 but for the plasma slab two times thinner. The transverse electric field before and after interaction are given in one plot.

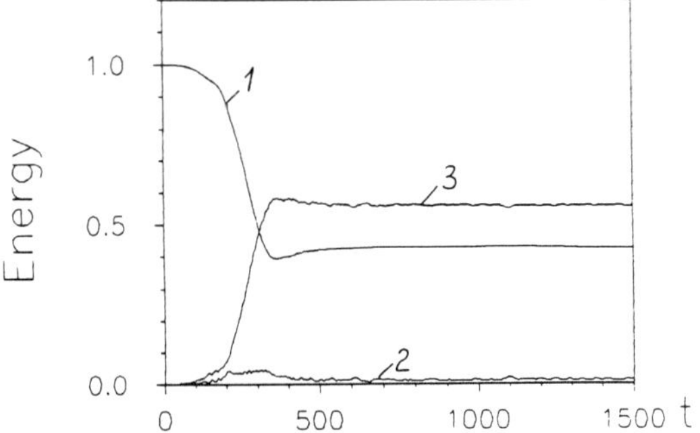


**Fig.12** Same as Fig.8 but for the plasma slab two times thinner.

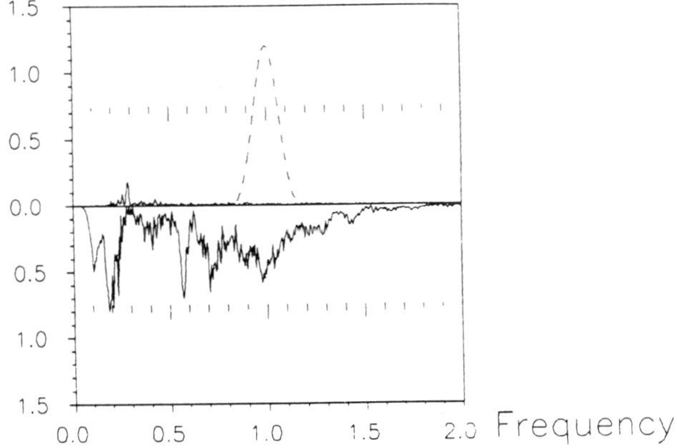


**Fig.13** Same as Fig.9 but for the plasma slab two times thinner.

the plasma wave by the pulse with the adiabatically slow rise of intensity should not be important. Still, the possibility of the pulse distortion could be connected with the backward SRS. The results of the numerical simulation for this case are shown in the Figs.14,15.

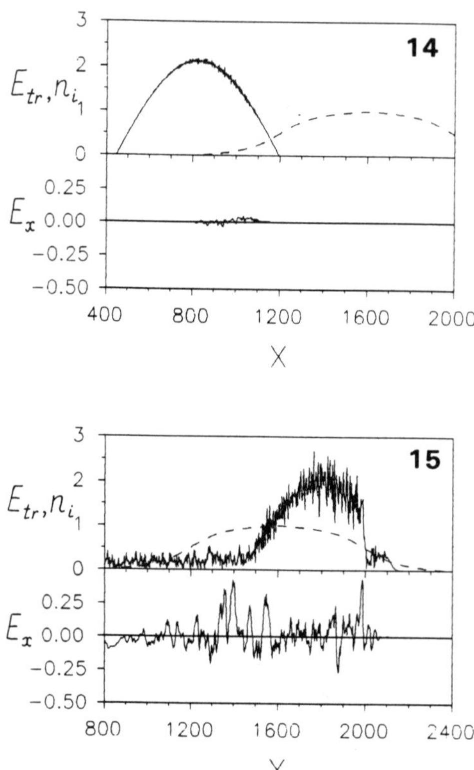


**Fig.14,15** Long laser pulse interaction with the plasma for $t = 400\omega_0^{-1}$ (Fig.14) and $t = 1400\omega_0^{-1}$ (Fig.15). The notations are the same as for top plots of Figs.4-7.

In the early time $t$=400 $\omega_0^{-1}$ (Fig.14) the laser pulse does not in fact excite fast plasma wave. The more precise analysis of the spectrum of the plasma electron density perturbations shows that in the region where the pulse amplitude is about unity ($|a| \simeq 1$) the density perturbations with the wave number two time greater than the wave number of the laser pulse exist. This fact pursue us that in this region the backward SRS develops. The same suggestion is as well supported by the presence of radiation with the frequency close to the initial one in the spectrum of the reflected radiation.

The depletion of the pulse energy in the region of $|a| \simeq 1$ due to the intense backward SRS results in the erosion of the leading front and formation of the jump of amplitude which could be seen in Fig.15 for $t$=1400 $\omega_0^{-1}$. For this moment no considerable electron density perturbations with $k_p \simeq 2k_0$ could be registered in a spectrum. It is probably since the width of the region of the maximum backward SRS growth rate has become small. Further on, in accordance with the previous consideration, the jump on the leading edge of the laser pulse can lead to the fast pulse depletion via the intense plasma wave excitation.

## VI. Conclusions and Discussion.

The interaction of ultra-intense laser radiation with underdense plasmas demonstrates a rich variety of the nonlinear phenomena. These include stimulated Raman scattering, process of self-modulation and absorption of the radiation energy through the plasma wave generation. Being responsible for the laser field evolution, all these processes are well understood only for fairly long laser pulses or for the ones of moderate intensity. We have turned to the very intense and at the same time ultrashort laser pulses to highlight new interesting features of the interaction and to study the relative role of different processes for this case. The most impressive effect among discussed in the present paper seems to be the extremely fast depletion of the pulse in the underdense plasma which is transparent for radiation of moderate intensity.

Our results have demonstrated that the steepening of the leading front of respectively long pulse ( which initially does not excite the plasma wave ) occurs and gives start to plasma wave excitation. The physical mechanism responsible for the formation of a jump on the laser pulse leading front and initiating the fast pulse depletion was identified as backward stimulated Raman scattering. Energy absorption via intense plasma wake excitation results in acceleration and heating of plasma electrons. For the shorter duration of laser pulses ( initially capable of the intense wakefield generation ) the backward Raman scattering does not manifest itself as a dominating process.

The frequency downshift of the laser field gives an opportunity for the spatially locked solitone-like transverse electromagnetic modes to appear. Being relativistically strong these modes have the same polarization as the initial one of the laser pulse and the frequency below the local plasma electron frequency.

We underline as well that even the plasma, which is fully transparent for the moderate radiation power, appears to be nontransparent for strongly relativistic pulse due to the processes considered in the present paper . In addition, the time scale of the absorption process is so short in comparison with that of ion motion that any significant ablation of the heated plasma hardly could be expected.

The nonlinear self-focusing [15,23] and the possibility of the optical guiding [6] of the laser pulse are out of the scope of the present paper though we believe them to be ones of primary importance in laser plasma interaction.

Further development of the theory of interaction of laser pulses with plasmas and solid targets for the pulses with very high power density and very short duration can enable us to extend further our understanding of the basic physics of the interaction of light with matter.

## References


1. Ebery J.H., Maine P. Strickland D., Mourou G. Laser Focus, N10, 84 (1987)
2. Maine P., Stricland D., Bado P., Pessot M., and Mourou G. IEEE J.Quan.Elec., **24**, 398 (1988)
3. Hutchinson M.H.R. Contemporary Phys., **30**, 355 (1989)
4. Tajima T., Dawson J.M. Phys. Rev. Letters, **34**, 267 (1979)
5. Gorbunov L.M., Kirsanov V.I. Zh. Eksp. Teor. Fiz., **93**, 509 (1987) (Sov. Phys. JETP, **66**, 290 (1987))
6. Esarey E., Ting A., Sprangle P., Joyce G. Comments on Plasma Phys., **12**, 291 (1989)
7. Bulanov S.V., Kirsanov V.I., Sakharov A.S. Pis'ma Zh. Eksp. Teor. Fiz., **50**, 176 (1989) (Sov. JETP Letters, **50**, 198 (1989))
8. Wilson E.J.N. Physica Scripta., **T30**, 69 (1990)
9. Chen F.F. Physica Scripta, **T30**, 24 (1990)
10. Katsouleas T. Phys. Rev. A, **33**, 2056 (1986)
11. Gorbunov L.M., Kirsanov V.I. Mtingva S.K. Kratk. Soobsh. po Fizike, N10, 27 (1989) (Sov. Phys. Lebedev Inst. Reps., N10, Allerton Press Inc., N.Y. (1989))

12. Rozenzweig J. IEEE Trans. on Plasma Sci., **PS-15**, 186 (1987)
13. Berezhiani V.I., Murusidze I.J. *"Proc. of the IV Int. Workshop on Nonlinear and Turbulent Processes in Physics, Kiev, USSR, Oct. 9-22, 1989"*. Kiev: Naukova Dumka, 235 (1989)
14. Sprangle P., Esarey E., Ting A. Phys. Rev. A., **41**, 4463 (1990)
15. Bulanov S.V., Kirsanov V.I., Sakharov A.S. Fiz. Plazmy, **16**, 935 (1990) (Sov. J. Plasma Phys., **16**, No.8 (1990))
16. Gorbunov L.M., Kirsanov V.I. *Nonlinear Theory of Strong Electromagnetic Wave-Plasma Interaction*, Proc. P.N.Lebedev Phys. Inst. of the USSR Ac. Sci., **219** (1991)
17. Kirsanov V.I. Kratk. Soobsh. po Fizike., N8, 36 (1987) (Sov. Phys. Lebedev Inst. Reps. N8, 49, Allerton Press Inc., N.Y. (1987))
18. Batha S.H., McKinstrie C.J. IEEE Trans. on Plasma Sci., **PS-15**, 131 (1987)
19. Horton W., Tajima T. Phys. Rev. A., **35**, 4110 (1986)
20. Kruer W.L. *The Physics of Laser PLasma Interaction*. Addison-Wesley (1988).
21. Kozlov V.A., Litvak A.G., Suvorov E.V. Zh. Eksp. Teor. Fiz., **76**, 148 (1979)
22. Yu M.Y., Shukla P.K., Tsintsadze N.L. Phys. Fluids, **25**, 1049 (1982)
23. Tsintsadze N.L. Physica Scripta, **T30**, 41 (1990)
24. Bulanov S.V., Kirsanov V.I., Sakharov A.S. Preprint of General Phys. Inst. of the USSR Acad. Sci., N86, Moscow, 1990

**Research Trends in Physics: Coherent Radiation Generation and Particle Acceleration**
Editorial Board: J.M. Buzzi, A. Prokhorov (Editor-in-Chief), P. Sprangle, and K. Wille
*La Jolla International School of Physics*, The Institute for Advanced Physics Studies, La Jolla, California

# Cyclotron Autoresonance MASER Amplifiers*

**B.G. Danly, G. Bekefi, C. Chen, A.C. DiRienzo,** W.L. Menninger, K.D. Pendergast,*** R.J. Temkin, and J.S. Wurtele**

Plasma Fusion Center
Massachusetts Institute of Technology
Cambridge, MA 02139

## ABSTRACT

Experimental and theoretical studies of cyclotron autoresonance maser (CARM) amplifiers are presented. The CARM, which has been shown theoretically to be capable of high efficiency and high power microwave generation with good rf phase stability, is a promising RF source for the next generation linear collider. Recently, an rf power of ~ 12 MW has been measured from a 35 GHz CARM amplifier energized by a MARX accelerator, corresponding to an electronic efficiency of ~ 6%. Detailed comparisons are made between the experimental results and a three-dimensional nonlinear theory that describes the self-consistent interaction of the electron beam and the electromagnetic wave. A 17 GHz CARM amplifier employing the MIT/SRL SNOMAD II linear induction accelerator is discussed.

* Research supported in part by the Department of Energy, Office of Basic Energy Sciences, under contract DE-FG02-89-ER14052, and in part by the Air Force Office of Scientific Research.

** Present Address: HQDA, OASA(RDA), Attn: SARD-DOV, Washington, D.C. 20310-0103.

*** Present Address: Institute of Defense Analysis, Washington, D.C.

## I. INTRODUCTION

Since its conception in the mid-70's,[1] the cyclotron autoresonance maser, or CARM, has received continued interest as a source of high power microwave and millimeter wave radiation. The CARM interaction occurs when a relativistic electron beam undergoing cyclotron motion in a uniform magnetic field $B_0\ \vec{e}_z$ interacts with a copropagating electromagnetic wave $(\omega,\ \vec{k})$. The cyclotron resonance condition is

$$\omega = k_z v_z + l\Omega_0/\gamma,$$

where $v_z$ and $\gamma$ are, respectively, the axial velocity and the relativistic mass factor of the beam electrons. The harmonic number is $l$, $\Omega_0 = eB_0/m_0c$ is the nonrelativistic cyclotron frequency, $m_0$ and $-e$ are the electron mass and charge, and $c$ is the speed of light in vacuum. Theoretical work on the CARM has concerned issues such as oscillator and amplifier nonlinear efficiency,[2,3] efficiency enhancement by tapering of the guide magnetic field and wave phase velocity,[4,5] multiple mode effects,[6,7] stability of the RF phase,[8] space-charge wave effects,[9,10] and stability of the amplifier to absolute instabilities.[11,12]

The experimental investigation of CARMs has also been carried out at a number of research centers. Several experiments were carried out in the early 1980's at the Institute of Applied Physics (IAP) in N. Novgorod, Russia.[13,14] More recently, experiments at IAP have produced radiation at 50 GHz with 10% efficiency.[15] Experiments in the U.S. which have observed CARM operation have been carried out at the Massachusetts Institute of Technology (MIT),[16–18] at the University of Michigan,[19,20] at the Naval Research Laboratory[21] and at LLNL.[22]

In this paper, the CARM theory is reviewed, and results obtained on the MIT CARM amplifier experiments are discussed. Our studies of the CARM amplifier are motivated by the need for high-power, high-frequency (10-40 GHz) RF sources for driving high-gradient RF accelerators. The MIT experiments are the first published CARM amplifier results. The measurements were carried out at 35 GHz using a relativistic electron beam (1.5 MeV). A linear gain of 50 dB/m was achieved with a saturated efficiency of 6.3% and power output of 12 MW. The status of our 17 GHz CARM amplifier experiments is reported.

## II. THEORY

In this section, we review the basic theory of the CARM. In particular, we discuss a linear kinetic theory and a three-dimensional, self-consistent, nonlinear theory of the CARM amplifier, based on a simple model which assumes a single transverse-electric (TE) mode in a cylindrical waveguide configuration. A general treatment of the CARM interaction which includes the coupling of the electron beam with an arbitrary number of TE and transverse-magnetic (TM) waveguide modes has been discussed elsewhere.[6,7]

For a CARM amplifier operating with a single $\mathrm{TE}_{mn}$ mode, the radiation field can be expressed in cgs units as

$$\vec{E}_t\,(r,\theta,z,t) = \frac{1}{2}A(z)\ \vec{e}_z \times \nabla_t J_m(k_\perp r) e^{i(m\theta-\omega t)} + \text{c.c.}\ , \qquad (1)$$

$$\vec{B}_t\,(r,\theta,z,t) = \frac{1}{2}\left(\frac{ic}{\omega}\right)\frac{dA(z)}{dz}\nabla_t J_m(k_\perp r)e^{i(m\theta-\omega t)} + \text{c.c.}\;, \tag{2}$$

$$B_z(r,\theta,z,t) = \frac{1}{2}\left(\frac{ick_\perp^2}{\omega}\right)A(z)J_m(k_\perp r)e^{i(m\theta-\omega t)} + \text{c.c.}\;, \tag{3}$$

where $\nabla_t = \vec{e}_r\,\partial/\partial r + (\vec{e}_\theta\,/r)\partial/\partial\theta$, $\omega = 2\pi f$ is the (angular) operating frequency, $A(z)$ describes the axial dependence of the mode, $J_m(x)$ is the first-kind Bessel function of order $m$, $k_\perp = \nu/r_w$ is the transverse wave number associated with the $\text{TE}_{mn}$ mode, $r_w$ is the waveguide radius, and $\nu$ is the $n$th zero of $J'_m(x) = dJ_m(x)/dx$. Let

$$A(z) = \frac{m_0c^2}{e}\left(\frac{\omega}{ck_\perp}\right)a(z)e^{i[k_z z+\delta(z)]}\;, \tag{4}$$

where $k_z = (\omega^2/c^2 - k_\perp^2)^{1/2}$ is the axial wave number of the vacuum $\text{TE}_{mn}$ mode. Both the wave phase shift $\delta(z)$ and the normalized wave amplitude $a(z)$ vary slowly with respect to the interaction length $z$ due to the CARM interaction.

## A. Linear Kinetic Theory

The linear regime of the CARM interaction has been investigated by several authors[3,6,7,23] using the Vlasov-Maxwell equations. The results of our analysis are summarized below for the equilibrium distribution function of the form

$$f_0(\vec{x},\vec{p}) = G(r_g)F(p_z,p_\perp)\;, \tag{5}$$

where $p_\perp = (p_x^2 + p_y^2)^{1/2}$, and $r_g$ is the electron guiding-center radius. This distribution function describes a class of azimuthally symmetric electron beam equilibria, provided that the equilibrium self-electric and self-magnetic fields are negligibly small.

Under circumstances in which the condition $k_\perp r_g \ll 1$ holds and there is no spread in either the electron energy or the electron axial momentum, it has been shown that to leading order in $c^2k_\perp^2/(\omega - l\Omega_0/\gamma - k_zv_z)^2$, the Laplace transform of the wave amplitude, denoted by $\hat{A}(s)$, is given by[6,7]

$$\left[s^2 - k_\perp^2 + \frac{\omega^2}{c^2} + \frac{\varepsilon k_\perp^2(\omega^2 + c^2s^2)}{(\omega - l\Omega_0/\gamma + iv_zs)^2}\right]\hat{A}(s) = \left[s + \frac{i\varepsilon k_\perp^2 v_z\omega}{(\omega - l\Omega_0/\gamma + iv_zs)^2}\right]A(0)\;, \tag{6}$$

where $dA(0)/dz = 0$ has been assumed, $l$ is the harmonic number, and $s = ik_z$ is the Laplace transform variable. The dimensionless coupling constant is defined by

$$\varepsilon = \frac{4\beta_\perp^2}{\gamma\beta_z}\left(\frac{I_b}{I_A}\right)\frac{X^2(r_L,r_g)}{(\nu^2 - m^2)J_m^2(\nu)}\;, \tag{7}$$

where $\gamma = (1 + |\,\vec{p}\,|^2/m_0^2c^2)^{1/2}$, $\beta_\perp = p_\perp/\gamma m_0c$, $\beta_z = p_z/\gamma m_0c$, $I_b$ is the beam current, $I_A = m_0c^3/e \simeq 17$ kA is the Alfvén current, $r_L = p_\perp/m_0\Omega_0$ is the electron Larmor radius, and

$$X(r_L,r_g) = J_{l-m}(k_\perp r_g)J'_l(k_\perp r_L) \tag{8}$$

is a geometric factor.

The linear gain in RF power can be obtained by the inverse Laplace transform of $\hat{A}(s)$, i.e.,

$$\frac{P(z)}{P(0)} = \left|\frac{A(z)}{A(0)}\right|^2 = \left|\frac{1}{2\pi i}\int_{\sigma-i\infty}^{\sigma+i\infty}\frac{\hat{A}(s)}{A(0)}e^{sz}ds\right|^2 . \tag{9}$$

This Laplace transform formalism enables us to evaluate launching losses which are important in predicting the maser performance. Note that the poles in the integrand in Eq. (9) correspond to the solutions of the dispersion equation

$$k_z^2 + k_\perp^2 - \frac{\omega^2}{c^2} = \frac{\varepsilon k_\perp^2(\omega^2 - c^2k_z^2)}{(\omega - l\Omega_0/\gamma - k_z v_z)^2} , \tag{10}$$

which has four $k_z$ roots for a real value of $\omega$.

## B. Nonlinear Theory

The system of equations describing the self-consistent evolution of the beam electrons and the wave amplitude and phase shift, $a(z)$ and $\delta(z)$, can be derived from the Maxwell equations and the Lorentz force equations. Within the eikonal approximations $|da/dz| \ll |k_z a|$ and $|d\delta/dz| \ll |k_z\delta|$ and the approximation of constant transverse guiding-center variables $r_g$ and $\theta_g$, the self-consistent CARM amplifier equations can be expressed in the following dimensionless form[6,7]

$$\frac{d\gamma}{d\hat{z}} = -\frac{\hat{p}_\perp}{\hat{p}_z}X(r_L, r_g)a\cos\psi , \tag{11}$$

$$\frac{d\hat{p}_z}{d\hat{z}} = -\frac{\hat{p}_\perp}{\hat{p}_z}X(r_L, r_g)\left[\left(\frac{1}{\beta_{ph}} + \frac{d\delta}{d\hat{z}}\right)a\cos\psi + \frac{da}{d\hat{z}}\sin\psi\right] , \tag{12}$$

$$\frac{d\psi}{d\hat{z}} = \frac{1}{\beta_{ph}} + \frac{d\delta}{d\hat{z}} - \frac{\gamma}{\hat{p}_z} + \frac{l\hat{\Omega}_0}{\hat{p}_z}$$
$$+ \frac{l}{\hat{p}_z\hat{p}_\perp}W(r_L, r_g)\left\{\left[\gamma - \hat{p}_z\left(\frac{1}{\beta_{ph}} + \frac{d\delta}{d\hat{z}}\right)\right]a\sin\psi + \hat{p}_z\frac{da}{d\hat{z}}\cos\psi\right\} , \tag{13}$$

$$\frac{da}{d\hat{z}} = g\left\langle X(r_L, r_g)\frac{\hat{p}_\perp}{\hat{p}_z}\cos\psi\right\rangle , \tag{14}$$

$$\frac{d\delta}{d\hat{z}} = -\frac{g}{a}\left\langle X(r_L, r_g)\frac{\hat{p}_\perp}{\hat{p}_z}\sin\psi\right\rangle , \tag{15}$$

where

$$\psi = k_z z + \delta(z) - \omega t + l\tan^{-1}(p_y/p_x) - (l - m)\theta_g + (l - 2m)\pi/2 , \tag{16}$$

$$g = \frac{4(\beta_{ph}^2 - 1)}{\beta_{ph}(\nu^2 - m^2)J_m^2(\nu)}\left(\frac{I_b}{I_A}\right) , \tag{17}$$

$$W(r_L, r_g) = lJ_{l-m}(k_\perp r_g)J_l(k_\perp r_L)/k_\perp r_L . \tag{18}$$

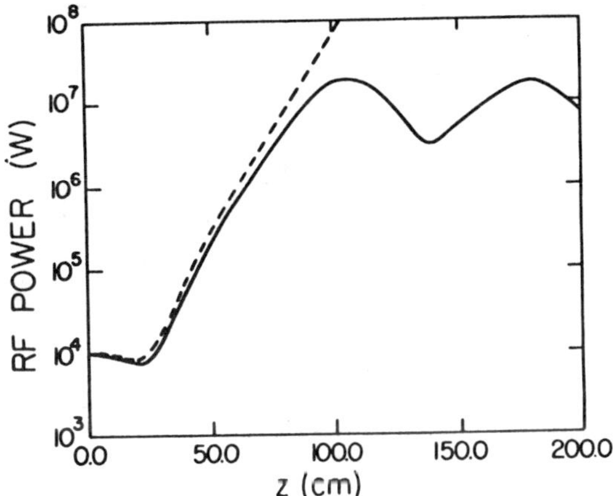


Figure 1: RF power as a function of the interaction length $z$, obtained from linear theory (dashed line) and computer simulation (solid line) with $\gamma = 3.96$, $I_b = 128$ A, $\Delta\gamma_z/\gamma_z = 0$, $\alpha = \beta_\perp/\beta_z = 0.27$, $r_g = 0$, $B_z = 5.62$ kG, and $f = 34.7$ GHz.

The normalized variables and parameters are defined by

$$\hat{z} = \omega z/c \ , \quad \hat{\Omega}_0 = \Omega_0/\omega \ , \quad \hat{p}_z = p_z/m_0c = \gamma\beta_z \ , \quad \hat{p}_\perp = p_\perp/m_0c = \gamma\beta_\perp \ , \quad \beta_{ph} = \omega/ck_z \ . \tag{19}$$

Finally, $\langle f \rangle = N^{-1}\sum_{i=1}^{N} f_i$ in Eqs. (14) and (15) denotes the ensemble averaging over the particle distribution, and typical number of macroparticles used in our simulations is $N = 1024$. The CARM amplifier equations (11)-(15) consist of $3N + 2$ first-order ordinary differential equations.

The RF power flow over the cross section of the waveguide is related to the normalized wave amplitude $a(z)$,

$$P(\hat{z}) = \frac{1}{8}\left(\frac{m_0^2c^5}{e^2}\right)\frac{\beta_{ph}(\nu^2 - m^2)J_m^2(\nu)}{(\beta_{ph}^2 - 1)}a^2(\hat{z}) \ , \tag{20}$$

where $m_0^2c^5/e^2 \simeq 8.7$ GW. It is readily shown from Eqs. (11), (14), and (20) that

$$P + \frac{I_b}{e}\langle\gamma\rangle m_0c^2 = \text{const.} \ , \tag{21}$$

corresponding to the conservation of total power flow through the waveguide.

We have developed a three-dimensional self-consistent code[4,7] for simulation studies of CARM amplifiers. The code solves a set of the CARM amplifier equations with multiple waveguide modes and has been benchmarked against our linear kinetic theory. It can model a single TE/TM mode, multiple TE and/or TM modes, cyclotron harmonics, magnetic field tapering, momentum and energy spread, waveguide losses, and various beam loading options. The results of the single-mode analysis are summarized in Figs. 1 and 2.

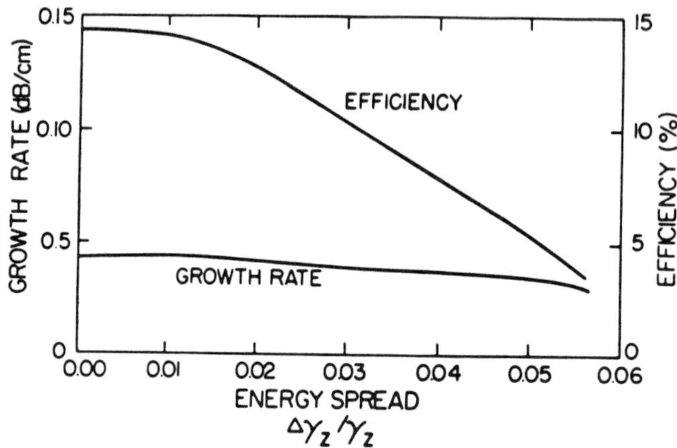


Figure 2: Linear growth rate and saturated efficiency as a function of (axial) energy spread. Here, $\gamma$ = 3.96, $I_b$ = 128 A, $\alpha$ = 0.27, $r_g$ = 0.16 cm, $B_z$ = 5.4 kG, and $f$ = 34.7 GHz.

Figure 1 depicts typical dependence of the RF power on the interaction length $z$ for the $TE_{11}$ mode obtained from linear theory discussed in Sec. II.A and the numerical simulation using Eqs. (11)-(15). The solid curve shows the simulation result while the dashed curve is obtained analytically from Eq. (9). It is evident that there is good agreement between simulation and theory in the linear regime.

Figure 2 illustrates the dependence of the saturated efficiency and the linear growth rate on energy spread $\Delta\gamma_z/\gamma_z$ as obtained from our simulation. When the energy spread is 0.044, typical of our CARM, the predicted efficiency is seen to be $\sim$ 6% and the growth rate $\sim$ 45 dB/m, values that are in good agreement with measurements described in Sec. III.A.

# III. EXPERIMENTS

## A. 35 GHz CARM Experiments

We have carried out extensive experimental and theoretical studies of a CARM amplifier at 35 GHz.[16,17] After describing briefly the experiments, we show the comparisons between experiment and theory. Detailed studies of the maser have been reported in our earlier papers.[16,17]

### Experimental Setup and Electron Beam Transport

A schematic of the CARM is shown in Fig. 3(a). The accelerating potential for the maser is supplied by a Marx generator (Physics International Pulserad 110A) capable of supplying a 1.5 MeV, 20 kA, 40 ns pulse to a 75 Ω matched load. The

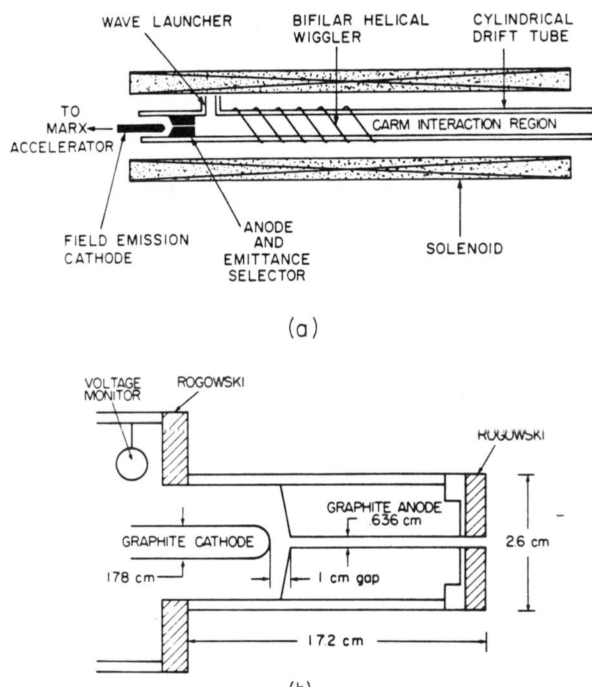


Figure 3: Experimental arrangement showing (a) the overall system, and (b) details of the electron gun.

electron beam is generated by a field emission (explosive emission) gun shown in Fig. 3(b) composed of a hemispherical graphite cathode and a conical anode[24] which also acts as an emittance selector. The entire 2 m long system (including the gun) is immersed in a uniform axial magnetic field $B_0$ of up to 8 kG. The axial magnetic field is generated by a solenoid energized by a capacitor bank that delivers a sinusoidal current pulse of $\sim$ 16 ms duration.

A measurement of the transmitted current $I_b$ as a function of the axial magnetic field $B_0$ at the emittance selector permits a determination of the normalized beam brightness $B_n$ and the normalized rms beam emittance $\varepsilon_n$ (rms) by means of the relations,[25]

$$B_n = \frac{\pi^2 I_b}{(\gamma_z \beta_z)^2 (\delta^4 V)} = \frac{2 I_b}{9 \varepsilon_n^2 (rms)} \ . \tag{22}$$

Here, the phase space volume $\delta^4 V$ is given by

$$\delta^4 V = \pi^2 b^4 \Omega_0^2 / 6 c^2 \gamma^2 \ , \tag{23}$$

with $b$ as the radius of the emittance selector. Equations (22) and (23) apply at sufficiently low magnetic fields for which $r_L \leq b/2$, where $r_L$ is the election Larmor radius. In this regime of magnetic fields, Eqs. (22) and (23) show that for a constant brightness gun, the transmitted current $I_b$ is proportional to $B_z^2$, a result which is in agreement with our measurements.

Table I illustrates the results of such measurements for the case of emittance selectors having radii of 0.076 cm and 0.318 cm, respectively. In addition to $B_n$ and $\varepsilon_n$ (rms) the table also lists the normalized axial momentum spread $\sigma_{pz} = \Delta p_z/m_0c$ and the corresponding energy spread $\Delta\gamma_z/\gamma_z$ of our beam, as derived from $\varepsilon_n$. It is noteworthy that the smaller radius beam is of high quality, albeit of low current. Being desirous of high power output from the CARM, we chose the higher current and larger emittance selector, with the poorer beam quality in all the measurements described.

Downstream from the emittance selector, a bifilar helical magnetic wiggler [see Fig. 3(a)] imparts transverse velocity $v_\perp$ to the electron beam. The wiggler consists of current carrying bifilar helical windings and is energized by current from a capacitor bank with time constant $\sim$ 100 $\mu s$. The wiggler field is gradually tapered from zero field to its desired maximum value and then abruptly terminated by a copper shorting ring. The tapering is achieved[26,27] by means of nichrome wire resistive rings, two per period, and extending over the full six-period wiggler length. The measured field profile is illustrated in Fig. 4(a). The wiggler characteristics are summarized in Table II. We have used two wigglers of different periodicity. This allowed us to spin up the beam in the so-called Group-I and Group-II regimes and thereby compare their merits.

Prior to taking radiation measurements, an axially moveable current collecting probe is used to measure the electron beam current at any axial position $z$ in the wiggler and CARM interaction regions. A 30% current loss is observed as the beam traverses the wave launcher and wiggler regions (see Fig. 3). However, no significant current loss occurs within the CARM interaction region itself.

A piece of thermal paper attached to the current collector acts as a witness plate. It is used in two ways: (1) align the beam concentrically in the evacuated cylindrical drift tube and solenoid; (2) observe the helical precession of the beam when the magnetic wiggler field is turned on. Figure 4(b) illustrates the burn marks produced by the beam as the probe is moved axially within the wiggler field. The radial excursion of the beam from the axis is a measure of the transverse velocity imparted by the wiggler; typically, $\alpha \equiv v_\perp/v_z \simeq 0.3$ for our experiments.

Table I. Electron Beam Characteristics

| Beam Radius $r_b$ (cm) | 0.076 | 0.318 |
|---|---|---|
| Current $I_b$ (A) | 5.0 | 195 |
| Normalized Brightness $B_n$ ($Acm^{-2}rad^{-2}$) | $3.2 \times 10^4$ | $5 \times 10^3$ |
| Normalized Emittance $\varepsilon_n$ (cm - rad) | $5.9 \times 10^{-3}$ | $9.4 \times 10^{-2}$ |
| $\Delta p_z/m_0c$ | $1.6 \times 10^{-3}$ | $2.4 \times 10^{-2}$ |
| $\Delta\gamma_z/\gamma_z$ | 0.003 | 0.044 |

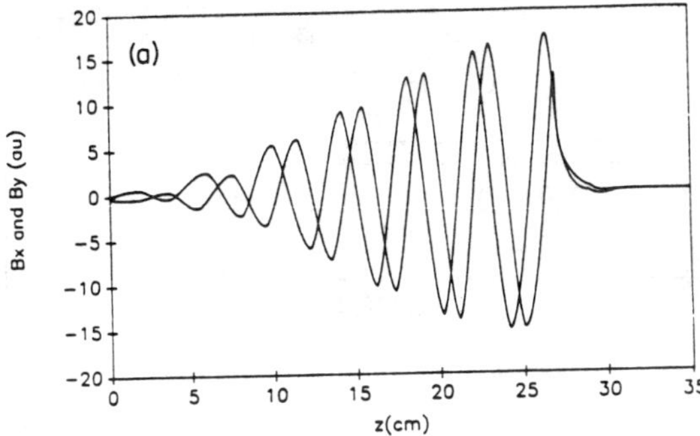


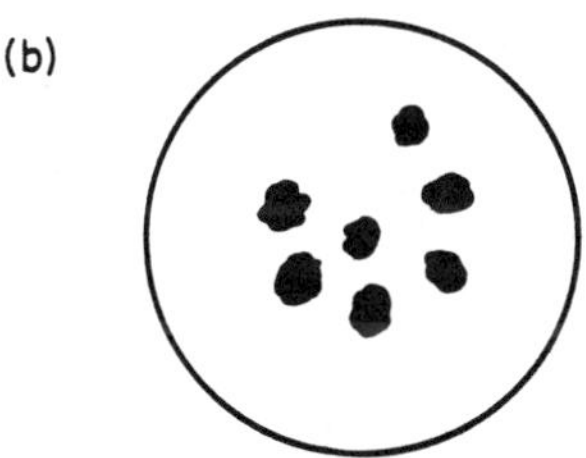


Figure 4: (a) Measurements of the wiggler field strength as a function of axial position. (b) Marks left on an axially movable witness plate within the wiggler, showing coherent beam precession. (The central spot is for the case when the wiggler field is tuned off.)

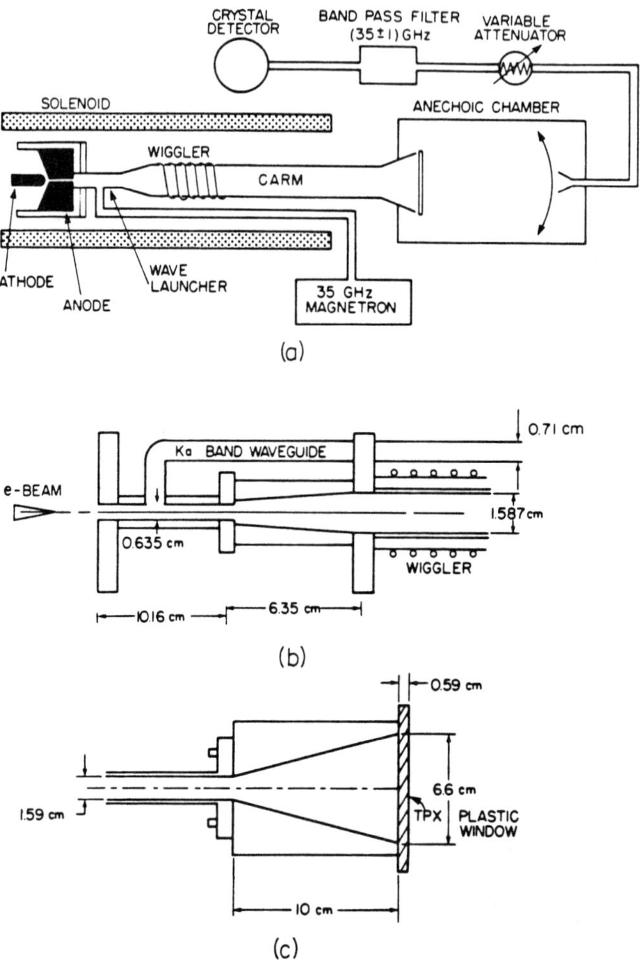


Figure 5: The 35 GHz wave system showing (a) the overall view, (b) details of the wave launcher, and (c) details of the wave transmitter.

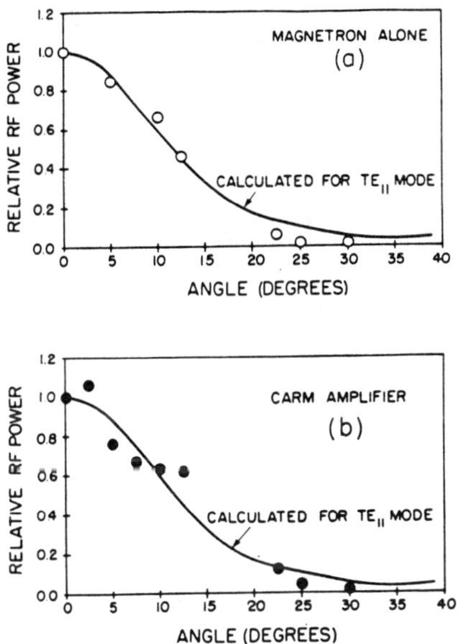


Figure 6: The measured far field radiation pattern for (a) the magnetron alone and (b) for the operating CARM amplifier.

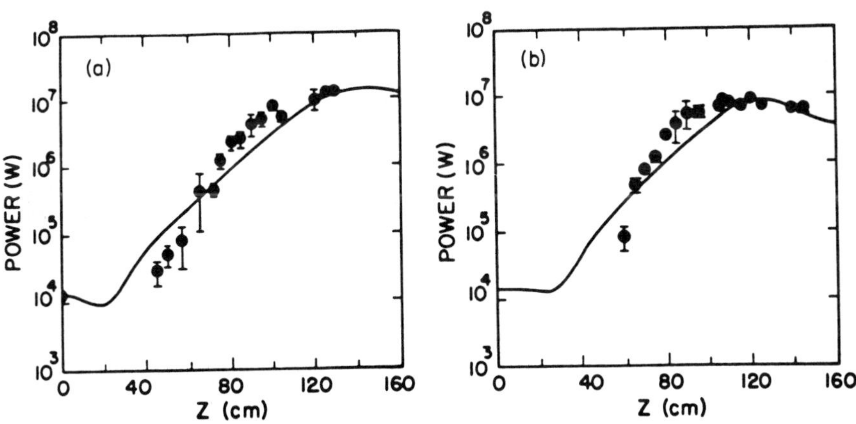


Figure 7: Comparison of the measured and computed RF output power as a function of length of the CARM interaction region for (a) the type I wiggler and (b) the type II wiggler.

The transport code, TRAJ,[28] has been used to study the particle dynamics in the field configuration consisting of an uniform axial guide field and a tapered wiggler field. In the numerical computations, the Biot-Savart law is used to generate the magnetic fields from the actual currents in the windings. The beam is treated as a group of up to 1024 macroparticles with a various possible initial phase space distributions. In this manner one obtains information about the average perpendicular energy of the macroparticles as defined by the parameter $\alpha = \beta_\perp/\beta_z$, the average pitch angle spread specified in terms of $\Delta\gamma_z/\gamma_z$, and the beam profile at any position $z$ within the wiggler or CARM interaction regions. The simulation shows that beam body rotation is minimal, and that the average electron Larmor radius is $r_L = 0.35$ cm. There is an undesirable guiding-center offset ($r_g = 0.16$ cm), which requires a slight repositioning of the cylindrical waveguide containing the beam.

## Amplifier Studies

The stainless steel drift tube has an internal radius of 0.79 cm and acts as a cylindrical waveguide whose fundamental $TE_{11}$ mode has a cutoff frequency of 11.16 GHz. A schematic diagram of the entire transport and detection system is given in Fig. 5(a).

A high power magnetron (~ 50 kW) operating at 34.73 GHz is the input power source for the CARM amplifier. The launcher [see Fig. 5(b)] consists of a section of circular waveguide of radius 0.32 cm into which the RF power is coupled from a standard Ka-band rectangular waveguide. This section of circular waveguide supports only the fundamental $TE_{11}$ mode at our operating frequency. Its radius is then adiabatically uptapered to the radius of the drift tube. A linearly polarized wave is thereby injected into the interaction region.

The output power from the CARM is sent by means of a conical horn [see Fig. 5(c)] into a reflection free "anechoic chamber". The vacuum interface between the transmitting horn and chamber is provided by a TPX plastic window[29] whose power reflectivity is measured to be $10^{-3}$. A small fraction of the radiated power is then collected by a receiving rectangular horn placed in the far (Fraunhoffer) field of the transmitter. Subsequently, the power is further reduced by means of precision calibrated attenuators and injected into a narrow band pass ($\pm 0.75$ GHz) filter. The power level is finally determined from the response of a calibrated crystal detector. The absolute CARM output power is obtained by a substitution method: the wiggler is turned off and the transmitted power from the device (in the absence of the CARM amplification process) is determined in terms of the known input power from the magnetron. The measurement is then repeated with the CARM interaction in place.

In order to assure ourselves that the radiation is predominantly in the $TE_{11}$ mode, the far field radiation pattern of the conical horn has been determined both in the absence and presence of the CARM interaction. The results of these measurements are illustrated in Fig. 6. The reasonably good agreement with a Kirchhoff-type diffraction theory[30] shown by the solid curves suggests that it is mostly the fundamental $TE_{11}$ mode that is being excited both in the absence and presence of the CARM interaction. Our measuring technique is not good enough to detect small admixtures of higher modes, if present.

The spatial growth rate of the electromagnetic wave is determined from the measurement of the output power as a function of the length of the interaction region. This length is varied by changing the distance that the electron beam is allowed to propagate in the drift tube. The application of a strong transverse magnetic field generated by a movable kicker magnet[26] is sufficient to deflect the electrons into the waveguide wall, and thus terminate the interaction at that point.

The RF launcher injects linearly polarized electromagnetic radiation, half of which, because of the wrong rotation of the RF electric field, does not participate in the CARM interaction. The remaining circularly polarized wave with the correct rotation is amplified and eventually emanates from the conical horn as circularly polarized radiation. This has been verified by rotating the pick up antenna 90° relative to the direction of polarization of the incident wave from the magnetron.

Table II. Magnetic Wiggler Characteristics

| Wiggler Type | Group I | Group II |
|---|---|---|
| Wiggler period (cm) | 4.06 | 7.00 |
| Wiggler length (cm) | 27 | 44 |
| Field strength $B_w$ (G) | 840 | 490 |
| Mean $\langle\alpha\rangle = \langle\beta_\perp/\beta_z\rangle$ (calculated) | 0.27 | 0.30 |

Table III. CARM Characteristics

| Wiggler Type | Group I | Group II |
|---|---|---|
| Beam Energy ($\gamma$) | 3.94 | 3.94 |
| Beam Current (A) | 128 | 128 |
| Axial magnetic field $B_0$ (kG) | 5.4 | 6.1 |
| RF frequency (GHz) | 34.73 | 34.73 |
| RF input power (kW) | 17 | 18 |
| RF output power (MW) | 12.2 | 8.7 |
| Saturated efficiency (%) | 6.3 | 4.5 |
| Linear growth rate (dB/m) | 50 | 62 |

The radiation output at the fixed magnetron input frequency of 34.73 GHz and fixed input power of 18 kW is optimized by varying both the guide and wiggler fields. Once the optimal values of $B_0$ and $B_w$ are determined, the radiation intensity is then measured as a function of the CARM interaction length $z$. This is done by a movable magnetic kicker magnet. The results of these measurements are depicted in Fig. 7 for both wigglers used (see Table II). The solid lines are from simulation using the single-mode version of the nonlinear model described by Eqs. (11)-(14). The overall agreement between experiment and theory is good.

## B. 17 GHz Induction-Linac-Driven CARM Amplifier

At the MIT Plasma Fusion Center a 17 GHz CARM amplifier which will utilize a 500 kV linear induction accelerator (LIA) is currently under investigation.[31,32] One of the principle applications of CARM amplifiers may be for driving future RF linear

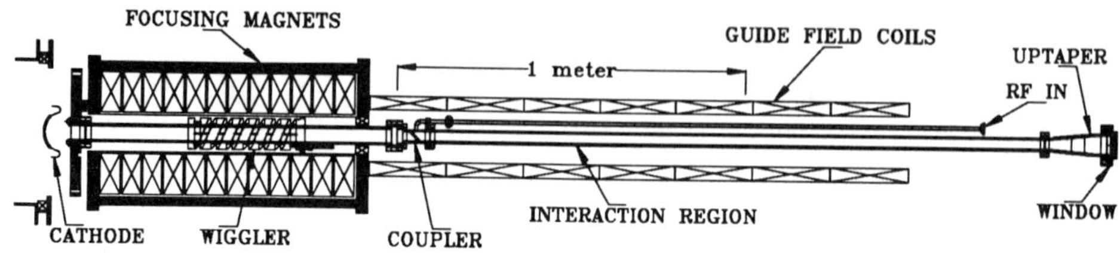


Figure 8: Schematic of induction-linac-driven 17 GHz CARM amplifier.

colliders. In future colliders, higher frequency (11.4- 35 GHz) structures will be required in order to achieve higher accelerating gradients without RF breakdown.[33,34]

At the present time, many different sources are under consideration for use in this capacity. The principle requirements for such a source are high gain, high efficiency, and good phase and power stability at high peak powers. The cyclotron autoresonance maser is a potentially promising candidate for the application of driving the next generation linear colliders.

The 17 GHz CARM amplifier experiment at MIT has been designed to run in the $TE_{11}$ mode. A schematic of the experiment is shown in Fig. 8. A three period bifilar helical wiggler with a wiggler wavelength of 9.21 cm and a field of up to 50 G will be used to spin-up the electron beam. Other parameters from the experiment are listed in Table IV. Initial operation will be carried out without magnetic guide field tapering. Improvements in the device efficiency should be realizable with the addition of tapering.

Fig. 9 shows how the efficiency of the CARM amplifier varies theoretically with detuning. This detuning is defined by $\Delta \equiv 2(1-\beta_{z0}/\beta_{ph})/(\beta_{\perp 0}^2(1-\beta_{ph}^{-2}))(1-\beta_{z0}/\beta_{ph}-\Omega_c/\gamma\omega)$. Different curves in the figure represent different values of coupling and of waveguide radius. The value of $\varepsilon/\varepsilon_c$ indicates how close the interaction is to the theoretical threshold for excitation of an absolute instability, with $\varepsilon/\varepsilon_c = 1$ corresponding to this threshold. In these expressions, the beam-wave coupling is defined in Eq. (7). The critical coupling which leads to absolute instability in an infinitely long system ($\varepsilon_c$) has been derived by Davies.[12]

For $\varepsilon/\varepsilon_c > 1$, the interaction is further prone to excite an absolute instability. Because of the short pulse length of the induction linac, it is unclear how much of a problem absolute instabilities will pose. The convective instability may still be able to dominate the absolute instability for values of $\varepsilon/\varepsilon_c > 1$. However, the CARM amplifier design is based on operation for values of $\varepsilon/\varepsilon_c$ just below 1. With such a coupling value, it is clear from Fig. 9 that higher detunings yield higher efficiencies, however Fig. 9 does not plot efficiencies for $\Delta > 0.4$ because the efficiency falls off rapidly.

An important consideration for any RF source to be used as a driver for linear colliders is that of phase and amplitude stability. We have demonstrated through simulations that a CARM amplifier that utilizes a bifilar helical wiggler to increase the pitch of the electron beam can be optimized to have excellent phase stability.[8] By setting the wiggler guide field and wiggle field to the right values, the correlation between $\gamma_0$ and $\beta_{\perp 0}$ runs exactly tangent to the zero phase shift curve for the CARM

| **Parameter** | **Design Value** |
|---|---|
| Beam Energy | 500 keV |
| Beam Current, $I_b$ | 500 A |
| Pulse Length | 30 ns |
| Beam Pitch, $\alpha_0 \equiv \beta_{\perp 0}/\beta_{z0}$ | 0.4 |
| Frequency, $\omega/2\pi$ | 17.136 GHz |
| Mode | $TE_{11}$ |
| Waveguide Radius, $r_w$ | 1.3 cm |
| Phase Velocity, $\beta_{ph}$ | 1.088 |
| Guide Field, $B_0$ | 3.06 kG |
| Detuning, $\Delta$ | 0.4 |
| Input Power, $P_{in}$ | 800 W |
| Est. Velocity Spread, $\sigma_{pz}/p_z$ | $< 1.6\%$ |
| Energy Spread, $\sigma_\gamma/\gamma$ | $< 1.6\%$ |
| Efficiency, $\eta$, untapered | 13.5% ($\sigma_{pz} = 0$) |
| | 9.3% ($\sigma_{pz} = 0.02$) |
| Output Power, $P_{sat}$ | 33.6 MW ($\sigma_{pz} = 0$) |
| | 23.3 MW ($\sigma_{pz} = 0.02$) |
| Saturation Length, $z_{sat}$ | 0.93 m ($\sigma_{pz} = 0$) |
| | 1.01 m ($\sigma_{pz} = 0.02$) |
| Gain | 46.2 dB ($\sigma_{pz} = 0$) |
| | 44.6 dB ($\sigma_{pz} = 0.02$) |

Table IV. 17 GHz CARM amplifier design parameters.

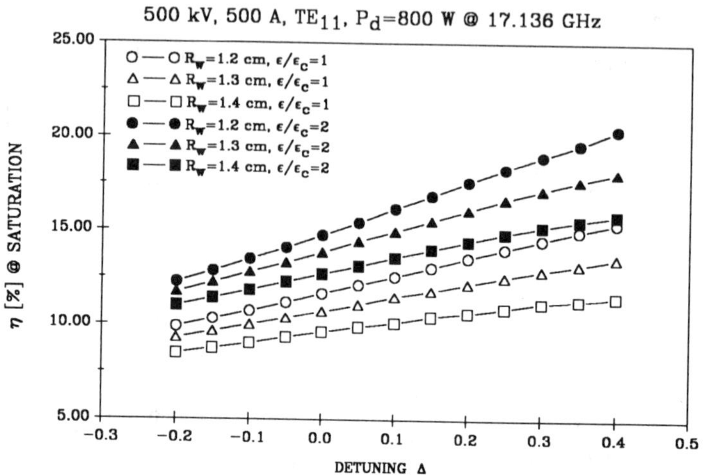


Figure 9: Theoretical CARM amplifier efficiency versus detuning $\Delta$.

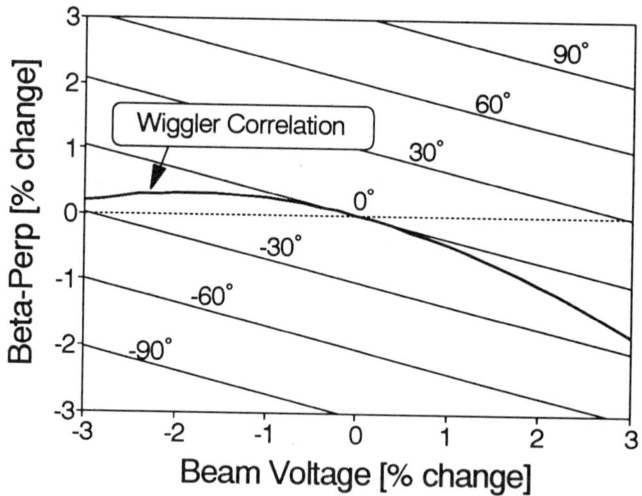


Figure 10: CARM amplifier constant phase curves over beam pitch and beam voltage. The optimal wiggler correlation is superimposed.

interaction, as shown in Fig. 10. This dramatically enhances the phase stability of the CARM amplifier. For the design parameters of Table IV, a phase stability of $\pm 0.8°$ is predicted over a voltage variation of $\pm 1\%$. This predicted phase stability remains to be verified experimentally.

## IV. CONCLUSIONS

The theory of CARM amplifiers has been reviewed, including the linear and nonlinear theory. A CARM amplifier operating at 35 GHz has been described, and an induction-linac-based CARM amplifier at 17 GHz has been discussed. The nonlinear efficiency of CARM amplifiers, while theoretically as high as 40–50% with magnetic field tapering,[4,5] remains to be demonstrated experimentally. Experimental efficiencies must be improved in order for CARMs to be considered as a viable candidate for an RF linear accelerator driver. Experimental issues for which significant work remains to be done include the improvement of beam quality and amplifier stability. These issues should not be problems in the long term for CARM development.

## References


[1]M.I. Petelin. On the theory of ultrarelativistic cyclotron self-resonance masers. *Radiophys. Quantum Elect.*, 17:686–690, 1974.

[2]V.L. Bratman, N.S. Ginzburg, G.S. Nusinovich, M.I. Petelin, and P.S. Strelkov. Relativistic gyrotrons and cyclotron autoresonance masers. *Int. J. Electron.*, 51:541–567, 1981.

[3]A.W. Fliflet. Linear and nonlinear theory of the Doppler shifted cyclotron resonance maser based on TE and TM waveguide modes. *Int. J. Electron.*, 61:1049, 1986.

[4]K.D. Pendergast, B.G. Danly, R.J. Temkin, and J.S. Wurtele. Self-consistent simulation of cyclotron autoresonance maser amplifiers. *IEEE Trans. Plasma Sci.*, PS-16:122–128, 1988.

[5]C. Chen and J.S. Wurtele. Efficiency enhancement in cyclotron autoresonance maser amplifiers by magnetic field tapering. *Phys. Rev. A*, 40:489, 1989.

[6]C. Chen and J.S. Wurtele. Multimode interactions in cyclotron autoresonance maser amplifiers. *Phys. Rev. Lett.*, 65:3389–3392, 1990.

[7]C. Chen and J.S. Wurtele. Linear and nonlinear theory of cyclotron autoresonance masers with multiple waveguide modes. *Phys. Fluids B*, B3, 1991.

[8]W.L. Menninger, B.G. Danly, and R.J. Temkin. Phase stability of cyclotron autoresonance maser amplifiers. 1992.

[9]A. Fruchtman and L. Friedland. Theory of a nonwiggler collective free-electron laser in uniform magnetic field. *IEEE J. Quantum Electron.*, 19, 1983.

[10]C. Chen, B.G. Danly, G. Shvets, and J.S. Wurtele. Effect of longitudinal space-charge waves of a helical relativistic electron beam on the cyclotron maser instability. *IEEE Trans. Plasma Sci.*, 1992. In Press.

[11]A.T. Lin, K.R. Chu, and A. Bromborsky. Stability and tunability of a CARM amplifier. *IEEE Trans. Electron Dev.*, 34:2621–2624, 1987.

[12]J.A. Davies. Conditions for absolute instability in cyclotron resonance maser. *Phys. Fluids*, B1:663–669, 1989.

[13]I.E. Botvinnik, V.L. Bratman, A.B. Volkov, N.S. Ginzburg, G.G. Denisov, B.D. Kol'chugin, M.M. Ofitserov, and M.I. Petelin. Free electron masers with distributed feedback. *JETP Lett.*, 35:516, 1982.

[14]I.E. Botvinnik, V.L. Bratman, A.B. Volkov, G.G. Denisov, B.D. Kol'chugin, and M.M. Ofitserov. The cyclotron autoresonance maser operated at a wavelength 2.4 mm. *Sov. Phys. Tech. Phys. Lett.*, 8:596, 1982.

[15]V.L. Bratman. Cyclotron autoresonance masers. In *Proceedings of the Workshop on Strong Microwaves in Plasmas, Suzdal, USSR*, Institute of Applied Physics, Gorky, 1990, 1990.

[16]G. Bekefi, A.C. DiRienzo, C. Leibovitch, and B.G. Danly. A 35 GHz cyclotron autoresonance maser (CARM) amplifier. *Appl. Phys. Lett.*, 54:1302–1304, 1989.

[17]A.C. DiRienzo, G. Bekefi, C. Chen, and J.S. Wurtele. Experimental and theoretical studies of a 35 GHz cyclotron autoresonance maser amplifier. *Phys. Fluids B*, 3:1755–1765, 1991.

[18]K.D. Pendergast, B.G. Danly, W.L. Menninger, and R.J. Temkin. A long-pulse CARM oscillator experiment. *Int. J. Electron.*, 1991. Accepted for Publication.

[19]J.G. Wang, R.M. Gilgenbach, J.J. Choi, C.A. Outten, and T.A. Spencer. Frequency-tunable, high-power microwave emission from cyclotron autoresonance maser oscillation and gyrotron interactions. *IEEE Trans. Plasma Sci.*, 17:906–908, 1989.

[20]J.J. Choi, R.M. Gilgenbach, and T.A. Spencer. High-Q bragg resonator CARM experiments on a long-pulse electron beam accelerator. In *Proc. Fifteenth International Conf. on Infrared and Millimeter Waves*, pages 143–145, SPIE, 1990.

[21]S.H. Gold and *et al.* High-voltage millimeter-wave gyro-traveling-wave-amplifier. *J. Appl. Phys.*, 69, 1991.

[22]B. Kulke, M. Caplan, D. Bubp, T. Houck, D. Rogers, D. Trimble, R. VanMaren, G. Westenskow, D.B. McDermott, N.C. Luhmann, and B.G. Danly. Test results from the LLNL 250 GHz CARM experiment. In *Proceedings of the 1991 Particle Accelerator Conference*, IEEE, 1991. Cat. No. 91CH3038-7.

[23]K.R. Chu and J.L. Hirshfield. Comparative study of the axial and azimuthal bunching mechanisms in electromagnetic cyclotron instabilities. *Phys. Fluids*, 21:461, 1978.

[24]R.H. Jackson and C.A. Sedlak. *Gyrotron Beam Generation With Helical Magnetic Fields*. Technical Report, Mission Research Corporation, 1983. MRC/WDC-R-061.

[25]This technique is described by D. Prosnitz and E.T. Scharlemann, Lawrence Livermore National Laboratory, ATA Note No. 229, Feb. 22, 1984.

[26]J. Fajans. *J. Appl. Phys.*, 55:43, 1984.

[27]J. Fajans, G. Bekefi, Y.Z. Yiu, and B. Lax. *Phys. Fluids*, 28:1995, 1985.

[28]K.D. Pendergast. *Theoretical and Experimental Research on a High-Power, High-Frequency Cyclotron Autoresonance Maser*. PhD thesis, MIT, 1991.

[29]A.C. DiRienzo. PhD thesis, MIT, 1991.

[30]M.N. Afsar. *IEEE Trans. Plasma Sci.*, 32:1598, 1984.

[31]D.L. Goodman, D.L. Birx, and B.G. Danly. Induction linac driven relativistic klystron and cyclotron autoresonance maser experiments. In H.E. Brandt, editor, *Intense Microwave and Particle Beams II*, pages 217–225, SPIE, 1991.

[32]W.L. Menninger, B.G. Danly, C. Chen, K.D. Pendergast, R.J. Temkin, D.L. Goodman, and D. Birx. Cyclotron autoresonance maser (CARM) amplifiers for RF accelerator drivers. In *Proceedings of the 1991 Particle Accelerator Conference*, IEEE, 1991.

[33]R.B. Palmer. *The Interdependence of Parameters for TeV Linear Colliders*. Technical Report SLAC-PUB-4295, Stanford Linear Accelerator Center, 1987.

[34]T.G. Lee, private communication, June, 1988.

**Research Trends in Physics: Coherent Radiation Generation and Particle Acceleration**

*La Jolla International School of Physics*, The Institute for Advanced Physics Studies, La Jolla, California

# Parameterizing the Free Electron Laser Physics

**G. Dattoli, H. Fang, L. Giannessi, L. Mezi, A. Torre, and R. Caloi**

ENEA, AREA INN, Centro di Frascati
00044 Frascati (Rome), Italy

*Abstract*

The Free Electron Laser is a rather complicated device, its performance depends indeed on a large number of parameters and on the interplay between them. In this note we show that an accurate choice of such parameters along with a suitable combination of them may be helpful to derive scaling laws relevant to e.g. the gain or output power. Such a view allows a kind of unitary description of the various effects occurring in Free Electron Laser physics. In particular, we show that the gain depression induced e.g. by the electron-beam energy spread is the same as that due to a finite width of the optical pulse.

(*) ENEA Guest

## 1. Introduction.

In a Free Electron Laser (FEL) a relativistic beam of electrons passing through the periodic magnetic field of an undulator couples to and amplifies a copropagating optical wave (fig. 1). The emitted wavelength is related to the undulator period $\lambda_U$ and to the electron energy $\gamma$ according to the approximate expression

$$\lambda_0 \cong \frac{\lambda_U}{2\gamma^2}, \tag{1.1}$$

the Lorentz contraction and the Doppler shift providing the frequency conversion factor $2\gamma^2$.

As is well known, the FEL can be configured as an amplifier for a signal, provided by an external source, or as an oscillator, if the radiation, spontaneously emitted by the electrons during their *wiggling* motion along the undulator, is stored in a resonator, thus giving rise to the necessary feedback mechanism.

The present paper refers to FELs operating with magnetic undulators, although the term *free electron laser* is presently used in a wider acceptation, denoting a great variety of devices, all involving *free electrons*, producing coherent radiation from the millimeter to the ultraviolet range.

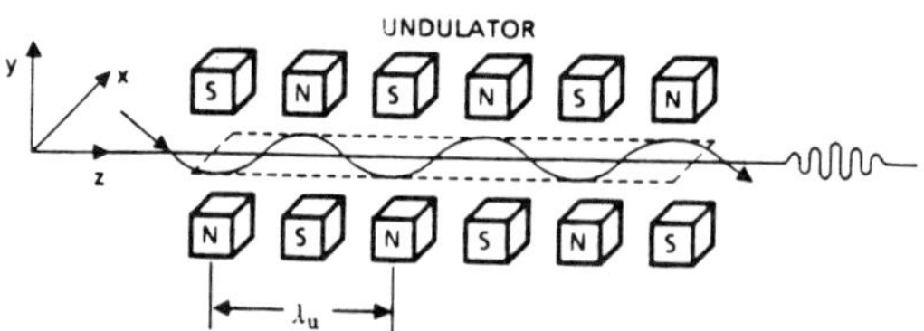


Fig. 1. Basic scheme of a Free Electron Laser.

Up to the present time about ten FELs have been successfully operated in the short wavelength region, as shown in fig. 2, where the electron energy is reported versus the operating wavelength [1]. As explicitly indicated, several types of accelerators have been used, including radio-frequency linear accelerators, storage rings, microtrons, induction linacs and electrostatic accelerators, the specific machine being chosen according to the wavelength of interest and to the requests on the characteristics of the e-beam.

Since the original demonstration of the possibility of operating a FEL as both amplifier and oscillator, realized by Madey and coworkers at Stanford University in 1977 [2], the performance of FELs has expanded in all directions. The shortest wavelength (240 nm) was recently achieved with the FEL operating on the storage ring VEPP-3 at Novosibirsk [3]. Correspondingly, the highest power (~ 1 GW) and gain (up to 40 dB) have been obtained at a very long wavelength (9 mm) with the induction linac ETA at Livermore [4]. In between lie a great variety of devices of intermediate power and wavelength. Furthermore, other devices are currently under design and construction.

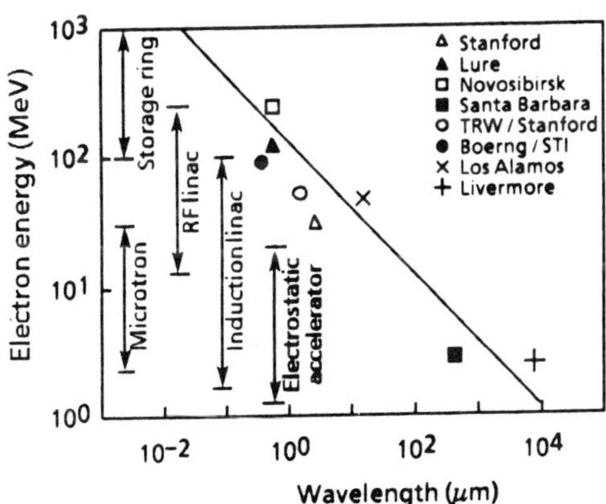


Fig. 2. FEL operating wavelentgh vs electron energy.

However, instead of the above quoted successful operations, FELs remain subtle and expensive devices, demanding for a very accurate design. In fact, FEL relies upon the combined action of the undulator magnetic field and of the electric and magnetic components of the optical field, which propagates *together* with the electron beam. The undulator magnetic field alone does not affect the electron energy, but forces the electron to follow a wiggle path, giving it a transverse velocity. Correspondently, the radiation field alone has no significant effect on either the electron trajectory or energy, since for relativistic motion the forces due to the electric and magnetic components nearly cancel. In combination, however, the undulator magnetic field guides the electrons through a transverse path, thus allowing the coupling of the electrons to the optical field.

It is therefore evident that the performance of the laser depends on the characteristics of the undulator, e.m. field and electron beam.

A brief discussion aimed at specifying the parameters, which influence the laser performance, as regards the undulator, the e.m. field and the electron beam, is presented in Sec. 2. Sec. 3 is devoted at briefly illustrating the results of the single-mode theory. The effect of the electron beam quality on the laser performance is discussed in Sec. 5, after introducing (Sec. 4) the so called $\mu$-parameters, accounting for the e-beam quality and emittances. Scaling laws relating the FEL gain in the small-gain weak-field regime are presented and commented. Deviations from the linear theory with increasing the gain coefficient are then discussed in Sec. 6, where simple and reliable formulae are given for the FEL gain versus the gain coefficient and $\mu$-parameters. The longitudinal dynamics is discussed in Sec. 7 within the context of the small signal regime. Accordingly, it is shown that an appropriate analytical approach to the FEL wave equation, ruling the pulse propagation in the weak field regime, allows to individuate specific parameters, which shortly account for the slippage mechanism and subsequent lethargy. Finally, a brief discussion of the transverse effects on the laser operation closes the paper.

## 2. Undulator, e.m. field and e-beam parameters.

The undulator parameters, which influence the FEL performance, can be summarized into

1) the undulator period $\lambda_U$,
2) the peak field strength $B_0$,

and

3) the undulator length $L_U$, or equivalently the number of undulator periods: $L_U = N\lambda_U$.

Typically, $B_0$ is several KG: $B_0 \sim 2-7KG$. The undulator period is usually in the range of a few centimeters: $\lambda_U \sim 2-10cm$, whereas the undulator length extends for a few meters: $L_U \sim 1-10m$, so that $N$ can vary from 20 to several hundred.

The undulators can be configured with a helical or linear polarization. In the first case, the direction of the magnetic field revolves around the undulator axis. On the other hand, in a linear undulator, the direction of the magnetic field changes sinusoidally, remaining however in one of the transverse directions, in the other one occurring the wiggle electron motion.

Linear undulators are most widely used, due to their flexibility. The undulator period $\lambda_U$ and $B_0$ combine in the dimensionless vector potential, denoted by $K$

$$K = \frac{eB_0\lambda_U}{2\pi m_0 c^2}, \tag{2.1}$$

whose value determines the features of the spontaneously emitted spectrum. FELs operate usually with $K \tilde{<} 1$, thus providing a first harmonic on the undulator axis, which dominates with respect to the other harmonics.

Finally, it is important to notice that the undulator magnetic field in not uniform in the transverse plane. The Maxwell 's equations demand indeed for the presence of a gradient, which contributes to the inhomogeneous broadening of the emission line and produces therefore a significant distortion of the FEL gain curve, as it will be seen in the following.

As already noticed, the e.m. field can be provided by an external source, a laser for instance, or by the spontaneously emitted radiation stored in a resonator. In both cases, the transverse profile of the field can be approximately described by a gaussian beam, which, as is well known, is characterized by the minimum spot-size $w_0$ (the *waist*) and by the Rayleigh distance $z_R$, the first giving the scale of variability of the field intensity in the transverse plane, the second being a measure of the divergence due to the natural diffraction.

Good coupling requires to mantain the overlap between the electron and optical beam cross section during the interaction. If the Rayleigh range is much smaller than the undulator length, the optical mode area becomes much larger than the e-beam cross area, and reduces the FEL gain in two ways: the optical field strength that bunches the electrons is smaller and only a small part of the wavefront is amplified. A Rayleigh range roughly equal to the undulator length seems to be a good choice, thus giving a mode area $\Sigma_L \sim \lambda_0 L_U$. This condition can be satisfied by means of an appropriate design of the optical resonator. However, it can be relaxed if the electron beam current is sufficiently high that the diffraction is balanced by the guiding effect due to the electron beam, which behaves as an optical fiber and guides the radiation.

These qualitative considerations apply to FELs operating in the optical range, operation in the millimeter or XUV range requiring different solutions for the resonator. In the first case, waveguides are usually required, whilst operation at very short wavelengths demands for amplified spontaneous emission regime, due to the lack of optical component with sufficiently high reflectivity and small absorption in the VUV and X-ray region.

On the other hand, the longitudinal structure of the optical field, which is relevant to both an amplifier working with a pulsed external source (rather than a c.w. one) and to the oscillator configuration, the pulsed structure of the optical field being due to the finite length of the electron beam, is basically determined by the slippage mechanism. Due to the difference in the velocities of the electron and optical beam, indeed, the latter slips ahead of the electron bunch one wavelength for each undulator period, thus giving a slippage distance $\Delta$ at the end of the undulator equal to [5]

$$\Delta = N\lambda_0. \tag{2.2}$$

It is therefore apparent that good coupling also demands for an appropriate timing between the optical and electron beam at the injection into the undulator.

Furthermore, the lethargic behaviour * resulting from the slippage mechanism, requires that, in the oscillator configuration, the resonator length be appropriately adjusted with respect to the nominal value, fixed by the time distance between two successive electron bunches.

Finally, the electron beam is usually characterized by

1) the energy spread,
2) the transverse size,

and

3) the angular divergence,

which are measured by the corresponding r.m.s. values: $\sigma_\epsilon$ for the energy spread, $\sigma_x$, $\sigma_y$ for the beam size in the radial and vertical direction, respectively, and $\sigma'_x$, $\sigma'_y$ for the angular divergence in

---

* By lethargy we mean that the optical pulse velocity slows down as a consequence of the slippage mechanism and of the gain.

the $x$ and $y$ direction.
In addition, as already noticed, the electron beam has a finite length, measured by the r.m.s. value $\sigma_z$, which dynamically combines with the longitudianl width $\sigma_E$ of the optical pulse and with the slippage distance $\Delta$ to model the field and hence to influence the laser gain.

## 3. Single mode theory.

The above qualitative discussion should clearly indicate that the gain of the FEL depends on a quite large number of parameters, relevant to the undulator, optical field and e-beam. Our work has been aimed at finding scaling laws, governing the FEL gain as a function of the above quoted parameters or their appropriate combinations, thus providing efficient tools for the design of FEL devices.

As a first step, let us briefly comment on the results of the single-mode theory, which is amenable for an analytical approach in the small-signal and small-gain regime [6].
Within that context, indeed, the gain curve is obtainable from the spontaneous spectrum, whose shape is given by the function

$$S(\nu)\propto\left(\frac{\sin\nu/2}{\nu/2}\right)^2. \tag{3.1}$$

The *detuning parameter* $\nu$, defined as

$$\nu=2\pi N\frac{\omega_0-\omega}{\omega_0},\quad \omega_0=\frac{2\pi c}{\lambda_0}, \tag{3.2}$$

measures the relative detuning of the operating frequency $\omega$ with respect to the resonance frequency $\omega_0$ in unity of the natural bandwidth, which is determined by the number of undulator periods as

$$\frac{\Delta\omega}{\omega_0}=\frac{1}{2N}. \tag{3.3}$$

The gain curve is then proportional to the derivative of the spontaneous spectrum with respect to $\nu$ , i.e.

$$G(\nu)=-\pi g_0\frac{d}{d\nu}\left[\frac{\sin(\nu/2)}{\nu/2}\right]^2, \tag{3.4}$$

and, as is well known, exhibits the familiar antisymmetric shape with the maximum value $G_{max}=0.85g_0$ at $\nu=2.6$ (see fig. 3).
The *small-gain coefficient* $g_0$ is a very important parameter, since it determines if the laser will reach the threshold and lase. Its explicit expression, reported here in the form holding for a linear undulator

$$g_0=2\pi N^3K^2\frac{\lambda_U^2 I}{\gamma^3 I_0\Sigma_E}[J_0(\xi)-J_1(\xi)]^2F, \tag{3.5}$$

clearly reveals the dependence of the FEL gain on the undulator parameters: $N$, $\lambda_U$, $K$, also contained in the argument $\xi$ of the Bessel function $J_0$ and $J_1$, which is $\xi=\frac{1}{4}\frac{K^2}{1+K^2/2}$. The effect of the electron beam is accounted for by the electron energy $\gamma$, the e-beam current $I$ and the cross section $\Sigma_E$. Furthermore, $F$ denotes the *filling factor*, which gives a qualitative account of the relative e-beam transverse cross section with respect to that of the optical field.
The gain curve, as given by the expression (3.4), the small-gain coefficient and the expression for the maximum tolerable width $\frac{\Delta\gamma}{\gamma}$

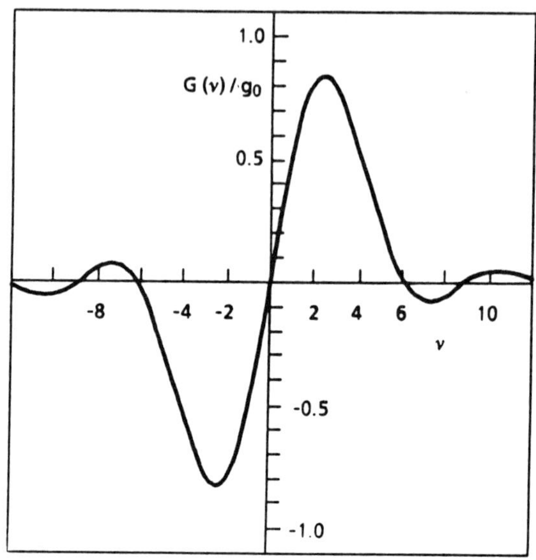


Fig. 3. Gain curve.

of the initial electron energy distribution, which is set on by the number of undulator periods as

$$\frac{\Delta\gamma}{\gamma} \tilde{<} \frac{1}{2N}, \tag{3.6}$$

represent the only tools provided by the one-dimensional theory for designing an FEL device.

They give useful indications, but do not take into account the effect of the previously introduced parameters, as the electron beam quality for instance. Furthermore, they work only for small values of $g_0$: $g_0 \tilde{<} 0.3$. For larger values of $g_0$, the linear theory does not apply. In addition, also within the context of the linear theory, significant deviations from its predictions are induced by the longitudinal and transverse effects as well.

As alternative to the recipes suggested by the single-mode theory, there are only numerical codes, which provide the most accurate and general description of FEL performance, but relatively less insight into the interplay of the various parameters, affecting the gain. As a consequence, it should be interesting and useful to develop simple formulae, which give reliable and accurate indications, obtainable furthermore from fast computations, on how the various parameters can influence the FEL operation.
We will firstly consider the effects of the e-beam quality.

## 4. Inhomogeneous broadening.

An immediate feeling of the effects caused by the energy and angular spread and by the transverse sizes of the electron beam on the FEL performance can be gained from the expression for the resonance frequency $\omega_0$:

$$\omega_0 = \frac{2\gamma^2 \omega_U}{1 + K^2 + \gamma^2\theta^2}, \quad \omega_U \equiv \frac{2\pi c}{\lambda_U} \tag{4.1}$$

reported in a more general form, which includes the possibility that the average trajectory of the electron be bent by the small angle $\theta$ with respect to the undulator axis.
It is evident that an energy and angular spread widens the emission line. Furthermore, as already noticed, the magnetic field exhibits a transverse gradient. So, electrons with different transverse coordinates experience different $K$-parameters, thus giving rise to a further broadening of the emission line, as a consequence of the e-beam transverse size.
An appropriate rehandling of (4.1) allows to rewrite the detuning parameter $\nu$ to include the e-beam parameters in the form

$$\nu = \nu_0\left(1 + \frac{\Delta\nu}{\nu_0}\right), \tag{4.2}$$

with $\nu_0$ denoting the detuning parameter relevant to an ideal electron beam (i.e. monochromatic and with zero emittance).

The deviations of $\nu$ from $\nu_0$ are measured by $\frac{\Delta\nu}{\nu_0}$, which can be significantly expressed as the sum of the relative shift in the resonance frequency induced by the

1) energy spread: $\frac{\delta\omega}{\omega_0}|_\epsilon$

2) transverse size: $\frac{\delta\omega}{\omega_0}|_{x,y}$

and

3) angular divergence: $\frac{\delta\omega}{\omega_0}|_{x',y'}$

compared to the homogeneous * bandwidth (3.3).

Averaging the expressions of the relative shifts $\frac{\delta\omega}{\omega_0}$ on the electron distribution with respect to the energy, transverse coordinates and angular divergences, which are assumed to be gaussian, allows to identify global parameters, the so called $\mu$-parameters, which account for the line broadening induced by the energy spread and finite emittance of the electron beam, compared with the homogeneous broadening.

In particular, for a linear undulator the parameters $\mu$ specialize as [7]

---

* *Homogeneous* means that the emission line broadening mechanism is due to the dynamics of the process (namely, the particle flight time along the undulator) and it is equal for all the electrons in the beam, independently of their own physical properties (i.e. momentum, energy and trajectory).

$$\mu_\epsilon = 4N\sigma_\epsilon, \quad \mu_{x,y} = \sqrt{|h_{x,y}|}\frac{2}{1+\frac{K^2}{2}}\frac{\gamma\epsilon_{x,y}}{\lambda_U}, \tag{4.3}$$

which explicitly displays the dependence on the e-beam energy spread ($\sigma_\epsilon$) and emittances ($\epsilon_x, \epsilon_y$) as well as on the undulator parameters, the transverse gradient being taken into account by $h_{x,y}$ linked to the sextupolar term in one of the transverse direction.
It is needless to say that the parameters (4.3) control the relevance of the inhomogeneous broadening effects. If $\mu \ll 1$, indeed, inhomogeneous effects should not create significant problems; conversely, if $\mu \tilde{>} 1$ they should be dominant.

## 5. Gain versus μ-parameters.

The parameters $\mu_\epsilon, \mu_x$ and $\mu_y$ introduced before play an important role within the context of the FEL dynamics, since they enter the FEL wave equation, which is the starting point of our analysis.
The FEL wave equation, originally developed by Colson within the context of the Maxwell-Lorentz theory of free electron laser dynamics [6], holds in the slowly-varying amplitude and phase approximation and in the small signal regime.
Its simplest form [8]

$$\frac{d}{d\tau}a(\tau) = -i\pi g_0 \int_0^\tau \tau' \frac{e^{i\nu_0\tau'} e^{-\pi^2\mu_\epsilon^2\tau'^2/2}}{(1+i\pi\mu_x\tau')(1+i\pi\mu_y\tau')} a(\tau-\tau')d\tau', \tag{5.1}$$

assumes the electron beam to be longitudinally and transversally uniform and does not account for the longitudinal and transverse structure of the optical field.
In the above equation, $a(\tau)$ is the dimensionless complex field amplitude, whose explicit expression is unessential within the

present context. The variable $\tau$ denotes the time normalized to the interaction time $L_U/c$, namely $\tau = L_U t/c$.

As a first step, let us consider the case of small gain: $g_0 \lesssim 0.3$, which should provide the corrections to the gain coefficient $g_0$ due to the $\mu$-parameters.

Assuming therefore the field amplitude $a(\tau)$ to remain almost unchanged during the interaction, i.e. $a(\tau) \sim a(0)$, eq. (5.1) gives for the field amplitude at $\tau = 1$ at the first order in $g_0$ the simple expression

$$a(1) = \left[1 + \frac{g_0}{2} G_1(\nu_0, \mu_\epsilon, \mu_x, \mu_y)\right] a(0). \tag{5.2}$$

The function $G_1(\nu_0, \mu_\epsilon, \mu_x, \mu_y)$, which writes

$$G_1(\nu_0, \mu_\epsilon, \mu_x, \mu_y) = -2\pi i \int_0^1 d\tau (1 - \tau) \frac{e^{i\nu_0 \tau} e^{\frac{-\mu_\epsilon^2 \tau^2}{2}}}{(1 + i\mu_x \tau)(1 + i\mu_y \tau)}, \tag{5.3}$$

can be regarded as a complex gain function accounting for the modifications of both the amplitude and phase of the optical field, due to the FEL interaction. It is indeed easy to see that the FEL gain in the small-signal regime is just given by the real part of $G_1$, namely

$$G(\nu_0, \mu_\epsilon, \mu_x, \mu_y) = g_0 \Re G_1(\nu_0, \mu_\epsilon, \mu_x, \mu_y). \tag{5.4}$$

Obviously for $\mu = 0$ (homogeneous broadening regime), the real part of $G_1$ reproduces the antisymmetric behaviour of the gain curve, obtained within the context of the one-dimensional theory.

The effect of the $\mu$-parameters can be easily recognized as giving rise to a broadening of the gain curve, a reduction of the value of the maximum gain and correspondingly a shift of $\nu$-value, where the gain is maximum, towards larger values (see fig. 4).

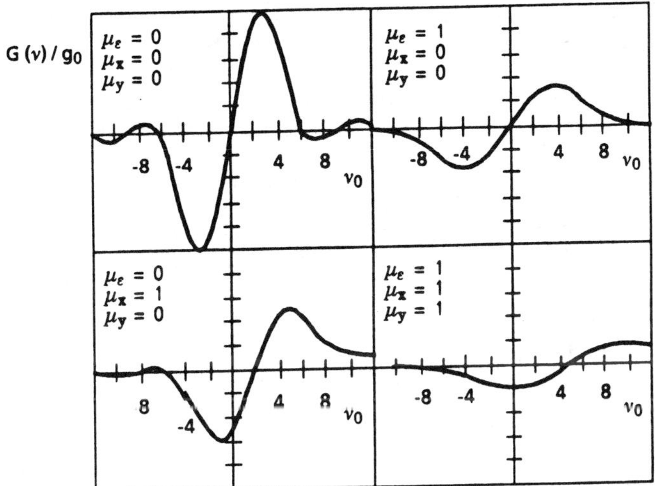


Fig. 4. Gain curves for different values of the parameters $\mu_\epsilon, \mu_x, \mu_y$.

A numerical analysis of the behaviour of the maximum of the real part of $G_1$ has allowed to deduce the following scaling law for the maximum gain as a function of $\mu_\epsilon$, $\mu_x$ and $\mu_y$ [9]:

$$G_{\max}(\mu_\epsilon, \mu_x, \mu_y) = \frac{0.85 g_0}{(1+1.7\mu_\epsilon^2)(1+\mu_x^2)(1+\mu_y^2)}, \qquad (5.5).$$

which holds for $0 \le \mu \le 1$. The above expression reproduces with good agreement the numerical data, as it is shown in fig. 5, where the maximum and minimum deviations between the numerical and fit curves are displayed as well.

For greater values of $\mu$: $\mu \le 3$, the contributions of the various parameters cannot be separated as in the above expression, thus leading to the more involved formula [10]

$$G_{\max}(\mu_\epsilon, \mu_x, \mu_y) = 0.85 g_0 \frac{g(\mu_{x,y}) e^{-\gamma \mu_\epsilon^2}}{(1+1.7\mu_\epsilon^2 g(\mu_{x,y}))}, \quad \gamma = 0.028 \qquad (5.6)$$

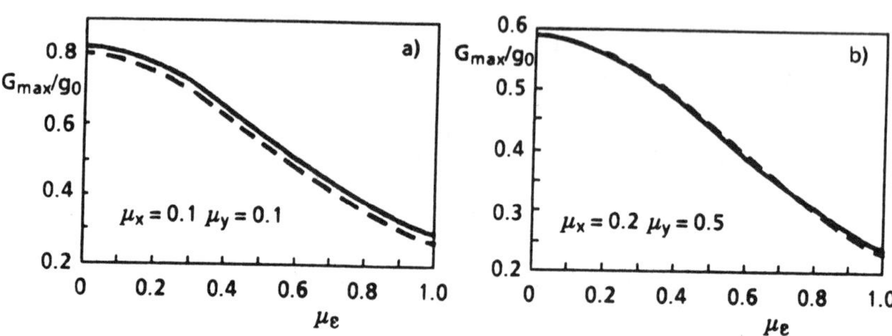


Fig. 5. Maximum gain vs. $\mu_\epsilon$ for different values of $\mu_x, \mu_y$; the dotted lines reproduce the scaling law (5.5).

with the function $g(\mu_{x,y})$ containing powers of $\mu_x, \mu_y$ according to

$$g(\mu_{x,y}) = \sum_{n=0}^{5} \sum_{m=0}^{n} \frac{\alpha_n \mu_x^m \mu_y^{n-m}}{(1+\mu_x^2)(1+\mu_y^2)}, \tag{5.7}$$

$$\alpha_1 = -0.155, \quad \alpha_2 = 0.068, \quad \alpha_3 = 0.070, \quad \alpha_4 = 0.022, \quad \alpha_5 = 0.009$$

## 6. Intermediate $g_0$ region.

The results illustrated in the previous section hold for small values of the gain coefficient $g_0$: $g_0 \leq 0.3$. For larger values of $g_0$, the linear theory does not correctly account for the FEL gain. Significant deviations of the gain curve as obtained from the linear theory with respect to that obtained solving the one-dimensional wave equation, can be revealed with increasing $g_0$, also in the homogeneous broadening regime, as undeniably displayed by the gain curves reported in fig. 6.

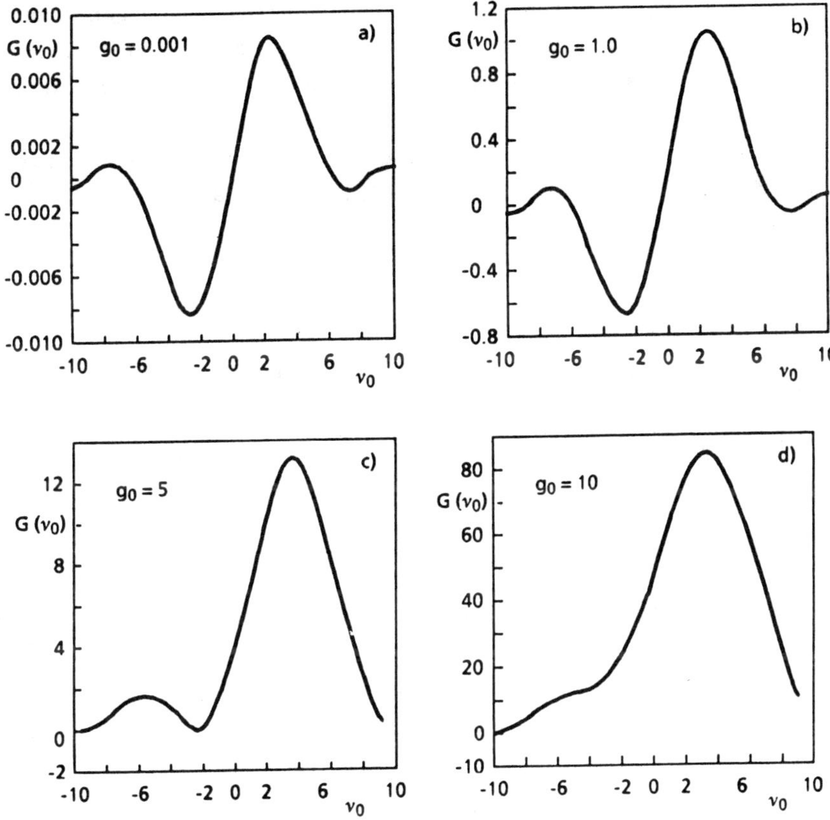


Fig. 6. Gain curves for different values of $g_0$.

A perturbative approach to the one-dimensional wave equation allows to write the gain in form of a series expansion in $g_0$ as [11]

$$G(\nu_0,\mu_\epsilon,\mu_x,\mu_y)=g_0g_1(\nu_0,\mu_\epsilon,\mu_x,\mu_y)+g_0^2g_2(\nu_0,\mu_\epsilon,\mu_x,\mu_y)+ g_0^3g_3(\nu_0,\mu_\epsilon,\mu_x,\mu_y), \qquad (6.1)$$

with the third power in the gain coefficient assuring a good agreement with the numerical curves up to $g_0\sim 10$.

The function $g_1$ is nothing but the real part of the function $G_1$ discussed before. The functions $g_2$ and $g_3$ account for the deviations from the linear behavior, including also the effect of the electron beam quality.

An analytical expression for the functions $g_1, g_2$ and $g_3$ can be worked out for $\mu = 0$, thus allowing to specialize (6.1) as

$$g_1(\nu_0) = \frac{2\pi}{\nu_0^3}[2(1-\cos\nu_0) - \nu_0 \sin\nu_0],$$

$$g_2(\nu_0) = \frac{\pi^2}{3\nu_0^6}[84(1-\cos\nu_0) - 60\nu_0\sin\nu_0 + 3\nu_0^2(1+5\cos\nu_0) + \nu_0^3\sin\nu_0], \tag{6.2}$$

$$g_3(\nu_0) = \frac{\pi^3}{60\nu_0^9}[11520(1-\cos\nu_0) - 9000\nu_0\sin\nu_0 + 360\nu_0^2(1+8\cos\nu_0) + 480\nu_0^3\sin\nu_0 - 20\nu_0^4(1+2\cos\nu_0) - \nu_0^5\sin\nu_0].$$

which reproduce remarkably well the gain curves up to $g_0 \leq 10$, as shown in fig. 7.

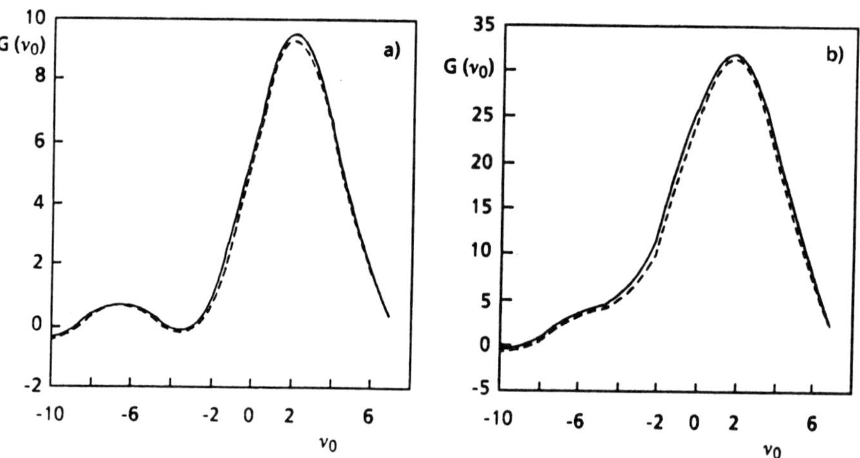


Fig. 7. Gain curves for $g_0 = 5$ (a) and $g_0 = 10$ (b); solid line: numerical curve, dotted line: fit curve.

Accordingly, the maximum gain can be given the manageable expression

$$G_{max} = 0.85 g_0 + 0.19 g_0^2 + 0.42 \times 10^{-2} g_0^3, \tag{6.3}$$

whose behaviour is visualized in fig. 8.
For $\mu \neq 0$, the functions $g_1, g_2$ and $g_3$ must be handled numerically, thus revealing that the parameters $\mu$ strongly reduce the value of $g_2$ and $g_3$, and hence strongly reduce the non-linear contributions to the FEL gain.
However, any fit of these functions has not been attempted. We have preforred to work directly with the wave equation, which once solved numerically has been used to calculate the maximum gain for values of $g_0$ up to 5 and values of $\mu$ up to 3. Scaling law have been then obtained from a fit procedure, thus leading to

$$G_{max}(\mu_\epsilon, \mu_x, \mu_y) = \frac{g_0(0.85 + 0.19 g_0) f(\mu_{x,y})}{1 + (1.7 + 0.30 g_0)\mu_\epsilon^2 f(\mu_{x,y})} e^{-0.06\mu_\epsilon^2}, \quad g_0 \lesssim 3, \quad 0 \lesssim \mu \lesssim 1.5 \tag{6.4}$$

where

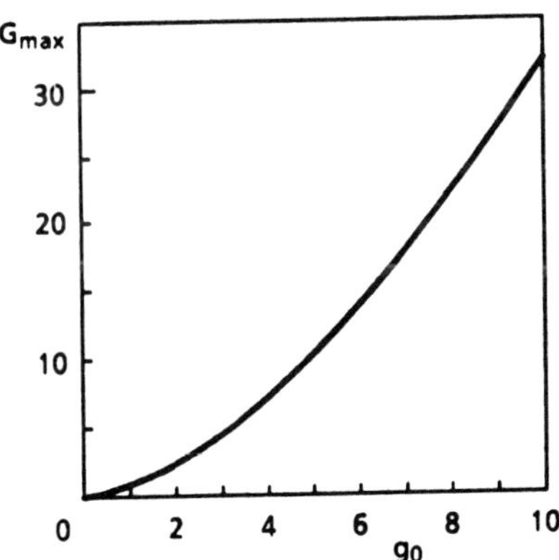


Fig. 8. Maximum gain vs. $g_0$ in the homogeneous broadening regime.

$$f(\mu_{x,y}) = \frac{1+0.06\mu_x\mu_y}{[1+(1.06+0.23g_0)\mu_x^{1.7}][1+(1.06+.23g_0)\mu_y^{1.7}].} \tag{6.5}$$

The expressions (5.5), (5.6) or (6.4) can be usefully exploited to design a FEL operating in the small signal regime, that is below saturation. In fact, although they assume a more involved form increasing $g_0$ and/or the parameters $\mu$, as clearly displayed by (6.4), they allow a fast evaluation of the higher order contributions in $g_0$ as well as of the degradation of the gain induced by the electron beam energy spread and emittances.

## 7. Longitudinal dynamics.

The propagation of an optical pulse in a FEL operating with a longitudinally shaped electron beam is ruled by the following wave equation [12]

$$\frac{\partial}{\partial\tau}a(\zeta,\tau) = -i\pi g_0 f(\zeta+\mu_E\tau)\int_0^\tau d\tau'\tau' e^{i\nu_0\tau'}a(\zeta+\mu_E\tau',\tau-\tau'), \tag{7.1}$$

where $a(\zeta,\tau)$ denotes the field complex amplitude, the longitudinal coordinate being normalized to the optical pulse width $\sigma_E$, i.e. $\zeta = z/\sigma_E$. Furthermore, $f(\zeta)$ accounts for the longitudinal distribution of the electron beam.

It is evident that due to the slipagge mechanism there is communication between adjacent wave front planes, as indicated by the shift in the longitudinal coordinate in both the electron distribution and the optical field amplitude, the parameter $\mu_E$ giving the slippage distance in unity of $\sigma_E$, i.e.

$$\mu_E = \frac{\Delta}{\sigma_E}. \tag{7.2}$$

As suggested by previous analyses *, let us expand the complex field amplitude $a(\zeta,\tau)$ in terms of Hermite functions $u_n(\zeta)$ as

$$a(\zeta,\tau)=\sum_{n=0}^{\infty}a_n(\tau)u_n(\zeta), \tag{7.3}$$

where $u_n(\zeta)$ is explicitly given by

$$u_n(\zeta)=\frac{1}{n!2^n\sqrt{\pi}}H_n(\zeta)e^{-\frac{\zeta^2}{2}}, \tag{7.4}$$

and the $\tau$-dependent coefficients $a_n(\tau)$ specify the contribution of $u_n(\zeta)$ to modelling the field.

Inserting the above expansion into the wave equation, we get a set of coupled equations for the $a_n$'s:

$$\frac{d}{d\tau}a_n(\tau)=-i\pi g_0\sum_{m=0}^{\infty}\int_0^{\tau}d\tau'\tau' e^{i\nu_0\tau'}f_{nm}(\tau,\tau')a_m(\tau-\tau'), \tag{7.5}$$

whose initial conditions are obviously specified by the overlap integrals between the optical envelope at the initial time $\tau=0$ and the Hermite function $u_n(\zeta)$.

The above equation resembles in some sense the one-dimensional wave equation, the significant difference, which accounts for the longitudinal dynamics, is in the presence of the summation over all the modes. As a consequence, also for an initially single-mode

* Previous analyses relevant to the dynamics of an FEL oscillator, operating with a bunched electron beam and in the small signal regime, have indicated that the optical pulse, which configures in the laser over many passes, can be approximately described by gaussian functions, thus naturally suggesting the present expansion on Hermite functions.

operation the FEL naturally operates over many modes, whose relative weight in modelling the optical field is determined by the strength of the coupling, specified by the matrix elements $f_{nm}$.
They are explicitly given by

$$f_{nm}(\tau,\tau')=\int d\zeta\, u_n(\zeta)f(\zeta+\mu_E\tau)u_m(\zeta+\mu_E\tau'), \tag{7.6}$$

and represent a kind of current form factor. Accordingly, they account for the coupling between the basis functions $u_n$ as a consequence of the slippage mechanism, the electron beam longitudinal distribution providing a further coupling source.
According to the analysis performed within the context of the one dimensional theory, we approach the solution to eq. (7.5) perturbatively, using the gain coefficient $g_0$ as perturbation parameter. We write indeed $a_n(\tau)$ in form of a series expansion in $g_0$ as

$$a_n(\tau)=\sum_{k=0}^{\infty} g_0^k a_n^{(k)}(\tau), \tag{7.7}$$

with the coefficients $a_n^{(k)}$ obeying recursive equations, whose explicit expressions are not relevant for the present discussion.
Skipping indeed all the details of the calculation, it can be easily recognized the possibility of expressing the FEL gain, which depends now on $\nu_0$, as before, and $\mu_E$ (apart from the specific electron beam distribution), in form of a series of powers of $g_0$, in full analogy with the expression obtained in the simple case of the single mode theory.
Within the present context, the corresponding functions $g_1, g_2$ and $g_3$ exhibit more complicated expressions, involving indeed rather intrigued combinations of the coefficients $a_n^{(k)}$, which in turn depend on the initial field distribution.

Presently, it is particularly interesting to consider the function $g_1$, that is the coefficient of $g_0$ in the gain expansion. It can be obtained as the real part of a complex function $G_1(\nu_0, \mu_E)$, which configures as the counterpart of the complex gain function (5.3), introduced within the context of the one dimensional theory. The function $G_1$ exhibits a specific expression for each mode $u_n$. Consequently, it is possible to identify for each mode $u_n$ a complex gain function, accounting for the modifications of the amplitude and phase of that mode, due to the FEl process.
It is evident that the explicit expression of the function $G_1(\nu_0, \mu_E)$ can give useful information on the parameters, which affect the gain, and can suggest a possible form for the dependence of the gain on those parameters.
Let us consider therefore as initial field distribution the simple gaussian function

$$a(\zeta, 0) = \pi^{-\frac{1}{4}} e^{-\frac{\zeta^2}{2}} \tag{7.8}$$

undergoing a FEL interaction with a longitudinally uniform electron beam, i.e. $f(\zeta) = 1$.
In that case the complex gain function assumes the simple and particularly enlightening expression

$$G_1(\nu_0, \mu_E) = -2i\pi \int_0^1 d\tau (1-\tau) e^{i\nu_0 \tau} e^{-\frac{\mu_E^2 \tau^2}{4}}, \tag{7.9}$$

which must be compared with the single mode gain function relevant to an electron beam, with an energy spread, accounting for by the parameter $\mu_\epsilon$. It is immediately recognized that $\mu_\epsilon$ enters the expression of the gain function in a similar way as $\mu_E$ in the above one. Consequently, the role of $\mu_E$ is completely equivalent to that of the parameter $\mu_\epsilon$, that is the effect of the electron beam energy spread is equivalent to that of the finite length of the optical pulse, as it should be expected.

This result is particularly interesting, since it allows to take advantage from the well established scaling laws of the gain versus the parameter $\mu_\epsilon$ (see eqs. (5.5-6)). As a consequence, we can say that the gain, at the first order in $g_0$ and for small values of $\mu_E$, scales with $\mu_E$ according to the simple formula

$$G_{max}(\mu_E) = \frac{0.85 g_0}{1+\mu_E^2/12}, \quad \mu_E \tilde{<} 2 \tag{7.10}$$

which has been numerically checked, as shown in fig. 9.

Finally let us briefly say that if the initial distribution is provided by a Hermite mode $u_n$, it is possible to identify a

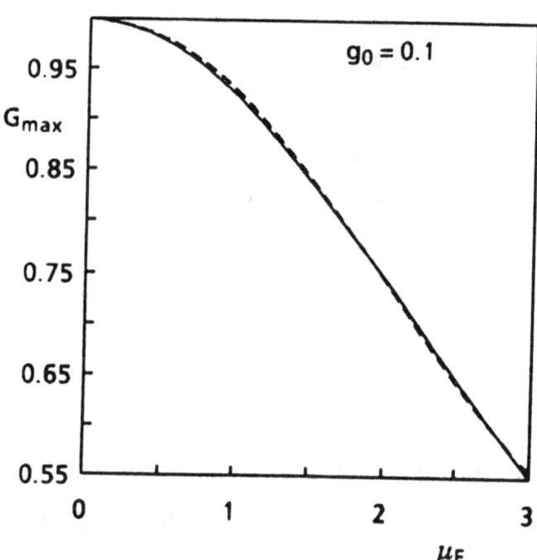


Fig. 9. Maximum gain vs. $\mu_E$ for $g_0 = 0.1$: comparison between the numerical (solid) and fit (dotted) curves.

corresponding $\mu_E^{(n)}$ parameter, given by the simple expression

$$\mu_E^{(n)} = (2n+1)\mu_E . \tag{7.11}$$

Accordingly, the maximum gain should scale as in (7.10).
Finally, let us stress that the present discussion **suggests** the way to take into account the effect of the e-beam quality on the FEl gain as well as the parameters characterizing the longitudinal dynamics. According to the expression (5.5), indeed, we can write down the following scaling law ruling the maximum gain for an initially gaussian shaped optical field:

$$G_{\max}(\mu_E) = \frac{0.85 g_0}{(1+1.7\mu_\epsilon^2 + 0.086\mu_E^2)(1+\mu_x^2)(1+\mu_y^2)}, \tag{7.12}$$

with the various parameters ranging as previously specified.
Before closing this section, we would briefly comment on the possibility of taking into account the transverse effects in a similar fashion. Approaching indeed the wave equation, which include the transverse distribution of the optical field as well as of the electron beam, within the context of the perturbative technique, illustrated before, it is possible to state that the beam waist $w_0$ plays the same role as the electron beam emittance. The parameter $\mu_T$, defined as [13]

$$\mu_T = \frac{1}{2w_0}, \tag{7.13}$$

enters the complex gain function in the same fashion as the parameters $\mu_x$ or $\mu_y$, accounting for the e-beam emittances.
As a consequence, it is immediate To generalize the expression (7.12) To include the transverse structure of the optical field.

## 8. Conclusions.

In this note we have tried to give a feeling of the complexity underlying the FEl physics and how, at the same time, some guide parameters, carefully chosen, can be exploited to derive e.g. scaling laws for the gain.
It is important to underline that the developed analysis is, in some sense, unifying and contains, in a very natural way, results already obtained within different contexts.
As it is well known, indeed, the theory of Super Modes [14] has allowed the derivation of the gain reduction, due to the finite length $\sigma_z$ of the electron pulse, in the form

$$G_{max} = \frac{0.85 g_0}{1+\mu_C/3}, \quad \mu_C \equiv \frac{\Delta}{\sigma_z}, \tag{8.1}$$

where $\mu_C$ is the so called *coupling* parameter. Equation (8.1) is consistent with eq. (7.10); in fact, since the width of the *fundamental* Super Mode, which can be approximately descrided with a gaussian, is

$$\sigma_E \cong 0.5\sqrt{\Delta\sigma_z}, \tag{8.2}$$

the relevant $\mu_E$ being therefore

$$\mu_E \cong 2\sqrt{\mu_C}, \tag{8.3}$$

the relation (7.10) turns into (8.1).
Such an example further supports the idea that the various parameters characterizing the FEL can be conveniently combined to get relatively simple expressions, which can provide simple relations for a *first sight* global view to the laser performance.

The topics we have discussed in this paper are relevant to an FEL amplifier rather than an oscillator. As already mentioned, the problems of a FEL oscillator operating with a bunched beam are associated with the lethargy. The gain is therefore also a function of the cavity mismatch $\delta L$ from the nominal value. A simple gain formula, accounting for the above effect and for the dependence on $\mu_c$, has been derived in ref. [15] and reads

$$G(\theta,\mu_c) \cong -0.85 g_0 \frac{\theta}{\theta_s}\{\ln[\frac{\theta}{\theta_s}\left(1+\frac{\mu_c}{3}\right)]-1\}, \tag{8.4}$$

where

$$\theta \equiv \frac{4\delta L}{g_0 \Delta}, \quad \theta_s = 0.456, \tag{8.5}$$

are usually referred to as the *cavity detuning parameter* (*c.d.p.*) and the *synchronous c.d.p.* respectively.

From the above equation we get that the maximum gain is located at

$$\theta^* = \frac{\theta_s}{1+\mu_c/3} \tag{8.6}$$

and is just given by eq. (8.1), as easily verified.

Equation (8.4) is the result of a numerical analysis and has been analytically checked within a perturbative procedure up to second order terms in $\mu_c$. Extension of eq. (8.4) also including energy spread and emittances have also been discussed, but we will not be reported here [15].

Finally, let us stress that the present note has been essentially devoted to discuss the FEL performance in the small signal regime and we did not mention any aspects of the saturation. It is perhaps worth concluding the paper with simple and effective considerations concerning this problem.

In conventional lasers the saturation intensity $I_S$ is a characteristic quantity, which corresponds to the intracavity intensity halving the small signal gain. Furthermore, the gain exhibits the following simple scaling law vs. the intensity

$$G \cong \frac{g}{1+\frac{I}{I_S}}, \tag{8.7}$$

with $g$ denoting the small signal gain.

A similar expression has been proved to hold for FEL, namely [16]

$$G \cong \frac{0.85 g_0}{1+\frac{I}{I_S}-0.14\frac{I}{I_S}\left(1-\frac{I}{I_S}\right)}. \tag{8.8}$$

The above equation reproduces the numerical data with an accuracy of few %. The expression for the FEL saturation intensity has been obtained as

$$I_S = \frac{m_0 c^3}{4\pi^2 r_0}\left(\frac{\gamma}{N}\right)^4 \frac{1}{[\lambda_U K(J_0(\xi)-J_1(\xi))]^2}, \tag{8.9}$$

where $r_0$ is the electron classical radius. Furthermore an important relation between the maximum output power $I_S$ and $g_0$ holds, namely [16]

$$I_{out} = \sim\frac{1}{2} g_0 I_S. \tag{8.10}$$

The link between the gain satured formula of the type (8.8) and those discussed in the previous section can be eventually found, but it will be not discussed here.

Our concluding remark is that a judicious integration of the analytical and numerical analysis may give the possibility of finding easily manageable relations characterizing the FEL in a wide region of parameters.

## References

(1) G. Dattoli, A. Renieri and A. Torre, ' Lectures notes on the theory of the Free Electron Laser and related topics', World Publ. Co. , to be published

(2) D. A. G. Deacon et al. , Phys. Rev. Lett. **38**, 892 (1977)

(3) G. N. Kulipanov et al., Nucl. Instr. Meth. **A296** , 1 (1990)

(4) T. J. Orzechowski et al., Phys. Rev. Lett. **54**, 889 (1985)

(5) W. B. Colson and A. Renieri, in Bendor Free Electron Laser Conf., J. Physique Colloque **C1**, Vol. 44 , eds. D.A.G. Deacon and M.Billardon, 11 (1983)

(6) W.B. Colson, Nucl. Instr. Meth. **A237**,1 (1985)

(7) G.Dattoli and A. Renieri, Laser Handbook, Vol. 4, eds. M.L. Stitch and M.S. Bass, North Holland (1985)

(8) J.C. Gallardo, L,R. Elias, G. Dattoli and A. Renieri, Phys. Rev. **A36**, 3222 (1987)

(9) G. Dattoli, T. Letardi, J.M.J. Madey and A. Renieri, IEEE J. Quantum Electron. **QE-20**, 637 (1984)

(10) G. Dattoli, H. Fang, L. Giannessi, M. Richetta, A. Torre and R. Caloi, Nucl. Instr. Meth. **A285**, 108 (1989)

(11) G. Dattoli, A. Torre, C. Centioli and M. Richetta, IEEE J. Quantum Electron. **QE-25**, 2327 (1989)

(12) G. Dattoli, H. Fang, M. Richetta and A. Torre, submitted for publication

(13) G. Dattoli, L. Giannessi, L. Mezi and A. Torre, in preparation

(14) G. Dattoli, J.C. Gallardo, T. Hermsen, A. Renieri and A. Torre, Part I, Phys. Rev. **A37**, 4326 (1988); G. Dattoli, T. Hermsen, L. Mezi, A. Renieri and A. Torre, Part II, Phys. Rev. **A37**, 4334 (1988)

(15) G. Dattoli, L. Giannessi, T. Hermsen, M. Richetta, A. Torre and R. Caloi, Proc. of 'New Laser Technologies and Applications', Olympia, 19-23 June 1988

(16) S. Cabrini, 'Teoria del Laser ad Elettroni Liberi', Thesis, Rome University (1991)

**Research Trends in Physics: Coherent Radiation Generation and Particle Acceleration**
Editorial Board: J.M. Buzzi, A. Prokhorov (Editor-in-Chief), P. Sprangle, and K. Wille
*La Jolla International School of Physics*, The Institute for Advanced Physics Studies, La Jolla, California

# Stimulated Backscattered Harmonic Radiation Generated by Intense Laser Interactions With Beams and Plasmas

**E. Esarey and P. Sprangle**

Naval Research Laboratory
Washington, DC 20375-5000

## Abstract

Stimulated backscattered harmonic radiation generated by the interaction of an intense pump laser field with an electron beam or plasma is analyzed using a nonlinear, relativistic, fluid theory valid to all orders in the pump laser amplitude. The backscattered radiation occurs at odd harmonics of the doppler shifted incident laser frequency. The growth rate and saturation level of the backscattered harmonics are calculated and thermal limitations are discussed. This mechanism may provide a practical method for producing coherent radiation in the XUV regime.

## I. Introduction

Recent technological advances have made possible compact terawatt laser systems with high intensities ($> 10^{18}$ W/cm$^2$), modest energies ($< 1$ J) and short pulses ($< 1$ ps).[1] These high intensities lead to a number of new laser–plasma and laser–electron beam interaction phenomena.[2] This paper discusses one such phenomenon, stimulated backscattered harmonic (SBH) radiation generated by the interaction of intense linearly polarized laser fields with electron beams or plasmas. The intense laser backscattering mechanism is essentially stimulated coherent scattering in the strong–pump regime. For sufficiently intense laser fields, the electron quiver velocity becomes highly relativistic. The high laser intensity, along with the induced nonlinear relativistic electron motion, results in the generation of SBH radiation at odd harmonics. In the interaction of an intense laser with a counterstreaming electron beam, SBH radiation is generated via a nonlinear free electron laser (FEL) mechanism,[3,4] resulting in a relativistic doppler frequency upshift. Hence, a laser–pumped FEL[5–9] may utilize both the harmonic and doppler upshifts to generate short wavelength coherent radiation. SBH radiation from stationary plasmas occurs at odd multiples of the incident laser frequency via a nonlinear Raman backscatter mechanism[10] in the strong–pump regime. Previous analyses[5–10] of stimulated backscattered radiation from electron beams or plasmas, due to the complex dynamics of the laser–electron interaction, have been limited to studies of the fundamental backscattered mode. The analysis presented in this paper includes the full nonlinear dynamics of the pump laser field and, hence, is capable of describing the generation of SBH radiation.

Significant radiation generation at high order harmonics is found to occur when the normalized pump laser amplitude exceeds unity, $a_0 \geq 1$, where $a_0 = |e|A_0/m_0c^2$ is the normalized vector potential amplitude of the pump laser field. The parameter $a_0$ is related to the power of a linearly polarized laser by $P_0[\text{GW}] = 21.5(a_0 r_0/\lambda_0)^2$, where $r_0$ is the spot size of the Gaussian profile and $\lambda_0$ is the laser wavelength. Physically, $a_0 \geq 1$ implies that the laser–induced electron quiver is highly relativistic, as may be seen from conservation of transverse canonical momentum in 1D which states $a_0 = \gamma v_\perp/c$, where $\gamma$ is the relativistic factor and $v_\perp$ is the quiver velocity. In terms of the laser intensity ($I_0 = 2P_0/\pi r_0^2$), $a_0 \simeq 10^{-9}\lambda_0[\mu\text{m}]I_0^{1/2}[\text{W/cm}^2]$. Relativistic electron motion, $a_0 \geq 1$, requires intensities $\geq 10^{18}$ W/cm$^2$ for wavelengths of $\sim 1\ \mu$m. Such intensities are currently available from compact laser systems.[1]

## II. Nonlinear Formulation

The 1D fields associated with the pump laser, backscattered radiation and the plasma fluid response are described by the normalized vector and scaler potentials, $\mathbf{a}(z,t) = |e|\mathbf{A}/m_0c^2$ and $\phi(z,t) = |e|\Phi/m_0c^2$. Coulomb gauge is assumed, $\nabla \cdot \mathbf{a} = 0$, which in 1D implies $a_z = 0$, where $z$ is along the axis of propagation of the SBH radiation. Introducing the independent variables $\eta = z + ct$ and $\xi = z - ct$, the transverse wave equation and Poisson's equation are given by

$$\left(\frac{\partial^2}{\partial\eta\partial\xi} - \frac{k_p^2}{4\gamma_0}\rho\right)\mathbf{a} = 0, \tag{1}$$

$$\left(\frac{\partial}{\partial\eta} + \frac{\partial}{\partial\xi}\right)^2 \phi = k_p^2\left(\frac{\gamma\rho}{\gamma_0} - 1\right), \tag{2}$$

where $k_p = \omega_p/c$ and $\omega_p = (4\pi|e|^2 n_0/m_0)^{1/2}$ is the plasma frequency evaluated at the ambient density $n_0$. The plasma response is determined by the fluid variable $\rho(\eta,\xi) = \gamma_0 n/\gamma n_0$, where $n(\eta,\xi)$ is the electron density, $\gamma(\eta,\xi) = (1-\beta^2)^{-1/2}$ is the relativistic factor, $\boldsymbol{\beta}(\eta,\xi) = \mathbf{v}/c$ is the normalized electron fluid velocity and $\gamma_0 = (1-\beta_0^2)^{-1/2}$ is a constant equal to the relativistic factor of the electrons prior to the interaction with the pump laser. In obtaining the transverse current in the wave equation, use is made of the fact that the transverse canonical momentum is invariant and that prior to the laser pulse interaction, the particle distribution is assumed to have no transverse velocity, i.e., $\boldsymbol{\beta}_\perp = \mathbf{a}/\gamma$.

The plasma fluid quantities are assumed to satisfy the cold, relativistic fluid equations (i.e., the axial momentum and continuity equations), which may be written in the form

$$\left[h^2\frac{\partial}{\partial\eta}-(1+a^2)\frac{\partial}{\partial\xi}\right]h=-h\frac{\partial}{\partial\xi}a^2+h^2\left(\frac{\partial}{\partial\eta}+\frac{\partial}{\partial\xi}\right)\phi, \tag{3}$$

$$\left[h^2\frac{\partial}{\partial\eta}-(1+a^2)\frac{\partial}{\partial\xi}\right]\rho=-\rho\frac{\partial}{\partial\eta}h^2+\rho h\left(\frac{\partial}{\partial\eta}+\frac{\partial}{\partial\xi}\right)\phi, \tag{4}$$

where $h(\eta,\xi)=\gamma(1+\beta_z)$. Thermal effects are neglected which assumes that the thermal velocity spread is sufficiently small so that electrons are not trapped in the plasma wave. Also, the ions are assumed to be stationary.

The radiation is assumed to consist of a large amplitude incident pump laser field (equilibrium) traveling towards the left, $\mathbf{a}^{(0)}$, and a small amplitude backscattered field (perturbation) traveling towards the right, $\mathbf{a}^{(1)}$, i.e., $\mathbf{a}=\mathbf{a}^{(0)}(\eta)+\mathbf{a}^{(1)}(\eta,\xi)$, where $|\mathbf{a}^{(0)}|\gg|\mathbf{a}^{(1)}|$. The pump laser field is assumed to be linearly polarized and a function only of $\eta$ (group and phase velocity equal to $c$, which assumes $k_p^2/k_0^2\ll 1$, where $k_0$ is the pump field wavenumber) with a constant amplitude $a_0$ (pump depletion effects are neglected), i.e., $a^{(0)}=a_0\cos k_0\eta$. Since the backscattered field may be temporally growing it is a function of both $\eta$ and $\xi$. The various quantities all have the general form $\mathbf{Q}=\mathbf{Q}^{(0)}(\eta)+\mathbf{Q}^{(1)}(\eta,\xi)$, where $\mathbf{Q}^{(0)}$ is the equilibrium quantity in the presence of only the pump field, $\mathbf{Q}^{(1)}$ is the perturbed contribution due to the backscattered field and $|\mathbf{Q}^{(1)}|\ll|\mathbf{Q}^{(0)}|$.

To determine the plasma equilibrium set up by the pump laser, the $\xi$ derivatives are neglected in Eqs. (2)-(4). In particular, two equilibrium constants of the motion exist, $c_1=h^{(0)}-\phi^{(0)}$ and $c_2=\rho^{(0)}h^{(0)}$. The behavior of $\phi^{(0)}$ must be determined from Poisson's equation and, in general, is highly nonlinear. However, simple solutions can be obtained in two relevant limits in which the characteristic temporal variation of the pump laser envelope, $\tau_L$ (typically the pump laser rise time), is (i) short and (ii) long compared to an effective plasma period, $\gamma_0^{3/2}/\omega_p$. For applications which utilized intense pump lasers with pulse lengths $\tau_L\sim 1$ ps, the short pulse limit is relevant to interactions with electron beams with densities $n_0/\gamma_0^3\ll 10^{16}$ cm$^{-3}$; whereas the long pulse limit is relevant to interactions with stationary plasmas with densities $n_0\gg 10^{16}$ cm$^{-3}$. In the short pulse (electron beam equilibrium) limit, $\tau_L\ll\gamma_0^{3/2}/\omega_p$, the effects of $\phi^{(0)}$ can be neglected[2] (i.e., $|\phi^{(0)}/h^{(0)}|\ll 1$), provided $a_0<2\gamma_0^{3/2}/\tau_L\omega_p$. Hence, $h^{(0)}=\gamma_0(1+\beta_0)$ and $\rho^{(0)}=1$. In the long pulse (plasma equilibrium) limit, $\tau_L\gg 1/\omega_p$, it can be shown[2] that $\phi^{(0)}=\gamma_{\perp 0}-1$ and, hence, $h^{(0)}=\gamma_{\perp 0}$ and $\rho^{(0)}=1/\gamma_{\perp 0}$, where $\gamma_{\perp 0}=(1+a_0^2/2)^{1/2}$. It is convenient to define the equilibrium parameters $h_0$ and $\rho_0$ such that $h^{(0)}=h_0$ and $\rho^{(0)}=\rho_0$, i.e, $h_0=\gamma_0(1+\beta_0)$ and $\rho_0=1$ for the electron beam equilibrium (short pulse limit); and $h_0 = \gamma_{\perp 0}$ and $\rho_0 = \gamma_{\perp 0}$ for the plasma equilibrium (long pulse limit).

To analyze the SBH radiation, Eqs. (1)-(4) are expanded about the equilibrium state to first order in the perturbations, $\mathbf{Q}^{(1)}$. Furthermore, the SBH radiation is analyzed in the strong-pump limit, i.e., the effects of the perturbed electrostatic potential, $\phi^{(1)}$, are neglected. This is valid provided[3,4,10] the temporal growth rate of the SBH radiation is much greater than the relativistically corrected plasma frequency, $\omega_p/\gamma_0^{3/2}$. Representing the perturbed quantities by the form $\mathbf{Q}^{(1)} = \mathbf{Q}_1(\eta)e^{ik\xi}/2 + c.c.$ and letting $\mathbf{a}_1 = \hat{\mathbf{a}}_1 e^{i\Delta k\eta}$, where $k$ and $\Delta k$ are complex, the linearized (with respect to $\mathbf{Q}^{(1)}$) versions of Eqs. (3)-(4) may be solved giving

$$h_1 \simeq \sum_{\ell,n=-\infty}^{\infty} \frac{k a_0 \hat{a}_1 (-1)^\ell J_n (J_\ell - J_{\ell+1}) \exp i\theta_{\ell,n}}{h_0 \left[\bar{k} - (1+2\ell)k_0 - \Delta k\right]}, \tag{5}$$

$$\frac{\rho_1}{\rho_0} \simeq \sum_{\ell,n=-\infty}^{\infty} \frac{2k\bar{k} a_0 \hat{a}_1 (-1)^\ell J_n (J_\ell - J_{\ell+1}) \exp i\theta_{\ell,n}}{h_0^2 \left[\bar{k} - (1+2\ell)k_0 - \Delta k\right]^2}, \tag{6}$$

where $\bar{k} = k\gamma_{\perp 0}^2/h_0^2$, $\theta_{\ell,n} = [(1+2\ell+2n)k_0 + \Delta k]\eta$, $J_n = J_n(b)$ are Bessel functions with argument $b = ka_0^2/4k_0h_0^2$ and where $\mathbf{a}_0 \cdot \hat{\mathbf{a}}_1 = a_0\hat{a}_1$ has been assumed.

The dispersion relation for the SBH modes may be obtained by substituting $\rho^{(1)}$, Eq. (6), into the perturbed version of the wave equation, Eq. (1). Requiring both sides of the wave equation to have the same $\eta$ dependences yields

$$\Delta k \simeq \sum_{\ell=-\infty}^{\infty} \frac{\rho_0 k_p^2 \bar{k} a_0^2 (J_\ell - J_{\ell+1})^2}{4\gamma_0 h_0^2 \left[\bar{k} - (1+2\ell)k_0 - \Delta k\right]^2}. \tag{7}$$

The frequency, $\omega$, and the temporal growth rate, $\Gamma$, may be determined by setting $k = \omega/c - \Delta k$. A particular SBH mode has the phase dependence $a^{(1)} \sim \exp(i\omega\xi/c + i2ct\Delta k)$, where $\Gamma = -2c\mathrm{Im}(\Delta k)$. The resonant denominator $D_{k,k_0} = \bar{k} - (1+2\ell)k_0 - \Delta k$, on the right of Eq. (7), indicates that the frequency of the $\ell^{th}$ SBH mode is $\omega = NM_0\omega_0$, where $\omega_0 = ck_0$ is the pump laser frequency, $N = 1+2\ell$ is the harmonic number and $M_0 = h_0^2/\gamma_{\perp 0}^2$ is the frequency multiplication factor (arising from the relativistic doppler upshift), i.e., $M_0 = \gamma_0^2(1+\beta_0)^2/\gamma_{\perp 0}^2$ for an electron beam and $M_0 = 1$ for a plasma.

Assuming $|\omega/c| \gg |\Delta k| \gg k_p/\gamma_0^{3/2}$, the growth rate of the $\ell^{th}$ mode, via Eq. (7), is given by

$$\Gamma \simeq \sqrt{3}\left[\frac{\rho_0\omega_p^2\omega_0 M_0 F_\ell}{\gamma_0(1+M_0)^2}\right]^{1/3}, \tag{8}$$

where $F_\ell = b_\ell\left[J_\ell(b_\ell) - J_{\ell+1}(b_\ell)\right]^2$ is the harmonic coupling function[4] with $b_\ell = Na_0^2/4\gamma_{\perp 0}^2$. A plot of $F_\ell^{1/3}$ versus $b_0 = a_0^2/4\gamma_{\perp 0}^2$ for $N = 1, 3, 5...19$ is shown in Fig. 1. For a relativistic electron beam with $h_0 \simeq 2\gamma_0 \gg \gamma_{\perp 0}$, $\Gamma \simeq \sqrt{3}(\omega_p^2\omega_0\gamma_{\perp 0}^2 F_\ell/4\gamma_0^3)^{1/3}$. For a plasma, $\Gamma \simeq \sqrt{3}(\omega_p^2\omega_0 F_\ell/4\gamma_{\perp 0})^{1/3}$. The asymptotic value of $F_\ell$ for $\ell \gg 1$ and $a_0^2 \gg 1$ is given by $F_\ell \simeq 0.169\ell^{-1/3}$. Hence, asymptotically, $\Gamma \sim \ell^{-1/9}$. For small arguments where $b_\ell < 1$ and $a_0^2 \ll 1$, $F_\ell \simeq 2(Na_0^2/8)^N(\ell!)^{-2}$. In particular, notice that for $\ell = 0$ and $a_0^2 \ll 1$, the growth rate of the fundamental backscattered radiation from a plasma is $\Gamma \simeq \sqrt{3}(\omega_p^2\omega_0 a_0^2/16)^{1/3}$, which is the standard result for the Raman backscatter instability in the strong–pump regime for a linearly polarized laser.[10]

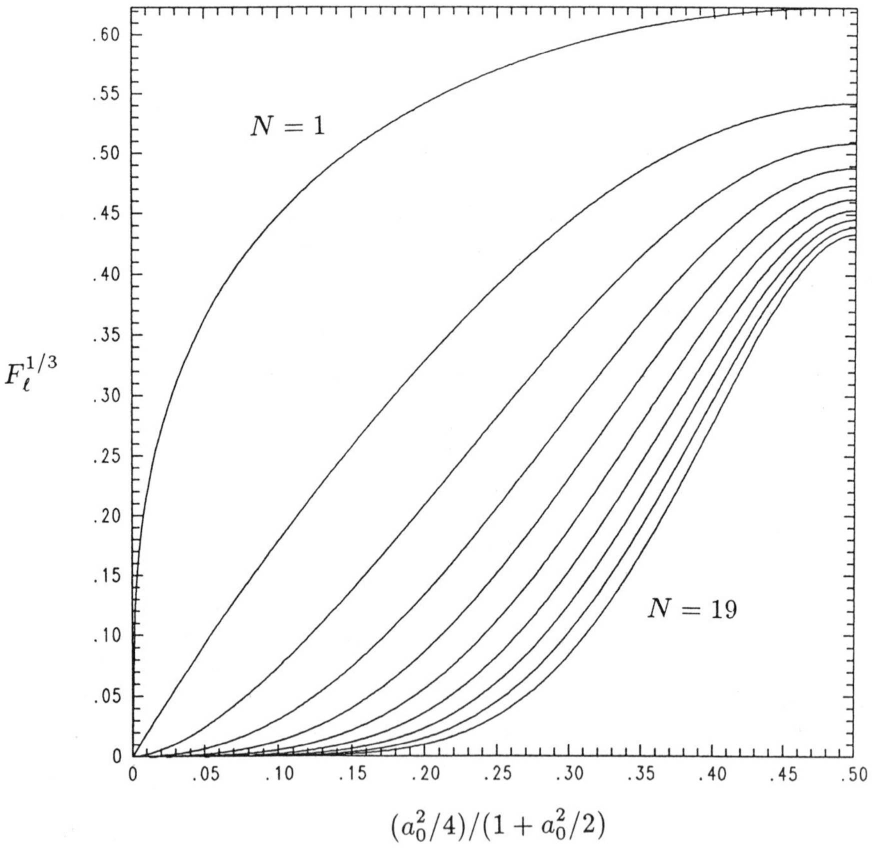


FIG. 1: The function $F_\ell^{1/3}$ (proportional to the growth rate $\Gamma$) versus the parameter $b_0 = (a_0^2/4)/(1 + a_0^2/2)$ for the harmonics $N = 1 + 2\ell = 1, 3, 5...19$.

## III. Saturation

Growth of the SBH radiation will continue until nonlinear effects (i.e., particle trapping) limit the amplitude. The saturation value of the SBH radiation may be determined from arguments based on electron trapping in the ponderomotive wave. Trapping occurs when the amplitude of the axial fluid velocity of the wave, $|\beta_z|$, becomes equal to the magnitude of the phase velocity of the wave, $|\beta_p|$. The axial fluid velocity is given by $\beta_z = (h^2 - \gamma_\perp^2)/(h^2 + \gamma_\perp^2)$, where $\gamma_\perp = (1 + a^2)^{1/2}$. In terms of equilibrium and perturbed quantities, $\beta_z = \beta_z^{(0)} + \beta_z^{(1)}$, where the average value of $\beta_z^{(0)}$ is $\langle \beta_z^{(0)} \rangle \simeq (M_0 - 1)/(M_0 + 1)$ and the leading order contribution to $\beta_z^{(1)}$ at resonance is $\beta_z^{(1)} \simeq 4M_0 h^{(1)}/h_0(1 + M_0)^2$. For a particular $\ell^{th}$ resonance, $h^{(1)}$ is composed of a collection of $n$ modes where the phase of the $(\ell, n)$ mode is given by $h_{\ell,n}^{(1)} \sim \exp(i\theta_{\ell,n} + ik\xi)$, as indicated by Eq. (5). The phase velocity of the $(\ell, n)$ mode at resonance is

$$\beta_p(\ell, n) = \frac{(M_0 - 1)Nk_0 - 2nk_0 - 2\Delta k}{(M_0 + 1)Nk_0 + 2nk_0}. \tag{9}$$

Trapping occurs first for the $n = 0$ mode, since it is the mode with phase velocity closest to the equilibrium fluid velocity, $\beta_p(\ell, 0) = \langle \beta_z^{(0)} \rangle + \delta\beta_p$, where $\delta\beta_p = -2\Delta k/(M_0 + 1)Nk_0$. The saturated amplitude of the SBH radiation, $|\hat{a}_1|$, may be obtained from the condition for deeply trapped electrons, $|\beta_z^{(1)}| \simeq 2|\delta\beta_p|$. This implies, using Eq. (5), that the ratio of power in the $\ell^{th}$ harmonic to that in the pump laser, $P_\ell/P_0 = N^2 M_0^2 |\hat{a}_1|^2/a_0^2$, is

$$\frac{P_\ell}{P_0} \simeq \left[\frac{\rho_0 \omega_p^2 (1 + M_0)}{\gamma_0 \omega_0^2 M_0^{1/2}}\right]^{4/3} \frac{F_\ell^{1/3}}{16 b_\ell J_0^2}. \tag{10}$$

Asymptotically ($\ell \gg 1$, $a_0^2 \gg 1$), $P_\ell/P_0 \sim \ell^{-10/9} J_0^{-2}$. Furthermore, the efficiency may be substantially increased by operating in a regime where $J_0(b_\ell) \simeq 0$. This corresponds to minimizing the perturbed fluid velocity $\beta_z^{(1)}(\ell, 0)$. At $J_0(b_\ell) = 0$, Eq. (10) no longer applies and saturation via particle trapping must be determined by consideration of the higher $n$ modes. In general, Eq. (10) is valid provided $P_\ell/P_0 < 1$. If this is violated, then saturation will be caused by some mechanism other than particle trapping, i.e., plasma thermalization or pump depletion.

## IV. Thermal Effects

The above results assumed a cold electron distribution. For a sufficiently thermal electron distribution, however, the axial velocity spread may become large enough to degrade the resonant interaction between the backscattered wave and the electrons. This thermal interaction regime corresponds to a weak resonant instability in which the growth rate is greatly reduced. The maximum thermal velocity spread which can be tolerated before the resonant interaction is degraded may be determined by considering the resonant denominator, $D_{k,k_0}$. The effects of an axial thermal velocity spread, $\beta_{th}$, on the resonance may be estimated by letting $\beta_0 \rightarrow \beta_0 + \beta_{th}$ in the expression for $D_{k,k_0}$, where $|\beta_{th}/(1+\beta_0)| \ll 1$. At resonance, this gives $D_{k,k_0} \simeq -(d_1+d_2)$, where $d_1 = (1+M_0^{-1})\Delta k$ and $d_2 = 2\gamma_0^2 N k_0 \beta_{th}$. Hence, in order for the effects of $\beta_{th}$ to be unimportant, it is necessary that $|d_2| \ll |d_1|$, which gives

$$\beta_{th} \ll \frac{F_\ell^{1/3}}{2\gamma_0^2 N}\left[\frac{\rho_o \omega_p^2 (1+M_0)}{\gamma_0 \omega_0^2 M_0^2}\right]^{1/3} . \tag{11}$$

Asymptotically ($\ell \gg 1$, $a_0^2 \gg 1$), the right side of Eq. (11) scales as $\ell^{-10/9}$. For an electron beam with $\gamma_0 \gg 1$, the normalized energy spread is given by $\Delta\gamma_{th}/\gamma_0 = \gamma_0^2 \beta_{th}$. For a stationary plasma, the thermal energy is given by $\Delta E_{th} = m_0 c^2 \beta_{th}^2/2$.

## V. Examples

Consider amplification of SBH radiation using a plasma with $n_0 = 10^{19}$ cm$^{-3}$. The pump laser parameters are $\lambda_0 = 1$ $\mu$m, $I_0 = 1.2 \times 10^{19}$ W/cm$^2$ ($a_0 = 3.0$), $r_0 = 5$ $\mu$m and $P_0 = 4.8$ TW. The interaction length is approximately two Rayleigh lengths, $2Z_R = 2\pi r_0^2/\lambda_0 = 150$ $\mu$m. Hence, a minimum pulse length of $4Z_R/c = 1$ ps is required. As an example, consider the third harmonic (and the fifth harmonic) at a wavelength of $\lambda_\ell = 3300$ Å (2000 Å). The e–folding length is $c/\Gamma = 1.8$ $\mu$m (2.0 $\mu$m). At saturation, $P_\ell/P_0 = 9.3 \times 10^{-5}$ ($5.6 \times 10^{-4}$), hence $P_\ell = 0.45$ GW (2.7 GW). The thermal requirement is $E_{th} < 75$ eV (22 eV). Plasma with sufficiently cold axial temperatures may be produced by laser–induced ionization.[11]

Consider amplification of SBH radiation using an electron beam with a current of 15 A, a radius of 10 $\mu$m and an energy of 250 keV ($\gamma_0 = 1.5$). The pump laser parameters are $\lambda_0 = 1$ $\mu$m, $I_0 = 5.5 \times 10^{18}$ W/cm$^2$ ($a_0 = 2.0$), $r_0 = 10$ $\mu$m and $P_0 = 8.6$ TW. The interaction length is either $2Z_R = 630$ $\mu$m or one–half the laser pulse length, whichever is shorter. As an example, consider the third harmonic at a wavelength of $\lambda_\ell = 1500$ Å. The e–folding length is $c/\Gamma = 35$ $\mu$m. At saturation, $P_\ell/P_0 = 1.1 \times 10^{-9}$, hence $P_\ell = 9.1$ kW. The electronic efficiency (the ratio of the saturated power to the initial electron beam power) is 0.47%. The thermal requirement is $\Delta\gamma_{th}/(\gamma_0 - 1) <$ 0.14%.

## VI. Conclusion

The generation of SBH radiation from electron beams and plasmas has been analyzed using a nonlinear fluid theory valid to all orders in the pump laser amplitude, $a_0$. The resonant frequency and growth rate of the various harmonics were determined in the strong–pump regime and the saturated power was obtained from particle trapping arguments. These results are summarized in Table I. The most stringent constraint on the generation of SBH radiation is the restriction of the axial velocity spread, which implies that cold axial electron distributions are necessary. The expressions for the growth rate, saturated power and thermal velocity spread indicate that high electron densities are required. This favors the use of dense plasmas over that of relativistic electron beams. However, the use of electron beams has an advantage in the ability to tune the frequency of the SBH radiation by adjusting the electron beam energy or the pump laser amplitude. As the capability for producing dense plasmas and electron beams with small thermal spreads improves, along with advancements in ultra–high power laser technology, SBH generation may provide a practical method for producing coherent radiation in the XUV regime.

## Acknowledgments

The authors wish to acknowledge the numerical assistance of T. Swyden. This work was supported by the Office of Naval Research and the Department of Energy.

## Table I.

## Growth Rates, Efficiencies and Thermal Requirements for Stimulated Backscattered Harmonic Generation†

| | Laser-Plasma ($M_0 = 1,\ \gamma_0 = 1$) | | Laser-Electron Beam ($M_0 \gg 1,\ \gamma_0 \gg 1$) | |
|---|---|---|---|---|
| | Arbitrary $a_0,\ N$ | $a_0 \ll 1,$ $N = 1$ | Arbitrary $a_0,\ N$ | $a_0 \ll 1,$ $N = 1$ |
| Growth Rate $\Gamma/\omega_0$ | $\sqrt{3}\left(\frac{\omega_p^2 F_\ell}{4\omega_0^2\gamma_{\perp 0}}\right)^{1/3}$ | $\sqrt{3}\left(\frac{\omega_p a_0}{4\omega_0}\right)^{2/3}$ | $\frac{\sqrt{3}}{\gamma_0}\left(\frac{\omega_p^2\gamma_{\perp 0}^2 F_\ell}{4\omega_0^2}\right)^{1/3}$ | $\frac{\sqrt{3}}{\gamma_0}\left(\frac{\omega_p a_0}{4\omega_0}\right)^{2/3}$ |
| Laser Eff.‡ $P_\ell/P_0$ | $\frac{\omega_p^2\gamma_{\perp 0}\Gamma/N}{\sqrt{3}\omega_0^2 a_0^2 J_0^2}$ | $\left(\frac{\omega_p^2}{2\omega_0^2 a_0}\right)^{4/3}$ | $\frac{\omega_p^2\gamma_0\Gamma/N}{\sqrt{3}\omega_0^2 a_0^2 J_0^2}$ | $\left(\frac{\omega_p^2}{2\omega_0^2 a_0}\right)^{4/3}$ |
| Electron. Eff. $\eta_e$ | — — | — — | $\frac{\Gamma/N}{2\sqrt{3}\omega_0 J_0^2}$ | $\frac{1}{2\gamma_0}\left(\frac{\omega_p a_0}{4\omega_0}\right)^{2/3}$ |
| Thermal Spread $\Delta E_{th}/m_0c^2 \ll$ | $\frac{1}{6}\left(\frac{\Gamma}{\omega_0 N}\right)^2$ | $\frac{1}{2}\left(\frac{\omega_p a_0}{4\omega_0}\right)^{4/3}$ | — — | — — |
| Energy Spread $\Delta\gamma_{th}/\gamma_0 \ll$ | — — | — — | $\frac{\Gamma/N}{2\sqrt{3}\omega_0}$ | $\frac{1}{2\gamma_0}\left(\frac{\omega_p a_0}{4\omega_0}\right)^{2/3}$ |

† $\omega_0$ is the pump laser frequency, $\omega_p$ is the plasma frequency, $\gamma_0$ is the initial relativistic factor, $a_0 = |e|A_0/m_0c^2$ is the normalized pump laser amplitude, $\gamma_{\perp 0} = (1 + a_0^2/2)^{1/2}$, $N = (2\ell + 1)$ is the harmonic number, $F_\ell = b\left[J_\ell(b) - J_{\ell+1}(b)\right]^2$ is the harmonic coupling function, $b = Na_0^2/4\gamma_{\perp 0}^2$ and $M_0$ is the frequency amplification factor, $\omega = NM_0\omega_0$.

‡ Formulas valid provided $a_0 > \omega_p^2/2\omega_0^2$.

## References


1. P. Maine, D. Strickland, P. Bado, M. Pessot and G. Mourou, IEEE J. Quantum Electron. **QE-24**, 398 (1988); M. Pessot, J.A. Squire, G.A. Mourou and D.J. Harter, Opt. Lett. **14**, 797 (1989); M. Ferray, L.A. Lompre, O. Gobert, A. L'Huillier, G. Mainfray, C. Manus and A. Sanchez, Opt. Commun. **75**, 278 (1990); M.D. Perry, F.G. Patterson and J. Weston, Opt. Lett. **15**, 1400 (1990); C. Sauteret, D. Husson, G. Thiell, S. Seznec, S. Gary, A Migus and G. Mourou, Opt. Lett. **16**, 238 (1991); T.S. Luk, A. McPherson, G. Gibson, K. Boyer and C.K. Rhodes, Opt. Lett. **14**, 1113 (1989).
2. P. Sprangle, E. Esarey and A. Ting, Phys. Rev. Lett. **64**, 2011 (1990); Phys. Rev. A **41**, 4463 (1990); A. Ting, E. Esarey and P. Sprangle, Phys. Fluids B **2**, 1390 (1990).
3. C. Roberson and P. Sprangle, Phys. Fluids B **1**, 3 (1989).
4. R.C. Davidson, Phys. Fluids **29**, 267 (1986).
5. A. Hasegawa, K. Mima, P. Sprangle, H.H. Szu and V.L. Granatstein, Appl. Phys. Lett. **29**, 542 (1976); A. Gover, C.M. Tang and P. Sprangle, J. Appl. Phys. **53**, 124 (1982); Y. Carmel, V.L. Granatstein and A. Gover, Phys. Rev. Lett. **51**, 566 (1983).
6. L.R. Elias, Phys. Rev. Lett. **42**, 977 (1979); I. Kimel, L. Elias and G. Ramiar, Nucl. Instrum. Methods **A250**, 320 (1986).
7. B.G. Danly, G. Bekefi, R.C. Davidson, R.J. Tempkin, T.M. Tran and J.S. Wurtele, IEEE J. Quantum Electron. **QE-23**, 103 (1987); J. Gea–Banacloche, G.T. Moore, R.R. Schlicher, M.O. Scully and H. Walther, IEEE J. Quantum Electron. **QE-23**, 1558 (1987); J.C. Gallardo, R.C. Fernow, R. Palmer and C. Pellegrini, IEEE J. Quantum Electron. **QE-24**, 1557 (1988).
8. K. Mima, Y. Kitawaga, T. Akiba, K. Imasaki, S. Kuruma, N. Ohigashi, S. Miyamoto, S. Fujita, S. Nakayama, Y. Tsunayaki, H. Motz, T. Taguchi, S. Nakai and C. Yamanaka, Nucl. Instrum. Methods **A272**, 106 (1988).
9. Y. Seo, Phys. Fluids B **3**, 797 (1991).
10. J.F. Drake, P.K. Kaw, Y.C. Lee, G. Schmidt, C.S. Liu and M.N. Rosenbluth, Phys. Fluids **17**, 778 (1974); D.W. Forslund, J.M. Kindel and E.L. Lindman, Phys. Fliuds **18**, 1002 (1975); W.L. Kruer, *The Physics of Laser Plasma Interactions* (Addison–Wesley, Reading, MA, 1988).
11. P.B. Corkum, N.H. Burnett and F. Brunel, Phys. Rev. Lett. **62**, 1259 (1989); N.H. Burnett and P.B. Corkum, J. Opt. Soc. Am. **B6**, 1195 (1989).

**Research Trends in Physics: Coherent Radiation Generation and Particle Acceleration**
Editorial Board: J.M. Buzzi, A. Prokhorov (Editor-in-Chief), P. Sprangle, and K. Wille
*La Jolla International School of Physics*, The Institute for Advanced Physics Studies, La Jolla, California

# Accelerator Technology For Bright Radiation Beam*

**Kwang-Je Kim**

Lawrence Berkeley Laboratory
University of California
Berkeley, CA 94720

## Abstract

We review the current and future accelerator technologies for generation of high brightness radiation beam.

## 1. Introduction

The most promising way known at present time to generate intense, bright radiation in short wavelength range is to pass relativistic electron beams through periodic magnetic structure called undulator. The spontaneous radiation from the periodic transverse acceleration in undulator is referred to as the undulator radiation. The high brightness of the undulator radiation is the main reason why several "third generation" synchrotron radiation facilities are being built around the world. Free electron lasers (FELs), which can be regarded as a further development of the undulator radiation, will produce fully coherent, intense, tunable radiation both in the infrared and in the ultraviolet and shorter wave length regions. An optimum operation of these radiation devices requires that certain conditions on the quality of the electron beam are satisfied. Here we review those conditions and the current and future accelerator technology that will provide electron beams of requisite quality.

*This work was supported by the Director, Office of Energy Research, Office of Basic Energy Sciences, Material Sciences Division of the U.S. Department of Energy under Contract No. DE-AC03-76SF00098.

## 2. Electron beam qualities for Undulators and Free Electron Lasers

Beam quality is mainly described by three quantities: beam current, energy spread and emittance. The emittance is a measure of transverse spread and is given by the area of the transverse phase space occupied by the electron beam. The current divided by the emittance is known as the brightness.

The requirement on emittance is that it be less than about λ, the wavelength of the radiation. To be quantitative, we introduce the rms emittance as follows:

$$\varepsilon_x^o = \sqrt{\langle x^2\rangle\langle x'^2\rangle - \langle x \cdot x'\rangle^2} \tag{1}$$

where x and x' are the transverse position and angle, respectively, and the brackets denote the operation of taking the ensemble average. The emittance defined in the above is sometimes referred to as the "unnormalized" emittance, since it is not invariant under adiabatic acceleration. It is also convenient to introduce the normalized emittance $\varepsilon_x$ as follows:

$$\varepsilon_x = \gamma\, \varepsilon_x^o \quad , \tag{2}$$

where $\gamma = E/mc^2$, E = electron energy, m = electron mass, c = speed of light. The normalized emittance is invariant under adiabatic acceleration as well as under beam transport consisting of linear focussing elements and drifts.

We can also associate phase space with a radiation field, and calculate its emittance [1]. The radiation emittance $\varepsilon_R$ is minimum for the Gaussian $TEM_{00}$ mode with the value

$$\varepsilon_R = \lambda/4\pi \tag{3}$$

To generate a transversely coherent radiation beam, it is clearly necessary that the electron beam emittance be smaller than the radiation emittance,i.e.,

$$\varepsilon_x^o \lesssim \frac{\lambda}{4\pi} \tag{4}$$

The transverse coherence is a necessary condition for FEL operation. For undulator radiation, the condition is desirable for high brightness. As an example, consider an FEL at "water window", $\lambda = 30$ Å. With a 1 GeV electron beam, the required normalized emittance is

$$\varepsilon_x \lesssim 0.5 \text{ mm-mrad} \tag{5}$$

The requirement on the relative energy spread for undulator radiation and FELs is that it be smaller than the typical bandwidth of the radiation process, i.e.,

$$\frac{\Delta\gamma}{\gamma} \lesssim \frac{1}{2N} \tag{6}$$

where N is the number of undulator periods. Typically this requirement leads to the condition that the relative beam energy spread $\Delta\gamma/\gamma$ be less than 0.1%.

The required current is typically several hundred Amperes.

The discussion in the above applies to undulator radiation and low gain FELs. For high gain FELs in the exponential gain regime [2], the emittance restriction Eq. (4) remains basically the same but N in Eq. (6) should be interpreted as the number of periods in one gain length. Rigorous discussion of beam quality effect in the high gain regime was discussed recently based on detailed FEL theory [3], [4].

## 3. Accelerator technology

Different considerations are applicable for storage rings and linacs as a source of bright electron beams. In the following, we consider them in turn.

### 3.1 Storage rings

Electron storage rings are promising as a source of bright electron beams because the unique damping mechanism improves and maintains the beam qualities in these machines [5].

In an ideal storage ring consisting of M achromatic bends, the minimum achievable emittance is

$$\varepsilon_x^o = \left(7.7 \times 10^{-13}\ \text{m-rad}\right) \frac{\gamma^2}{M^2} \tag{7}$$

This equation implies that storage rings for low emittance must have large M and thus are big: typical storage rings for state-of-the-art light source have a circumference on the order of a few hundred meters and larger.

There are several instabilities in storage rings that limit the beam quality [6]. Among these, the most significant for the present discussion is the so-called microwave instability [7], as a result of which the achievable peak current I is limited by the energy spread as follows:

$$I \leq 2\pi\alpha \frac{E}{e} \frac{1}{Z_n/n} \left(\frac{\Delta\gamma}{\gamma}\right)^2 \tag{8}$$

where e is the electron charge, $Z_n/n$, known as the broad band impedance, characterizes the interaction of the beam with the vacuum chamber environment, and $\alpha$ is a quantity known as the momentum compaction which is the coefficient relating the change in the orbit frequency to the change in momentum. For a large I, it is desirable to minimize the impedance and maximize $\alpha$. For modern storage rings, the achievable value of $Z_n/n$ is limited to about 1 Ohms. On the other hand,the momentum compaction is proportional to $1/M^2$, and therefore cannot be arbitrarily increased without compromising the emittance requirement because of Eq. (7).

The accelerator community has gained considerable experience recently in the arts of building high brightness electron storage rings in connection with the synchrotron radiation facility construction projects at several places around the world. The unnormalized beam emittance and the energy spread in these storage rings is typically about $10^{-8}$ mm-mrad and 0.1%, respectively. Undulators placed in the straight sections of these machines produce tunable, high brightness, short wavelength radiation beam that cannot be obtained by any other method. The performance of such devices are summarized in Fig. 1.

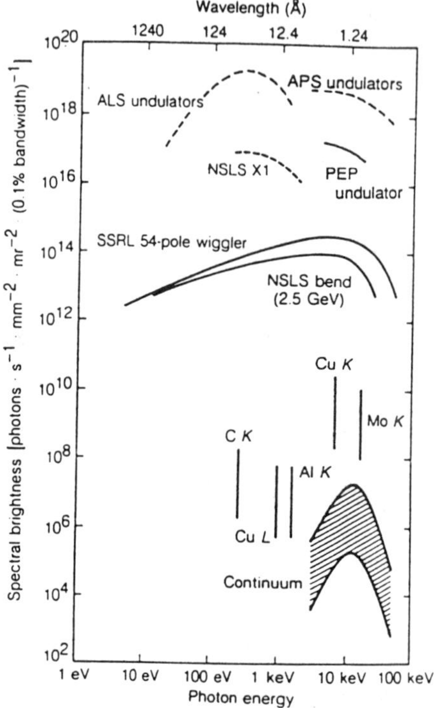


**Figure 1.**

Average spectral brightness within a 0.1% bandwidths, as a function of photon energy, for a variety of synchrotron radiation sources.

An interesting idea to obtain a higher peak current in a storage ring is to design the lattice so that the momentum compaction vanishes, i.e., the concept of the isochronous storage ring [8], [9]. In such a ring, the microwave instability develops so slowly that it becomes irrelevant. A more detailed analytical can experimental investigation is necessary to understand the stability of the isochronous ring.

### 3.2 RF Linacs

In RF linacs, it is necessary to start out with a good emittance beam, since there is no damping mechanism. The most commonly used electron source is the thermionic gun based on emission from heated cathode surface. Although electron beam from thermionic guns have a low emittance (the normalized rms emittance is less than 1 mm-mrad) the current is rather modest, being about one Amperes. In order to obtain several hundred Amperes, it is therefore necessary to bunch the long pulses from the gun to shorter pulses of higher current. This "bunching" process involves nonlinear mixing in phase space, causing a significant emittance degradation. Thus, the emittance the 10 nC pulse for the SLC gun after bunching is about 300 mm-mrad [10]. Another example is the emittance of the 1 nC, 10 psec pulse for the ALS injector, which is about 30 mm-mrad [11].

Brighter beams appear to be possible with the recent development of the laser driven RF guns[ 12]. In this gun, a photo-emissive surface, the photo-cathode, is placed in an RF cavity and is illuminated by intense laser beams to knock out the electrons from the surface. The advantage over the thermionic gun is that the current from photo-cathode is high so that bunching is not necessary. The time structure of the electron beam is controlled by that of the laser beam, and can be tailored to match to the requirements of the RF accelerating sections.

The emittance of RF photo-cathode gun is larger than the intrinsic emittance of the photo-emission process due to the time variation of the RF field and due to the action of the space charge force while the beam is accelerated to a relativistic energy [13]. Of these, the latter effect is usually the more important and gives rise to the following emittance:

$$\varepsilon_x^{sc} = \frac{\pi}{4}\left(\frac{2mc^2}{E_{acc}}\right)\frac{I}{I_A}\frac{1}{(3\sigma_x/\sigma_z + 5)} \tag{9}$$

where $E_{acc}$ is the peak electric field of acceleration, $I_A$=17,000 A is the Alfven current, $\sigma_x$ and $\sigma_z$ is respectively the rms value of the transverse and the longitudinal beam sizes.

With the parameters of the BNL photo-cathode gun [14], Eq. (9) gives $\varepsilon_x$=4 mm-mrad, which , although smaller by an order of magnitude than that produced by thermionic guns, is still an order of magnitude larger than that required for an FEL at water window, Eq. (5). It is possible to improve the emittance of the RF photo-cathode further by correcting the correlated part of the emittance growth [15],[16]. However, it would be difficult to obtain emittance values much smaller than 1 mm-mrad required for x-ray FELs. A possible way to over come this impasse is to introduce a correlation between the transverse distribution and energy spread by means of $TM_{210}$ mode of microwave cavities [17]. Such a "beam conditioning" will effectively remove the emittance restriction, Eq. (4).

The RF cavities for acceleration can be either of the room temperature type or of the superconducting type. The room temperature cavities must necessarily operated in a pulsed mode because of the large Ohmic loss at copper surfaces. On the other hand, the RF loss in superconducting cavities is negligible. Thus linacs using superconducting cavities can be operated in a CW mode, thus producing a very high average electron beam power and hence the radiation power. Another significant advantage of the CW operation is the possibility that various beam fluctuations can be controlled to a much lower level than is feasible in room temperature linacs. In addition, the RF frequency and cavity shape can also be optimized with the view to minimize the electron beam instabilities. The development of superconducting RF cavity technology benefits from nuclear physics projects and high energy physics collider projects.

## 4. Prospects for the future development in FELs

Although spontaneous radiation from undulators in storage rings provide intense radiation with some degree of coherence in hitherto inaccessible wavelength regions, it is the FEL technology that offers the opportunity for generation of truly coherent radiation. The FEL development in the future needs to be pursued in two separate directions. In the infrared region, where the technology for the accelerator and the optical cavities are available, the challenge is to construct user facilities. In the short wavelength region, the challenge is to develop the technology. The electromagnetic spectrum and the projected FEL capability is shown and compared with other sources in Fig. 2.

Infrared FELs have been built and operated, but these first devices were oriented toward learning about FELs rather than toward serving a community of users. The task in the future is to build an FEL that satisfies a unique set of criteria required for a user facility. An important such criterion is the stability of the FEL output,in wavelength, in intensity and in direction. Thus the choice of the accelerator system and design must be made with the view of ensuring the required stability. As an example, the IR FEL for the Combustion Dynamic Facility [18] is based on a 500 MHz superconducting RF accelerator with the fluctuation in the electron beam energy reduced to less than 0.01 %. Another important criterion is the ease of wavelength coverage and tuning. For

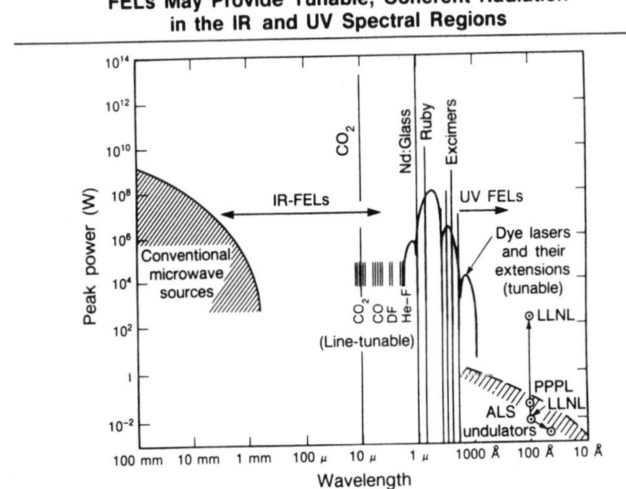


2. Performance of FELs compared to other sources in different wavelength regions.

this purpose, the FEL optical system that includes the optical cavity, outcoupling, and the optical transport must be properly designed.

The short wavelength record in FEL is 2400 Å obtained at Novosibirsk in 1988 [19]. Realizing an FEL operation at wavelength shorter than 1000 Å is more difficult because the electron beam requirements are more demanding, and perhaps more importantly, because high reflectivity mirrors required for optical cavity are currently not available. The use of mirrors can be avoided if FELs are run in the amplifier mode. As the input radiation, if available at all, is usually quite weak, in the short wavelength region, the gain of the FEL in the amplifier mode must necessarily be very high. In the case of extreme high gain (one million or larger), the FEL can amplify the initial noise signal( the undulator radiation) to intense coherent radiation [20]. Although operation in this so-called self-amplified spontaneous emission(SASE) regime requires neither mirrors nor coherent input signals, the requirements on the electron beam qualities and the undulator construction is very demanding. Despite these difficulties, several laboratories are pursing short wavelength FELs [21], [22], [23], [24] because of the potentially high scientific pay-off [25].

## References

[1] K-J. Kim, Nucl. Instr. Meth. **A246**,71 (1986).

[2] R. Bonifacio, C. Pellegrini and L. N. Narducci, Opt. Commun. **50**, 373 (1984).

[3] S. Krinsky, L.-H. Yu, and R. L. Gluckstern, Phys. Rev. Lett., **64**, 3011 (1990).

[4] Y. H. Chin, K.-J. Kim, M. Xie, LBL Preprint LBL-30673 (Aug. 1991).

[5] For a review, see M. Sands, SLAC preprint, SLAC-121 (1970).

[6] See for example, S. Chattopadhay, SPIE Proc. Vol. 1227, 160 (1990).

[7] D. Boussard, CERN LAB II/RF/INT/75-2 (1975).

[8] S. Chattopadhyay et. al., Proc. ICFA workshop on Low Emittance $e^-$-$e^+$ Beams, p. 76, BNL-52090 (1987).

[9] C. Pellegrini and D. Robin, Nucl. Instr. Meth., **A301**, 27 (91).

[10] M.B. James, J.E. Clendenin, S.D. Ecklund and R.H. Miller,IEEE Trans. Nucl. Sciences, Vol NS 30,2992 (1983).

[11] C. H. Kim, LBL preprint LBL-29227 (June 1990).

[12] J.S. Fraser, R.L. Sheffield, E.R. Gray and P.M. Giles, Proc. 1987 Particle Acc. Conf., IEEE Cat. No. 87 CH 2387-9, 1705 (March, 1987).

[13] K-J. Kim, Nucl. Instr. Meth., **A275**, 201 (1989).

[14] K. Batchelor, et. al., Proc. 1988 Linac Conf. CEBAF Report 89-001, 540 (1988).

[15] B. Carlsten, Nucl. Instr. Meth., **A285**, 313 (1989).

[16] J. C. Gallardo and R. B. Palmer, Nucl. Instr. Meth. **A304**, 345 (1991).

[17] A. M. Sessler, D. H. Wihittum and L.-H. Yu, Phys. Rev. Lett., **68**, 309 (1992).

[18] CDRL-FEL conceptual design report, LBL preprint (under preparation).

[19] G. N. Kulipanov, et. al., Nucl. Instr. Meth., **A296**, 1 (1990).

[20] K.-J. Kim, C. Pellegrini, AIP Conference Proc. No. 147, 17 (1986); K.-J. Kim, Phys. Rev. Lett., **57**, 1871 (1986).

[21] K.-J. Kim, J. J. Bisognano, A. A. Garren, K. Halbach and J. M. Peterson, Nucl. Instr. Meth., **A239**, 54 (1985).

[22] J. C. Goldstein, B. D. McVey and B. E. Newnam, SPIE Proc. Vol. 582, 350 (1986).

[23] J. E. La Sala, D. A. G. Deacon, J. M. J. Madey, Nucl. Instr. Meth., **A250**, 262 (1986).

[24] UV FEL Conceptual Design Report, BNL (December 1991).

[25] K.-J. Kim and A. M. Sessler, Science, **Vol. 250**, 88 (1991).

**Research Trends in Physics: Coherent Radiation Generation and Particle Acceleration**
Editorial Board: J.M. Buzzi, A. Prokhorov (Editor-in-Chief), P. Sprangle, and K. Wille
*La Jolla International School of Physics*, The Institute for Advanced Physics Studies, La Jolla, California

# Beam-Gap Interaction -- From Gigawatt to Nanowatt

**Y.Y. Lau**

Naval Research Laboratory
Beam Physics Branch
Plasma Physics Division
Washington, DC 20375-5000

Recent interests in nanoelectronics (< $10^{-9}$W) and ultra-high power microwave generation (> $10^{9}$W) have invited a reexamination of the interaction of an electron beam with a vacuum gap. In the former case, the dimension may become so small that a quantum mechanical treatment would be required. For the latter case, a classical (non-quantum) analysis suffices, but the intense space charge that necessarily accompanies a high current beam renders the conventional klystron theory invalid. Here, we summarize our recent studies of these two regimes. We pay special attention to the comparison with the conventional theories.

We use a one-dimensional model. The gap is formed by two parallel plates, and may be subject to a bias voltage $V_g$. When the (one-dimensional) electron beam enters the gap, it carries a current density J and electron energy E. In the studies of ultra-high power microwave generation, we allow J, E and $V_g$ to be modulated.

(a) Limiting Current in a Quantum Diode

In general, when the injected current is too high, the electrostatic potential in the self-fields of the beam becomes so high that the beam electrons cannot propagate across the gap. A steady state, non-relativistic, non-quantum mechanical treatment shows that the upper bound of the transmittable current density $J_c$ is proportional to $E^{3/2}$ [Child-

Langmuir Law[1]]. When the gap width D becomes too small, specifically, when the electron wavelength is of order (or greater than) D, quantum mechanical treatment would be required. The quantum version of Child-Langmuir Law[2] shows that the limiting current $J_q$ is proportional to $E^{1/2}$ for small E. Thus, for small E, $J_q >> J_c$. The underlying reason for this is that tunneling effects permit current transmission beyond the classical Child-Langmuir value. The smooth transition from $J_q$ to $J_c$ as E increases is shown in Ref. 2.

(b) Dynamical Limiting Current and Microwave Conversion Efficiency in a High Power Microwave Gap

Multi-gigawatts of coherent rf power (at 1.3 GHz) were recently generated in the relativistic klystron amplifier at the Naval Research Laboratory.[3] This device employs a modulated intense electron beam whose instantaneous current may reach the limiting value beyond which electrons begin to be reflected by a gap. Our analysis includes relativistic effects, but is otherwise classical. Numerical codes have been developed to extend the Child-Langmuir Law to include a time-varying gap voltage and a modulated incident beam.[4] In Ref. 4, the meaning of beam-loaded gap capacitance under dynamical conditions is also clarified. We then examine the frequency stability and optimal conversion efficiency when the gap is connected to an external load, modeled by an RLC circuit. After some study over a vast parameter space, we find that a fully modulated beam cannot deliver to the load, on the average, more than 57 per cent of its beam power because of nonlinear beam loading. We reach this conclusion using beam parameters similar to the NRL relativistic klystron amplifier experiments.[3] If the beam current is only 60 per cent modulated, the maximum power conversion efficiency from the beam to the load is found to be only 35 per cent, a value consistent with experimental observations.[3]

This work was supported by the Office of Naval Research and by the Strategic Defense Initiative Organization/Innovative Science and Technology Office, managed by the Harry Diamond Laboratory. My interest in high power diodes is sustained by extensive collaboration with my colleagues in Refs. 2-4; in particular, D. G. Colombant and M. Friedman.

## References

1. See, e.g., R. C. Davidson, Physics of Nonneutral Plasmas, (Addison-Wesley, Redwood City, CA, 1990), p. 463.

2. Y. Y. Lau, D. Chernin, D. G. Colombant and P.-T. Ho, Phys. Rev. Lett. 66, 1446 (1991).

3. M. Friedman, J. Krall, Y. Y. Lau and V. Serlin, J. Appl. Phys. 64, 3353 (1988); Rev. Sci. Instrum. 61, 171 (1990); IEEE Trans. PS-18, 553 (1990).

4. D. G. Colombant and Y. Y. Lau, Phys. Rev. Lett. 64, 2320 (1990); Also, in Proc. Soc. Photo-Optical Instrum. Eng. Vol. 1407, p. 13 (1991), Ed.: H. E. Brandt.

Research Trends in Physics: Coherent Radiation Generation and Particle Acceleration

*La Jolla International School of Physics*, The Institute for Advanced Physics Studies, La Jolla, California

# Optical Guiding of Relativistically Strong Laser Pulses in Plasma

**A.G. Litvak, V.A. Mironov, and A.M. Sergeev**

Institute of Applied Physics, Russian Academy of Sciences
603600 Nizhny Novgorod, Russia

A promising method for the generation of strong electric fields in a plasma for the high-energy particle acceleration is to excite plasma waves using ultrashort laser pulses.

As applied to accelerators, this idea was first formulated back in [1] but a detailed consideration of the nonlinear processes in a plasma has been performed in recent years [2-6] when such experiments became feasible. The excitation of a plasma wave by a short (compared to the plasma wavelength) laser pulse is due to the ponderomotive force; physically, this process is similar to plasma wave emission by a moving discharge. The amplitude of the excited wave is maximum for pulse duration equal to the plasma wavelength $\ell \simeq c/\omega_p$ , when this amplitude reaches the magnitude of the order of the ponderomotive potential [4]. Therefore, to generate strong electric fields in a plasma it is necessary to use relativistically strong short laser pulses the propagation of which in a plasma is defined by a competition between the relativistic and the electrostriction nonlinearities. These are appropriate conditions for the laser beam relativistic self-focusing and optical guiding effects leading to the cumulative generation of a plasma wake field and to a noticeable increase of its longitudinal extension.

The self-focusing effects of electromagnetic wave beams in a plasma, that are due to the relativistic dependence of electron mass on their oscillatory energy, were first considered in [7] and later discussed in [8] and [9]. In those, as well as in many papers to follow (see, e.g. 10–13 ) it was shown that if the wave beam power P exceeds the so-called critical power of self-focusing $P_{cr}$, which is given by $P_{cr} = 15\,\omega^2/\omega_p^2$ GW for the case of relativistic nonlinearity, then the beam is focused to the scale that is determined by the effects of nonlinearity saturation and the power on the order of the critical magnitude is subsequently "trapped" into the regime of nonlinear optical guiding. In 4 it is shown, however, that this is true only for rather long pulses ( $\ell \gg c/\omega_p$) while the competition between the relativistic and the electrostriction nonlinearities, at which the plasma oscillation excitation along the pulse front attenuates the total nonlinearity of the plasma, is radically important for ultrashort pulses. The self-focusing in this case was considered qualitatively in [4], where it was shown that the front of the wave packet must propagate like in the linear case, i.e. with diffraction spreading, but the bulk of it can be trapped into the optical guiding regime. Nevertheless, a qualitative theory of the process has not been developed yet. In this paper an attempt is made to investigate qualitatively the self-focusing processes of an ultrashort laser pulse as it propagates in a rarefied plasma.

1. Basic equations. The excitation of a plasma wake by a short laser pulse, with the nonlinear deformation of its transverse and longitudinal spatial structure taken into account, can be investigated within the following set of equations:

$$-2ik\frac{\partial A}{\partial z}+\frac{\omega_p^2}{u^2\omega^2}\frac{\partial^2 A}{\partial \tau^2}+\Delta_\perp A+\frac{\omega_p^2}{c^2}\frac{\Phi}{1+\Phi}A=0 \qquad (1)$$

$$\frac{\partial^2\Phi}{\partial\tau^2}-\omega_p^2\frac{1+|A|^2-(1+\Phi)^2}{2(1+\Phi)^2} \qquad (2)$$

which is an elementary generalization of the equations which were used for the one-dimensional case in [4] and [5]. In these equations we have adopted the notation: $\Phi=e\varphi/m_0c^2$, the scalar potential of plasma oscillations; $A=eA/m_0c^2$, a slowly varying complex amplitude of the vector-potential of a circularly polarized electromagnetic wave; $\tau=t-z/u_0$, the local time; $u_0=k^2c^2/\omega^2$, the group velocity of the wave packet; $\Delta_\perp=\partial^2/\partial x^2+\partial^2/\partial y^2$, the transverse Laplacian; $\omega_p=(4\pi e^2N_0/m)^{1/2}$, the Langmuir frequency; $N_0$, the electron density in unperturbed plasma; $\omega$ and $k=\omega/c$, the frequency and the wavenumber of the electromagnetic wave, respectively. We suppose here that the wave beam radius $a$ is larger than the plasma wavelength, $a \gg c/\omega_p$.

Below we shall consider the propagation of a short (compared to the excited plasma wavelength $\lambda_p = 2\pi c/\omega_p$ ) laser pulse in a rarefied ( $\omega_p \ll \omega$ ) plasma along paths of limited length $z \ll c\omega^4/\omega_p^3$ , where we can neglect the dispersion spreading effects, i.e. omit the term with $\partial^2 A/\partial \tau^2$ in (1).

2. Weak nonlinearity. In the simplest case of weak nonlinearity ( $A^2 \ll 1$ , $\varphi \ll 1$ ) the amplitude of the plasma wave, generated by such a short pulse, is given by

$$\varphi_m = \omega_p \int_{-\infty}^{\infty} |A|^2 d\tau \tag{3}$$

Here the self-consistent distribution of the field A is described by

$$-i \frac{\partial A}{\partial z} + \Delta_\perp A - \varphi A = 0 \tag{4}$$

$$\frac{\partial^2 \varphi}{\partial \tau^2} = -|A|^2 \tag{5}$$

where new dimensionless variables are introduced:

$$z_{(n)} = \frac{\omega_p}{2\omega} k_p z, \quad \vec{r}_{(n)} = k_p \vec{r}_\perp, \quad \tau_{(n)} = \omega_p \tau, \quad k_p = \omega_p / c,$$

elsewhere below the index n is omitted.

Equations (4) and (5) have an ordinary integral:

$$\int |A(\vec{r}, z, \tau)|^2 d\vec{r} = W(\tau) \tag{6}$$

which corresponds to retaining the total energy flux in the beam:

$$P = \tilde{P} W(\tau), \quad \tilde{P} = \frac{m^2 c^5}{4\pi e^2} \cdot \frac{\omega^2}{\omega_p^2}$$

To analyse the solution we can make use of the methods [14] for the investigation of the spatio-temporal wave collapses in media with nonstationary nonlinearity. In the case of a rectangular pulse ($W(\tau)$ = const inside the pulse) the set (4)-(5) has a self-similar solution:

$$\begin{aligned} A &= (\exp \alpha\tau)\, u(\zeta) \exp(i\gamma z \exp 2\alpha\tau) \\ \varphi &= (\exp 2\alpha\tau)\, V(\zeta), \quad \zeta = r \exp \alpha\tau \end{aligned} \tag{7}$$

The self-similar functions $V(\zeta)$ and $u(\zeta)$ correspond to localized solutions of the set of equations:

$$\begin{aligned} &\Delta u + (V - \gamma) u = 0 \\ &\zeta^2 \alpha^2 \frac{\partial^2 V}{\partial \zeta^2} + 5\alpha^2 \zeta \frac{\partial V}{\partial \zeta} + 4\alpha^2 V = -u^2 \end{aligned} \tag{8}$$

The magnitude of the self-similarity parameter $\alpha$ can be determined by a parabolic approximation of potential distribution near the axis (r = 0). Consequently, $\alpha = \sqrt{W}/4$ and the vector-potential spatial distribution has a form

$$u = \frac{\sqrt{W}}{4} \exp - \zeta^2 / a_0^2 \tag{9}$$

where $a_0$ is the beam width at the forefront of the pulse, while $\tilde{P}$ has the dimension of power and differs from the critical self-focusing power by the numerical factor of the order of $2\pi$ in the case of stationary cubic nonlinearity.

Employing Eqs. (3) and (9) we obtain a relation for the amplitude of a plasma wave (in dimensional parameters):

$$\varphi_m = \frac{\omega^2}{\omega_p^2} \frac{\sqrt{P/\tilde{P}}}{K_0^2 r^2} \left[ \exp\left(-\frac{r^2}{a^2}\right) - \exp\left(-\frac{r^2}{a^2} \exp\sqrt{P/\tilde{P}}\, \omega_p \tau_0 / 2\right) \right] \tag{10}$$

where $\tau_0$ is the pulse duration.

The distribution of the plasma wave potential in the axial region ( $r \ll a$ ) is described by the following expression:

$$\varphi_m = \frac{\omega^2}{\omega_p^2} \frac{\sqrt{P/\tilde{P}}}{(Ka)^2} \left[ \left( \exp\frac{1}{2} \sqrt{P/\tilde{P}}\, \omega_p \tau_0 - 1 \right) - \frac{r^2}{2a^2} \left( \exp\sqrt{P/\tilde{P}}\, \omega_p \tau_0 - 1 \right) \right] \tag{11}$$

From this equation it follows that the nonlinear self-contraction effects can be neglected when $\sqrt{P/\tilde{P}}\, \omega_p \tau_0 \ll 1$ . The self-contraction effect is significant only when the power is greater than $\tilde{P}_{cr} = 4\tilde{P}\,(\omega_p \tau_0)^{-2}$ which is actually the critical power of self-focusing for the nonstationary case (a similar estimate was obtained in [4] for a one-dimensional model). When $P \gg \tilde{P}_{cr}$, the amplitude of the wake wave increases exponentially, $\sim \exp\left(\frac{1}{2}\sqrt{P/\tilde{P}}\,\omega_p \tau_0\right)$, and is accompanied by a decrease of the radius of the wake region.

3. The saturation of nonlinearity (the paraxial optics approximation). The self-similarity relations obtained for the case of weak nonlinearity should be used merely as reference points of the problem since they do not take into account the nonlinearity saturation effects which are significant for strong relativistic fields $A^2 \gg 1$ that are of practical importance for accelerators. The potentialities of nonlinear beam guiding for the extension of the plasma wake region should be considered using more complicated equations that can be represented as

$$-i\frac{\partial A}{\partial z} + \Delta_\perp A + \frac{\varphi}{1+\varphi} A = 0 \tag{12}$$

$$\frac{\partial^2 \varphi}{\partial \tau^2} = \frac{1 + |A|^2 - (1+\varphi)^2}{2(1+\varphi)^2} \tag{13}$$

Consider the set (12)-(13) in the paraxial optics approximation (also known as aberrationless approximation) [10] which assumes that the wave beam has a quasi-Gaussian shape

$$A = \frac{\sqrt{W}}{a_0 b} \exp -\frac{r^2}{2a_0^2 b^2} + i\,æ r^2 \tag{14}$$

of a relative width b that depends on $\chi$ and $\tau$ : $b = b(\chi, \tau)$, and the distribution of the refractive index of a plasma and of the potential $\varphi(r)$ in the axial region, determined by (13), can be approximated by a parabola:

$$n(r) = n_0 + n_2 r^2 = \frac{\varphi(r)}{1+\varphi(r)}, \quad \varphi(r) = \varphi_0 + \varphi_2 r^2 \tag{15}$$

Here $a_0$ is the initial width of the Gaussian beam at the plasma boundary. The resulting equation for the beam width

$$\frac{\partial^2 b}{\partial Z^2} = \frac{1}{b^3} + n_2 b, \quad n_2 = \frac{\varphi_2}{(1+\varphi_0)^2} \tag{16}$$

where a new variable $Z = \chi / a_0^2$ is introduced, is a standard usual equation for the paraxial approximation in the theory of self-focusing. The quantities $\varphi_0$ and $\varphi_2$ in (16) are solutions to the following equations:

$$\frac{\partial^2 \varphi_0}{\partial \tau^2} = -\frac{\varphi_0 (1+\varphi_0/2)}{(1+\varphi_0)^2} + \frac{W}{2a_0^2 b^2 (1+\varphi_0)^2},$$

$$\frac{\partial^2 \varphi_2}{\partial \tau^2} = -\frac{\varphi_2 (1+W/a_0^2 b^2)}{\cdot (1+\varphi_0)^3} + \frac{W}{2a_0^4 b^4 (1+\varphi_0)^2} \tag{17}$$

The constant $\varkappa$ in (14) is determined by the expression $\varkappa = -\frac{1}{2b}\frac{\partial b}{\partial Z}$.

The set of equations (16)-(17) is simpler than the basic one because it has fewer independent variables; in this set $b = b(Z, \tau)$, i.e. the dependence on the transverse coordinate r is excluded. Nevertheless, this set also needs a laborious numerical integration.

Numerical integration of the "exact" equations (12) and (13) and of the equations in the paraxial approximation was performed for the converging Gaussian beam of width $b_0$ and convergence $\frac{\partial b}{\partial z}(z=0)=b'_{0z}$ specified at the plasma boundary. The time dependence of the incident pulse power was represented in the form

$$W = W_0 \exp\left[-\frac{(\tau-T)^2}{\tau_0^2}\right]$$

In both cases the calculation results are in good qualitative and quantitative agreement.

Consider as an example the results of integration for the following parameter values: $a_0 = 10$, $\tau_0 = 1$, $T = 3$, $b_0 = \sqrt{2}$ and $b'_0 = -\frac{1}{\sqrt{2}}$. The value $b'_0 = -\frac{1}{\sqrt{2}}$ denotes the distance from the linear focus of the beam to the plasma boundary $z = 1$ and the length of the caustic in the linear case is $l = b_0^2 = 2$. Compare the results of numerical integration of the quasioptical equations (12) and (13) and the equation in the paraxial approximation for $W_0 = 10$ and $W_0 = 100$. Figures 1 and 2 show that the plasma wave amplitude is a little larger in the paraxial approximation. As the laser pulse power increases, both the amplitude of the wake plasma wave grows and the region of its

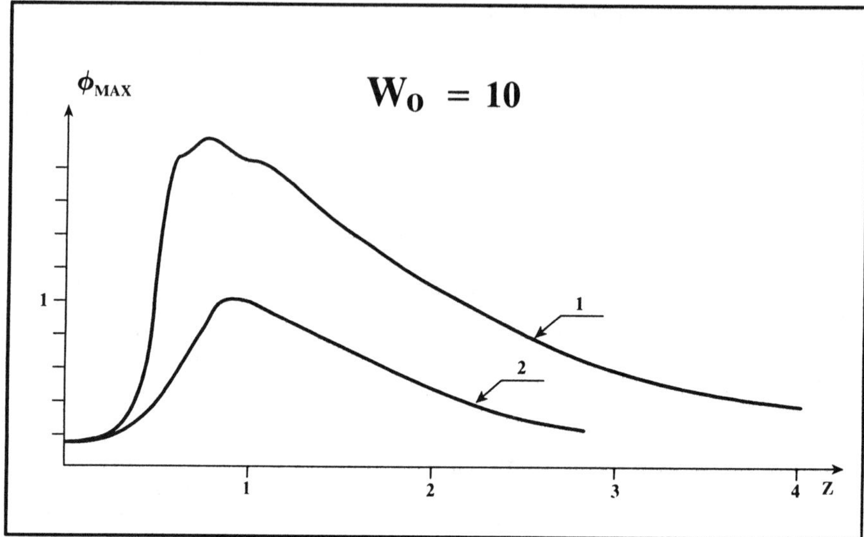


Fig. 1

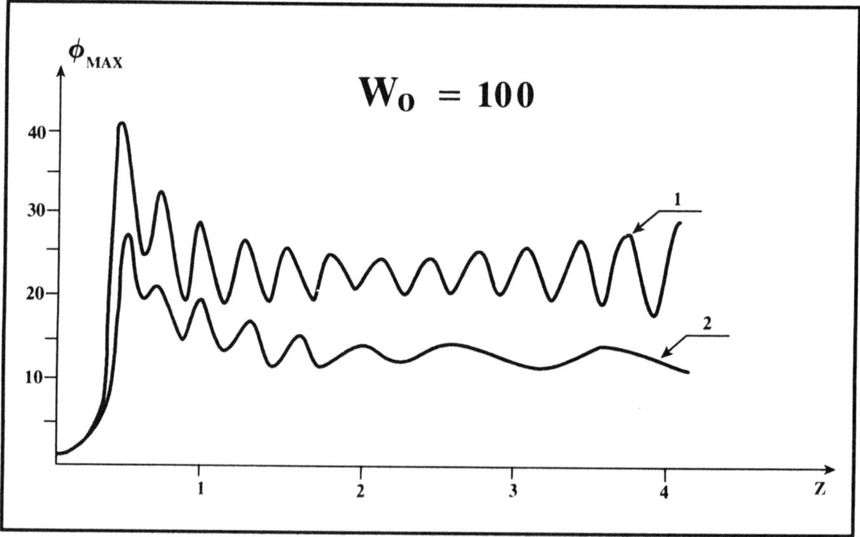


Fig. 2

generation broadens markedly. Apparently, this is caused by the effect of laser pulse self-focusing that leads to nonlinear guiding of radiation and, consequently, to the increase in the length of the nonlinear caustic.

The structural changes in the wave beam are described by the nonlinear amplification coefficient of the wave in a plasma: $S = |E(\tau, z)|^2 / W_0$, and are presented in Fig. 3,4,5 for $W_0 = 100$. The wave beam narrows in the region of the linear focus. The effect of nonlinear saturation leads to the trapping of the wave beam into the regime of optical guiding. As the radiation is transmitted through the nonlinear focal region, the pulse is self-modulated, which is accompanied by the formation of short (in time) regions of intense field and facilitates enhanced generation of a wake wave even when the spatial structure is restored.

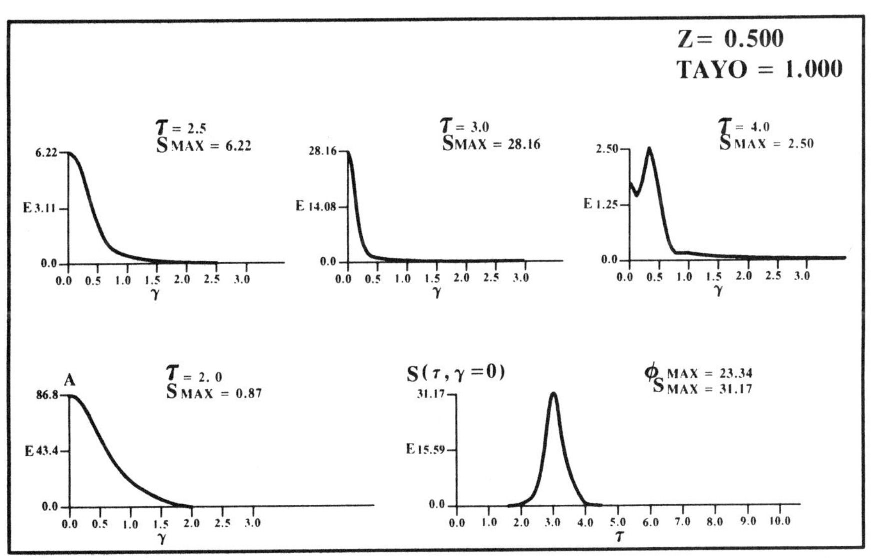


Fig.3

Z=0.980
TAYO = 1.000
τ = 2.5
SMAX = 2.53
τ = 3.0
SMAX = 8.32
τ = 4.0
SMAX = 2.84
τ = 2.0
SMAX = 2.11
S(τ,γ=0)
φ MAX = 14.11
S MAX = 25.53
γ
τ

Fig.4

Z= 1.710
TAYO = 1.000
τ = 2.5
SMAX = 3.02
τ = 3.0
SMAX = 22.61
τ = 4.0
SMAX = 4.47
τ = 2.0
SMAX = 1.22
S(τ,γ=0)
φ MAX = 13.17
S MAX = 23.80
γ
τ

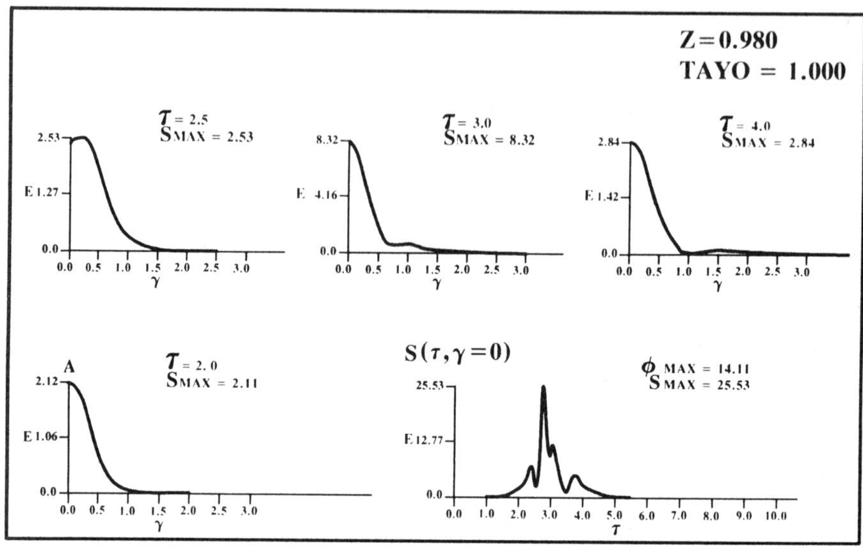

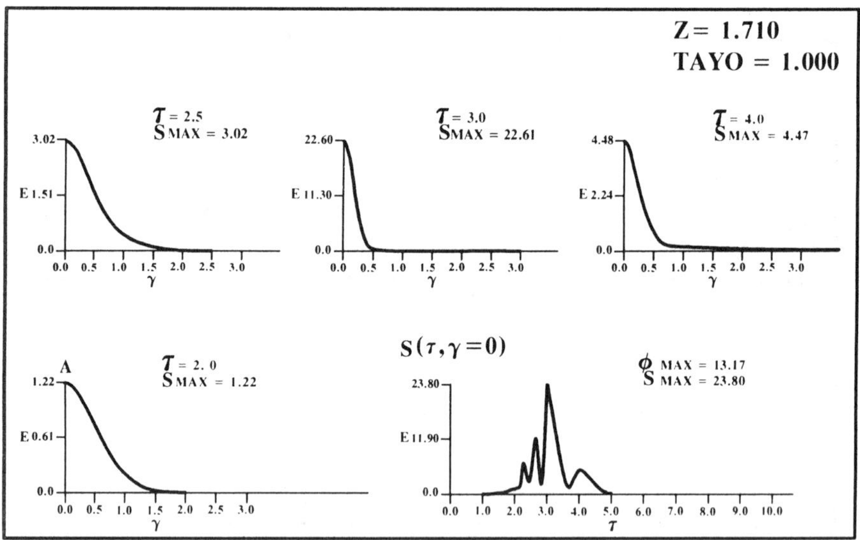

Fig.5

## R E F E R E N C E S


1. T.Tajima, J.Dawson. Phys.Rev.Lett., 43, 267 (1979).
2. L.M.Gorbunov, V.I.Kirsanov. ZhETF, 93, 509 (1987); Sov.Phys. JETP, 66, 290 (1987).
3. P.Sprangle, E.Esarey, A.Ting, G.Joyce. Appl.Phys.Lett., 53, 1246 (1988).
4. P.Sprangle, E.Esarey, A.Ting. Phys.Rev.Lett., 64, 2011 (1990).
5. S.Bulanov, V.Kirsanov, A.Sakharov. Pis'ma v ZhETF, 50, 176 (1989).
6. V.N.Tsytovich, U.De Angelis, R.Bingham. Comments Plasma Phys. Contr.Fusion, 12, 249 (1989).
7. A.G.Litvak. Doctor Thesis, Gorky University, USSR, 1967; ZhETF, 57, 629 (1969); Sov.Phys.JETP, 30, 344 (1970).
8. C.Max, J.Arons, A.B.Langdon. Phys.Rev.Lett., 33, 209 (1974).
9. G.Schmidt, W.Horton. Comments Plasma Phys., 9, 85 (1985).
10. A.G.Litvak. Voprosy Teorii Plazmy, v.10, p.164, Atomizdat, Moscow, 1980 (in Russian) in Reviews of Plasma Physics, v.10, ed. M.Leontovich, Consultants Bureau, N.Y.-London, 1986, p.293.
11. P.Sprangle, C.M.Tang, E.Esarey. IEEE Trans.Plasma Sci., 15, 2 (1987).
12. W.B.Mori, C.Joshi, J.M.Dawson, D.W.Forslund, J.M.Kindel. Phys.Rev.Lett., 60, 1298 (1988).
13. N.L.Tsintsadze. Phys.Scripta, T30, 41 (1990).
14. A.G.Litvak, V.A.Mironov, A.M.Sergeev. Phys.Scripta, T30, 57 (1990).

Research Trends in Physics: Coherent Radiation Generation and Particle Acceleration

*La Jolla International School of Physics*, The Institute for Advanced Physics Studies, La Jolla, California

# Chaos in a Free Electron Laser

**L. Michel, A. Bourdier, and J.M. Buzzi**

Centre National de la Recherche Scientifique, École Polytechnique
91128 Palaiseau Cedex France

## ABSTRACT

It is confirmed that, when the equilibrium self-fields are taken into account, the motion of an electron in a helical wiggler with guide field may be chaotic. Only two independant constants of motion in involution were found. Also, there is evidence of chaos from numerical calculations of Poincaré maps. The trajectory of an electron in a linearly polarized wiggler with guide field is, however, found to be nonintegrable. Resonances can be predicted from a one dimensional Hamiltonian perturbed by a small "time" dependant quantity.

## I. INTRODUCTION

The electron orbits in helical as well as linear wigglers are examined assuming the presence of a guide field in both. In addition for the helical wiggler case, one finds that when self-fields are included, chaos appears in the electron orbits. In this case, the first attempt was to find analytically a third constant of motion. Finally using Poincaré surface of section plots, agreement was found with the results for stochastic trajectories obtained by C. Chen and R.C. Davidson[1].

The nonintegrability of the electron motion is demonstrated by elaborate work on the computation of Poincaré maps. A trick[2] is used to find simply but very accurately the intersection of a numerically integrated trajectory with a surface of section.

The numerical results obtained are quite similar to those found by C. Chen and R.C. Davidson, although chaotic behaviour appears for higher currents.

In the linear wiggler case, stochastic trajectories are found when taking into account a guide field only. This situation is avoided in most experiments[3]. Still, it is interesting to consider for a very small radius beam[4]. The nonintegrability in this case is surprising as three constants of the motion can easily be found. One has to realize that they are not independant and in involution. Using a canonical transformation we simplify the problem by introducing a guiding center and perform Poincaré sections. At last, the equations of the motion can be derived from a one dimensional "time" dependent Hamiltonian. Resonances can then be predicted.

## II. HELICAL WIGGLER CASE

### II-1. THEORETICAL FORMULATION OF THE PROBLEM

Consider the motion of a relativistic electron in a uniform axial field $B_0\vec{e}_z$, a constant amplitude helical wiggler magnetic field $\vec{B}_\omega = -B_\omega(\vec{e}_x \cos k_\omega z + \vec{e}_y \sin k_\omega z)$, and the self-electric and self-magnetic fields produced by a relativistic nonneutral electron beam with radius $r_b$[1]. The beam is assumed to have a constant axial velocity $V_b\vec{e}_z$ ($V_b$ is the average velocity of the beam). A uniform density profile representing a possible state of equilibrium for the beam[5,3,6] is considered. The self fields can be expressed as[5]

$$\vec{E}_s = -\frac{m\omega_{pb}^2}{2e}\left(x\vec{e}_x + y\vec{e}_y\right), \tag{1}$$

$$\vec{B}_s = \frac{m\omega_{pb}^2\beta_b}{2ec}\left(y\vec{e}_x - x\vec{e}_y\right). \tag{2}$$

where $\omega_{pb}^2 = \frac{n_b e^2}{m\varepsilon_0}$, $\beta_b = \frac{V_b}{c}$, and $n_b$ is the electron density in the laboratory-frame.

The assumption of a constant velocity for the beam to calculate $\vec{E}_s$ and $\vec{B}_s$ will now be discussed. The velocity of the beam may satisfy the equations of the motion of one particle. Then, for a given density in the beam-frame, and when the velocity along the beam axis is relatively constant, a value of $\beta_b = \left\langle \frac{v_z}{c} \right\rangle$ is determined and the two models are compared. For not too high densities, the trajectories in the phase space have the same behaviour and are nearly identical. For example, when $\beta_b = 0.65$, $B_0 = 1.42$ T and $B_\omega =$ 710 G, we must have $n_b \lesssim 10^{13}$ $cm^{-3}$. Since the same results are obtained with different but realistic hypotheses, for low densities, it is reasonable to study the motion of one particle in a beam with constant velocity.

Consequently, the scalar and vector potentials, $\phi_s$ and $\vec{A}$ can be defined by[1]

$$\phi_s = m \frac{\omega_{pb}^2}{4e} \left(x^2 + y^2\right), \tag{3}$$

$$\vec{A} = B_0 x \vec{e}_y + A_\omega \left(\vec{e}_x \cos k_\omega z + \vec{e}_y \sin k_\omega z\right) + \frac{\beta_b}{c} \phi_s \vec{e}_z , \tag{4}$$

with $A_\omega = \frac{B_\omega}{k_\omega}$ and $k_\omega = \frac{2\pi}{\lambda_\omega}$ .

Next, let us point out that another term should be added to the vector potential. We apply an helical magnetic field

$$\vec{B}_{\omega 0} = - B_{\omega 0} \left[\vec{e}_x \cos k_\omega z + \vec{e}_y \sin k_\omega z \right], \tag{5}$$

which induces a helical motion of the beam. This motion creates a new transverse magnetic field. The total transverse field is given by

$$\vec{B}_\omega = \vec{B}_{\omega 0} + \vec{B}_{\omega i}. \tag{6}$$

We assume

$$\vec{B}_{\omega i} = \alpha \vec{B}_{\omega 0} , \tag{7}$$

and that $\vec{B}_{\omega i}$ satisfies

$$\vec{\nabla} \times \vec{B}_{\omega i} = \mu_0 \vec{J}_\perp = - \mu_0 n_b e \vec{v}_\perp , \tag{8}$$

where $\vec{J}_\perp$ and $\vec{v}_\perp$ are respectively the transverse current and particle velocity.

Since the self fields are assumed to be weak, one has

$$\vec{v}_\perp \approx \frac{\Omega_\perp v_{z0}}{\gamma k_\omega v_{z0} - \Omega_c} \left[\vec{e}_x \cos k_\omega z + \vec{e}_y \sin k_\omega z \right], \tag{9}$$

with $\Omega_\perp = \frac{eB_{\omega 0}}{m}\left(1 + \alpha\right)$ and $\Omega_c = \frac{eB_0}{m}$.

Using equation (5) for $B_{\omega 0}$ in equation (8) gives

$$\alpha = \frac{\omega_{pb}^2}{k_\omega^2 c^2} \frac{\left(1 + \alpha\right)\beta_{//}}{\left(\zeta_{//} - \beta_{//}\right)} , \tag{10}$$

with $\beta_{//} = \frac{v_{z0}}{c}$ and $\zeta_{//} = \frac{eB_0}{mc\gamma k_\omega}$.

Then, if $\zeta_\perp = \frac{\Omega_\perp}{k_0 c\gamma} = \frac{e\left(1 + \alpha\right) B_{\omega 0}}{mc\gamma k_\omega}$ and $\beta_\perp = \frac{v_\perp}{c}$ are introduced, the normalized constant transverse velocity is

$$\beta_\perp = \frac{\zeta_\perp \beta_{//}}{\left|\beta_{//} - \zeta_{//}\right|} . \tag{11}$$

This expression can be combined with

$$\gamma^{-2} = 1 - \beta_\perp^2 - \beta_{//}^2 = 1 - \beta^2 , \tag{12}$$

which leads to the following equation for $\beta_{//}$

$$\beta_{//}^4 - 2\zeta_{//}\beta_{//}^3 - \beta_{//}^2\left(\beta^2 - \zeta_{//}^2 - \zeta_\perp^2\right) + \beta_{//}\left(2\zeta_{//}\beta^2\right) - \zeta_{//}^2\beta^2 = 0. \tag{13}$$

The factor $\alpha$ in equation (7) can be calculated by iteration. If $\alpha = \alpha^{(n)}$, then $\zeta_\perp^{(n)} = \dfrac{e(1+\alpha^{(n)})B_{\omega 0}}{mc\gamma k_\omega}$ and equation (13) gives $\beta_{//}^{(n)} = f\left(B_0, B_{\omega 0}, k_\omega, \gamma, \alpha^{(n)}\right)$. Equation (10) becomes

$$\alpha^{(n+1)} = \frac{\omega_{pb}^2}{k_\omega^2 c^2} \frac{\left(1+\alpha^{(n)}\right)\beta_{//}^{(n)}}{\left(\zeta_{//} - \beta_{//}^{(n)}\right)}. \tag{14}$$

By this result, the modified group-I and group-II orbits have been determined .

Taking into account this transverse magnetic field corresponds to multiplying the wiggler field by a factor which is close to unity at low density and which is of the order of two for the highest currents we consider (a few kA). As a consequence $\vec{B}_{\omega i}$ does not modify the essence of the model and will not be included in the following discussions.

The Hamiltonian for one electron becomes

$$H = \left[\left(\vec{P} + e\vec{A}\right)^2 c^2 + m^2c^4\right]^{1/2} - e\phi_s \tag{15}$$

$$= \gamma mc^2 - e\phi_s \tag{16}$$

where $\gamma$ is the relativistic mass factor and $\vec{P}$ the canonical momentum.

Since H is independent of time, it is a constant of motion.

Following C. Chen and R.C. Davidson, we perform the canonical transformation

$$x = \left(\frac{2P_\varphi}{m\Omega_c}\right)^{1/2} \sin\left(\varphi + k_\omega z'\right) - \left(\frac{2P_\psi}{m\Omega_c}\right)^{1/2} \cos\left(\psi - k_\omega z'\right), \tag{17}$$

$$y = \left(\frac{2P_\psi}{m\Omega_c}\right)^{1/2} \sin\left(\psi - k_\omega z'\right) - \left(\frac{2P_\varphi}{m\Omega_c}\right)^{1/2} \cos\left(\varphi + k_\omega z'\right), \tag{18}$$

$$z = z', \tag{19}$$

$$P_x = \left(2m\Omega_c P_\varphi\right)^{1/2} \cos\left(\varphi + k_\omega z'\right), \tag{20}$$

$$P_y = \left(2m\Omega_c P_\psi\right)^{1/2} \cos\left(\psi - k_\omega z'\right), \tag{21}$$

$$P_z = P_{z'} - k_\omega P_\varphi + k_\omega P_\psi. \tag{22}$$

Introducing the following dimensionless parameters and variables

$$\widehat{\Omega}_c = \frac{\Omega_c}{ck_\omega}, \; a_\omega = \frac{eA_\omega}{mc}, \; \widehat{\phi}_s = \frac{e\phi_s}{mc^2}, \; \widehat{H} = \frac{H}{mc^2}, \; \widehat{P}_{z'} = \frac{P_z}{mc},$$
$$\widehat{P}_\varphi = \frac{k_\omega P_\varphi}{mc}, \; \widehat{P}_\psi = \frac{k_\omega P_\psi}{mc}, \; \widehat{z'} = k_\omega z', \; \tau = ck_\omega t, \; \widehat{\omega}_{pb}^2 = \frac{\omega_{pb}^2}{ck_\omega}. \tag{23}$$

The Hamiltonian becomes

$$\widehat{H}\left(\varphi,\psi,\widehat{P}_\varphi,\widehat{P}_\psi,\widehat{P}_{z'}\right) = -\widehat{\phi}_s + \left\{ 2\widehat{\Omega}_c\widehat{P}_\varphi + a_\omega^2 + 2a_\omega\left(2\widehat{\Omega}_c\widehat{P}_\varphi\right)^{1/2}\cos\varphi + \widehat{p}_z^2 + 1 \right\}^{1/2}, \tag{24}$$

with

$$\widehat{\phi}_s = 2\varepsilon\widehat{\Omega}_c\left[\widehat{P}_\varphi + \widehat{P}_\psi - 2\left(\widehat{P}_\varphi\widehat{P}_\psi\right)^{1/2}\sin\left(\varphi + \psi\right)\right], \tag{25}$$

and

$$\widehat{p}_z = \widehat{P}_{z'} - \widehat{P}_\varphi + \widehat{P}_\psi + \beta_b\widehat{\phi}_s, \tag{26}$$

which is the axial mechanical momentum. $\varepsilon = \dfrac{\widehat{\omega}_{pb}^2}{4\widehat{\Omega}_c^2}$ is a dimensionless parameter which specifies the strength of the self-fields.

Since $\hat{H}$ is independant of $\hat{z}'$, it follows that $\hat{P}_{z'}$ is another constant of motion.

As $\varepsilon \to 0$, $\hat{H}$ becomes independant of $\psi$. As a consequence $\hat{P}_\psi$ is a third constant of motion which makes the motion integrable[7].

In the integrable limit ($\varepsilon = 0$), the steady-state orbits are recovered, giving some physical meaning to the variables. With dimensional variables, the equations of the steady-state trajectories are[1]

$$x = \pm\left(\frac{2P_{\varphi 0}}{m\Omega_c}\right)^{1/2} \sin\left(k_\omega z_0 + k_\omega v_{z0} t\right) - \left(\frac{2P_{\psi 0}}{m\Omega_c}\right)^{1/2} \cos\psi_0\,, \tag{27}$$

$$y = \pm\left(\frac{2P_{\varphi 0}}{m\Omega_c}\right)^{1/2} \cos\left(k_\omega z_0 + k_\omega v_{z0} t\right) + \left(\frac{2P_{\psi 0}}{m\Omega_c}\right)^{1/2} \sin\psi_0\,, \tag{28}$$

$$z = z_0 + v_{z0} t\,, \tag{29}$$

for $\cos\varphi_0 = \pm 1$ .

These three equations describe helical trajectories.

Let us try now to find a new constant of motion (when $\varepsilon \neq 0$) using Noether's theorem, which states that if the Lagrangian L of a system is invariant under the infinitesimal transformation[8]

$$t \to t + \varepsilon'\psi(t,\vec{r})\,,\ \vec{r} \to \vec{r} + \varepsilon'\vec{\eta}\,(t,\vec{r})\,,$$

then

$$\frac{\partial L}{\partial \vec{v}}\vec{\eta} + \left(L - \vec{v}\frac{\partial L}{\partial \vec{v}}\right)\psi = \text{Const.} \tag{30}$$

For convenience, we use a new gauge

$$\vec{\vec{A}} = \left(-\,B_0\frac{y}{2} + A_\omega \cos k_\omega z\right)\vec{e}_x + \left(B_0\frac{x}{2} + A_\omega \sin k_\omega z\right)\vec{e}_y + \frac{\beta_b}{c}\phi_s\vec{e}_z\,, \tag{31}$$

and the following screw transformation $t \to t$, $x \to x + \varepsilon' k_\omega y$, $y \to y - \varepsilon' k_\omega x$, $z \to z - \varepsilon'$.

The Lagrangian of the system

$$L = - mc^2 \sqrt{1 - \frac{v^2}{c^2}} - e\tilde{\vec{A}}.\vec{v} + e\phi_s , \tag{32}$$

is invariant under the transformation when neglecting terms in $\varepsilon'^2$. Since $\psi = 0$ and $\vec{\eta} = \left(k_\omega y\ ,\ - k_\omega x\ ,\ - 1\right)$, equation (30) is equivalent to $\tilde{\vec{P}} . \vec{\eta}$ = Const, that is to say

$$\tilde{P}_x k_\omega y - \tilde{P}_y k_\omega x - \tilde{P}_z = \text{Const}\ , \tag{33}$$

where the $\tilde{P}_i$'s are the components of the canonical momentum in the new gauge.

To transform the constant of motion expressed by relation (33) in terms of the first gauge, the following invariance is used, namely

$$\vec{p} = \vec{P} + e\vec{A} = \tilde{\vec{P}} + e\tilde{\vec{A}} \tag{34}$$

then

$$\tilde{P}_x = P_x + \frac{1}{2} eB_0 y\ , \tag{35}$$

$$\tilde{P}_y = P_y + \frac{1}{2} eB_0 x\ , \tag{36}$$

$$\tilde{P}_z = P_z\ . \tag{37}$$

Using the canonical transformation defined by equations (17-22), one finds

$$-P_{z'} = \text{Const.} \tag{38}$$

This is just the second constant of motion already found, instead of a third new constant of motion.

## II-2. NUMERICAL INTEGRATION

At last, to check if the problem is integrable or not, we integrate numerically the equations of the motion obtained with the Hamiltonian defined by equation (24).

A fourth order Runge-Kutta method with adaptative stepsize has been used to integrate the differential equations for the electron trajectory. The Hamiltonian is conserved with a great accuracy. The fact that the numerical results are very reliable appears in figure 1 displaying two separatrices.

The initial condition for $\widehat{P}_\psi$ is fixed at the value $\left(\dfrac{2\widehat{P}_{\psi i}}{\widehat{\Omega}_c}\right)^{1/2} = \dfrac{1}{4}$.

The numerical computation of the Poincaré maps was improved by using Hénon's method[2]. The equations of the electron trajectory can be written in the following form

$$\begin{aligned} \frac{dx_1}{dt} &= f_1\left(x_1 \ldots x_n\right) \\ &\vdots \\ \frac{dx_n}{dt} &= f_n\left(x_1 \ldots x_n\right). \end{aligned} \tag{39}$$

The successive intersections of the corresponding trajectory with the surface $\Sigma$ defined by

$$x_n - a = 0\,, \tag{40}$$

where a is a constant, are considered.

Since $x_n$ is a dependent variable, it is not possible to specify in advance its variation over one integration step. The idea of the method is to divide the (n - 1) first equations by the last one, obtaining a new set of equations for which t is a dependant variable.

The system (39) has been integrated until a change of sign is detected for $\delta = x_n - a$, then one shifts to the new system for one step, taking $\Delta x_n = -\delta$ as an integration step. This brings us exactly on the surface of section.

This method and this trick have been applied to obtain accurately the intersection of a trajectory with a surface to the equation of motion obtained from the Hamiltonian defined in equation (24). This motion occurs in a three dimensional phase space $\left(\varphi, \psi, \widehat{P}_\psi\right)$

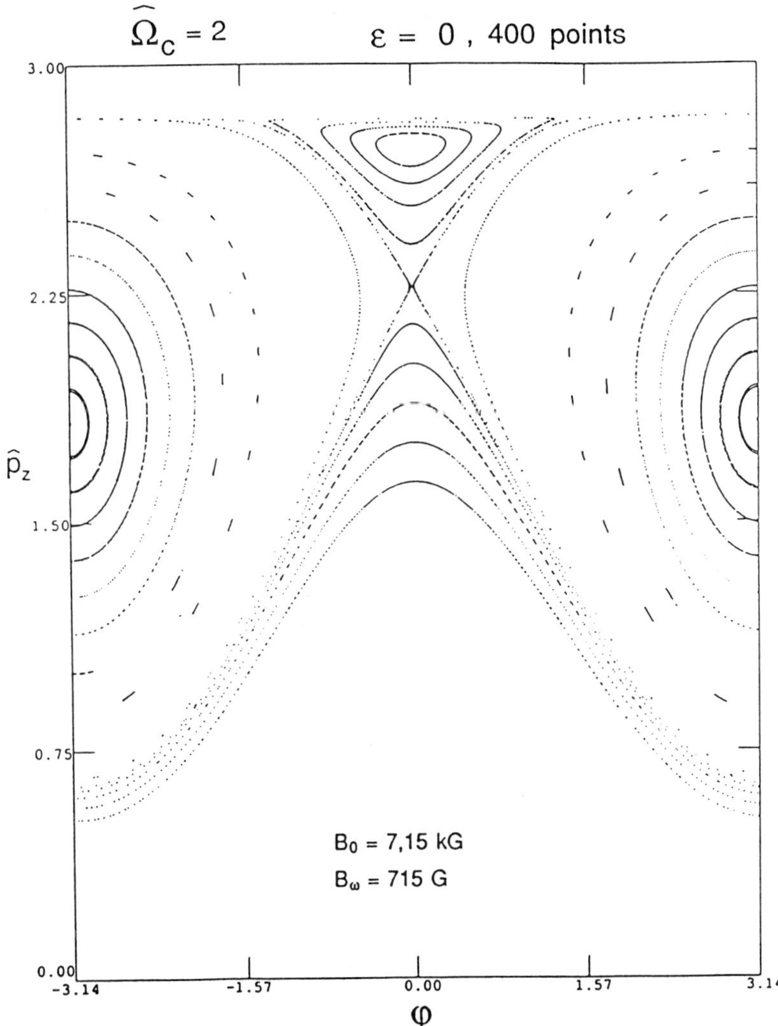


Fig 1 : Contour plots in the integrable phase plane $(\varphi, \widehat{p}_z)$ calculated for equation (24), for $\varepsilon = 0$, $\widehat{\Omega}_c = 2$, $\widehat{H} = 3$ and $a_\omega = 0.2$ .

The plane $(\varphi, \hat{p}_z)$ with $\psi = 0 \pmod{2\pi}$ is chosen to be the surface of section.

## II-3. DISCUSSION

The hypothesis of C. Chen and R.C. Davidson which consists of assuming a constant velocity for the beam seems valid for low intensities.

Another self-field should be added to this model. Taking this field into account implies multiplying $B_\omega$ by a factor which can be of the order of two for high currents. This correction could be taken into consideration in a more sophisticated theory, but that would not change the main result which is to predict stochastic electron trajectories.

Using Noether's theorem one finds the second constant of motion $P_{z'}$, which had already been found with the Hamiltonian method.

The Poincaré maps obtained by C. Chen and R.C. Davidson are found to be valid in the integrable limit ($\varepsilon = 0$). The Hamiltonian given by equation (24) is confirmed to be nonintegrable when $\varepsilon \neq 0$ (Fig. 2).

In supercritical cases (when there is only one elliptic fixed point), chaos appears for higher densities than had previously been predicted[1]. The results are in good agreement with previous resonance analysis. A period-two and a period-three island close to the two elliptic fixed points corresponding to Group-I and group-II orbits are found (Fig 3).

## III. LINEAR WIGGLER CASE

Here, the self-fields are neglected and the motion of one electron is considered in the following magnetic field

$$\vec{B} = B_0\vec{e}_z + B_\omega \sin k_\omega z\, \vec{e}_x, \tag{41}$$

the corresponding Hamiltonian is

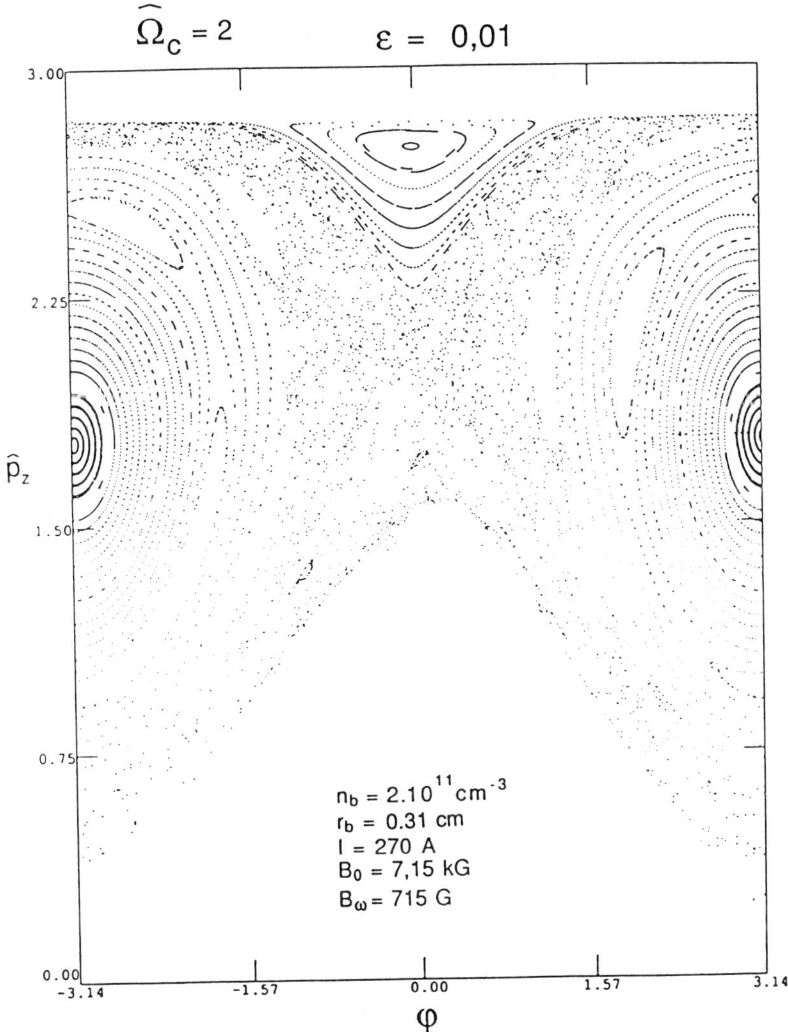


Fig 2 : Nonintegrable surface of section plots with $\psi = 0 \pmod{2\pi}$ for $\varepsilon = 0.01, \widehat{\Omega}_c = 2, \widehat{H} = 3, a_\omega = 0.2, \beta_b = 0.91$ . When $r_b = 3.1$ cm , the corresponding intensity is I = 270 A.

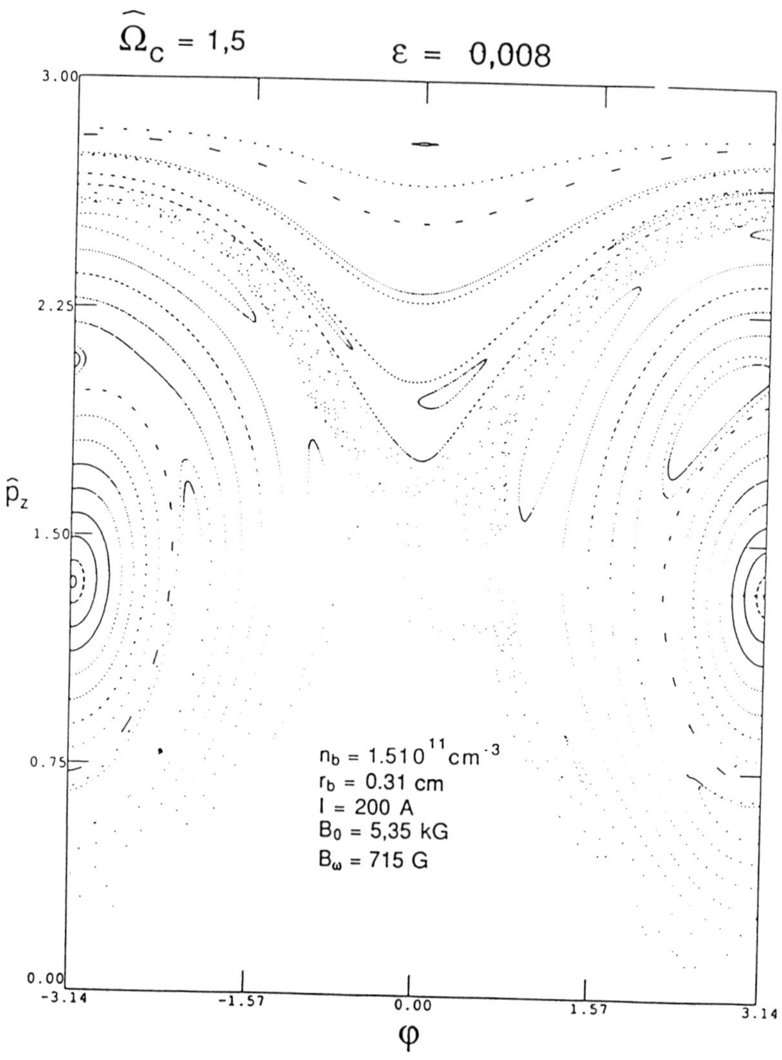


Fig 3 : Nonintegrable surface of section plots with $\psi = 0 \pmod{2\pi}$ for $\varepsilon = 0.008, \hat{\Omega}_c = 1.5, \hat{H} = 3, a_\omega = 0.2, \beta_b = 0.91$ .

$$H = \left\{ c^2 \left[ P_x^2 + \left( P_y + eB_0 x + \frac{eB_\omega}{k_\omega} \cos k_\omega z \right)^2 + P_z^2 \right] + m^2 c^4 \right\}^{1/2} . \tag{42}$$

Integrating the two first equations of Hamilton, one obtains simply two constants of motion

$$\begin{aligned} C_1 &= P_x + eB_0 y , \\ C_2 &= P_y . \end{aligned} \tag{43}$$

Unfortunately, they are not in involution

$$\{C_1, C_2\} = eB_0 . \tag{44}$$

Given the type 2 generating function

$$F_2(x,y,z,P_1,P_2,P_3) = (P_1 - eB_0 y)x + P_2\left(y - \frac{P_1}{eB_0}\right) + P_3 z , \tag{45}$$

yields the canonical transformation

$$\begin{aligned} x &= Q_1 + \frac{P_2}{eB_0} , \\ y &= Q_2 + \frac{P_1}{eB_0} , \\ z &= Q_3 , \\ P_x &= - eB_0 Q_2 , \\ P_y &= - eB_0 Q_1 , \\ P_z &= P_3 . \end{aligned} \tag{46}$$

In the new variables, one obtains

$$H = \left\{ c^2 \left[ e^2 B_0^2 Q_2^2 + \left( P_2 + \frac{eB_\omega}{k_\omega} \cos k_\omega Q_3 \right)^2 + P_3^2 \right] + m^2 c^4 \right\}^{1/2} . \tag{47}$$

H is independent of $Q_1$ and $P_1$, so both of these are invariants. We can come to the same conclusion transforming $-C_2/eB_0$ and $C_1$. They define a guiding center denoted by $X_G$ and $Y_G$

$$X_G = Q_1 \,,\; Y_G = \frac{P_1}{eB_0} \,.$$

As the Hamiltonian is not an explicit function of time, H is another constant of motion. At last two independent constants in involution are obtained. Having failed in finding a third constant, Poincaré maps have been plotted to demonstrate the nonintegrability of the motion. To do so, the following simple normalized equations of motion derived from equations (47) have been solved numerically

$$\begin{aligned} \dot{\hat{Q}}_2 &= \frac{1}{\gamma}\left(\hat{P}_2 + a_\omega \cos \hat{Q}_3\right), \\ \dot{\hat{Q}}_3 &= \frac{\hat{P}_3}{\gamma}, \\ \dot{\hat{P}}_2 &= -\frac{\hat{\Omega}_c^2}{\gamma}\hat{Q}_2, \\ \dot{\hat{P}}_3 &= a_\omega \dot{\hat{Q}}_2 \sin \hat{Q}_3 \,. \end{aligned} \tag{48}$$

The dimensionless variables defined by $\tau = ck_\omega t$, $\hat{Q}_i = k_\omega Q_i$, $\hat{P}_i = P_i/mc$ ··· have been introduced.

The motion occurs in a three dimensional space $\left(\hat{Q}_2, \hat{Q}_3, \hat{P}_2\right)$. The plane $\left(\hat{Q}_2, \hat{P}_2\right)$ with $\hat{Q}_3 = 0 \pmod{2\pi}$ is chosen to be the Poincaré surface of section. The numerical method used is the same as in the helical wiggler case.

When $B_\omega = 0$, $P_3$ is a third constant of motion, and the phase curves are ellipses with equation

$$\hat{\Omega}_c^2 \hat{Q}_2^2 + \hat{P}_2^2 = \hat{H}^2 - \hat{P}_3^2 - 1 \,. \tag{49}$$

When $B_\omega$ is small enough, equation (49) approximately describes the Poincaré maps (Fig 4).

One case for which the trajectories are chaotic is shown in figure 5. There seems to be resonances which may be predicted from a one dimensional Hamiltonian perturbed by a small quantity depending on z.

The Hamiltonian (42) leads to

$$\begin{aligned}
\dot{P}_x &= -\frac{eB_0}{m\gamma}\left(P_y + eB_0x + e\frac{B_\omega}{k_\omega}\cos k_\omega z\right), \\
\dot{P}_z &= \frac{eB_\omega}{m\gamma}\sin k_\omega z\left(P_y + eB_0x + e\frac{B_\omega}{k_\omega}\cos k_\omega z\right), \\
\dot{x} &= \frac{P_x}{m\gamma}, \\
\dot{z} &= \frac{P_z}{m\gamma}.
\end{aligned} \tag{50}$$

Let us consider z as an independent variable. Hence

$$\begin{aligned}
\frac{dx}{dz} &= \frac{\dot{x}}{\dot{z}} = \frac{P_x}{P_z}, \\
\frac{dP_x}{dz} &= -\frac{\dot{P}_x}{\dot{z}} = -\frac{eB_0}{P_z}\left(P_y + eB_0x + e\frac{B_\omega}{k_\omega}\cos k_\omega z\right).
\end{aligned} \tag{51}$$

Equation (42) gives

$$P_z = \left[\frac{H^2}{c^2} - P_x^2 - m^2c^2 - \left(P_y + eB_0x + e\frac{B_\omega}{k_\omega}\cos k_\omega z\right)^2\right]^{1/2}. \tag{52}$$

We note that $-P_z$ can also be interpreted as a one dimensional Hamiltonian for which z plays the part of time. Equations (51) can be derived from the following Hamilton's equations

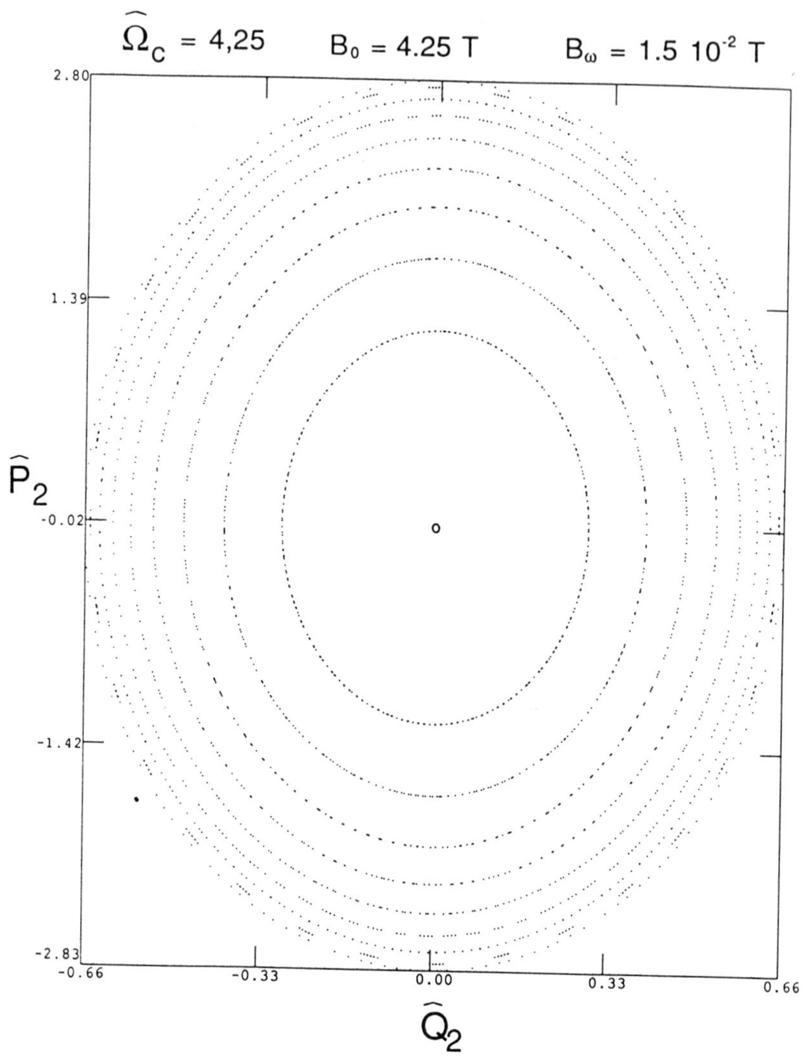


Fig 4 : Regular surface of section plots with $k_\omega z = 0 \pmod{2\pi}$ in the linear wiggler case, for $\widehat{H} = 3$, $B_0 = 4.25$ T, $B_\omega = 1.5\ 10^{-2}$ T and $\widehat{\Omega}_c = 4.25$ .

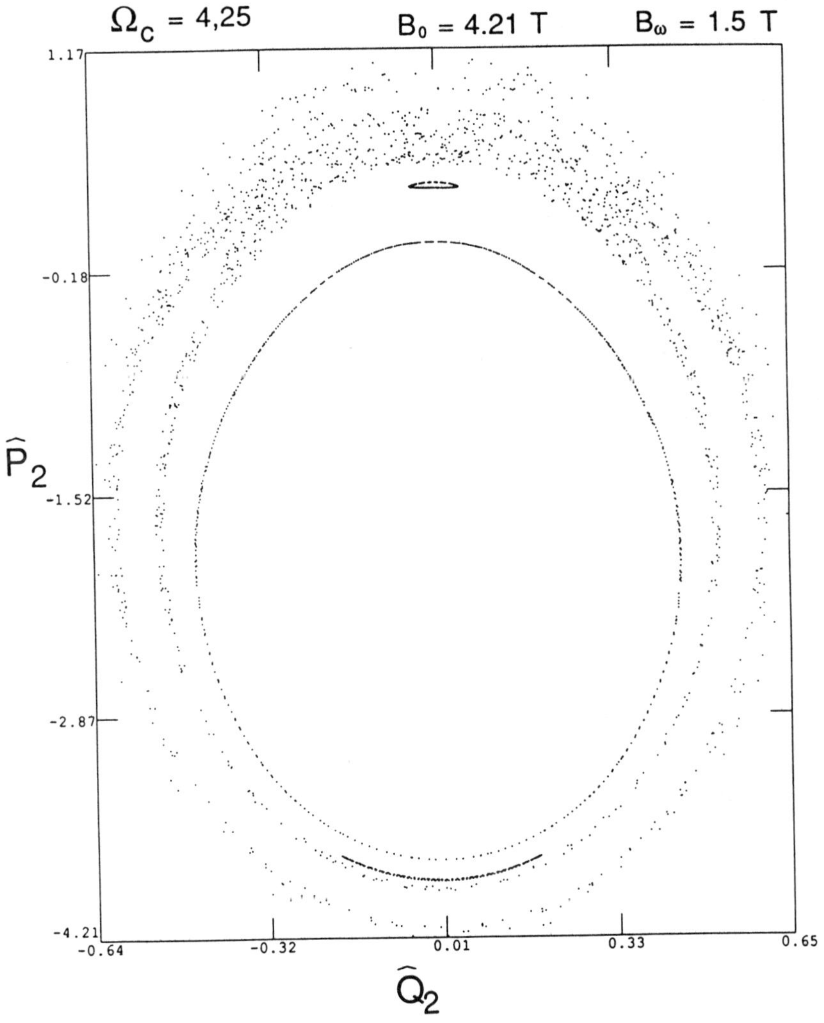

Fig 5 : Surface of section plots for chaotic trajectories in the linear wiggler case, with $Q_3 = 0 \pmod{2\pi}$, for $\widehat{H} = 3$, $B_0 = 4.21$ T, $B_\omega = 1.5$ T and $\widehat{\Omega}_c = 4.25$ .

$$\frac{dx}{dz} = \frac{\partial(-P_z)}{\partial P_x}$$
$$\frac{dP_x}{dz} = -\frac{\partial(-P_z)}{\partial x} \tag{53}$$

For a weak pump, the equations of motion can be linearized by setting

$$P_z = P_{z0} + P_{z1} + P_{z2} \ldots ,$$
$$x = x_0 + x_1 + x_2 \ldots , \tag{54}$$
$$P_x = P_{x0} + P_{x1} + P_{x2} \ldots ,$$

with

$$P_{z0} = \left[\frac{H^2}{c^2} - P_{x0}^2 - m^2c^2 - \left(P_y + eB_0x_0\right)^2\right]^{1/2} . \tag{55}$$

The different variables are evaluated under the condition that the frequency is far from the magnetoresonance[9]. Calculating $\dot{P}_{x1}$ and $\dot{x}_1$, neglecting second order terms, leads to

$$\ddot{x}_1 + \omega_0^2 x_1 = -\omega_0^2 \frac{B_\omega}{B_0 k_\omega} \cos k_\omega z , \tag{56}$$

with

$$\omega_0 = \frac{eB_0}{P_{z0}} . \tag{57}$$

This equation predicts one resonance for $k_\omega = \pm\omega_0$ which is not acceptable by our hypotheses.

Then, taking into account third order terms, we have

$$\ddot{x}_3 + \omega_0^2 x_3 = \omega_0 P_{z2}\left(\omega_0 x_1 + \frac{eB_\omega}{k_\omega P_{z0}}\cos k_\omega z\right) - \frac{P_{x1}\dot{P}_{z2}}{P_{z0}^2} - \frac{\dot{P}_{x1}P_{z2}}{P_{z0}^2} \tag{58}$$

where

$$P_{z2} = -\frac{P_{x1}^2}{2P_{z0}} - \frac{\omega_0^2 x_1^2}{2} P_{z0} - \frac{\omega_0^2}{k_\omega} P_{z0}\frac{B_\omega}{B_0} x_1 \cos k_\omega z - \frac{\omega_0^2 B_\omega^2}{2k_\omega B_0^2} P_{z0}\cos^2 k_\omega z \,. \tag{59}$$

There is a resonance whenever

$$k_\omega = n\,\omega_0 \,, \tag{60}$$

with $n = \pm\frac{1}{3}, \pm 3$.

The intersections between the trajectories and a surface of section were determined. To do so, dimensionless variables were introduced: $\widehat{P}_i = \frac{P_i}{mc}$, $\hat{z} = k_\omega z$, $\hat{x} = k_\omega x$, $\widehat{\Omega}_c = \frac{\Omega_c}{ck_\omega}$, $a_\omega = \frac{eA_\omega}{mc}$,

$\widehat{H} = \frac{H}{mc^2}$, $\tau = ck_\omega t$. The plane $\left(\hat{x}, \widehat{P}_x\right)$ with $\hat{z} = 0 \pmod{2\pi}$ is the Poincaré surface of section.

In figure 6 a period-three island appears corresponding to the $n = 3$ resonance condition.

Figure 7 shows that for other conditions, we have stochastic trajectories which confirm that the system is nonintegrable.

## III-2. DISCUSSION

By performing Poincaré sections, it has been found that the motion of one electron in a uniform magnetic field and in the field of a linear wiggler is nonintegrable. Such a problem can, indeed, in the case of a weak pump, be reduced to a one dimensional motion perturbed by a periodic "time" dependant source. As a result, resonances have been predicted.

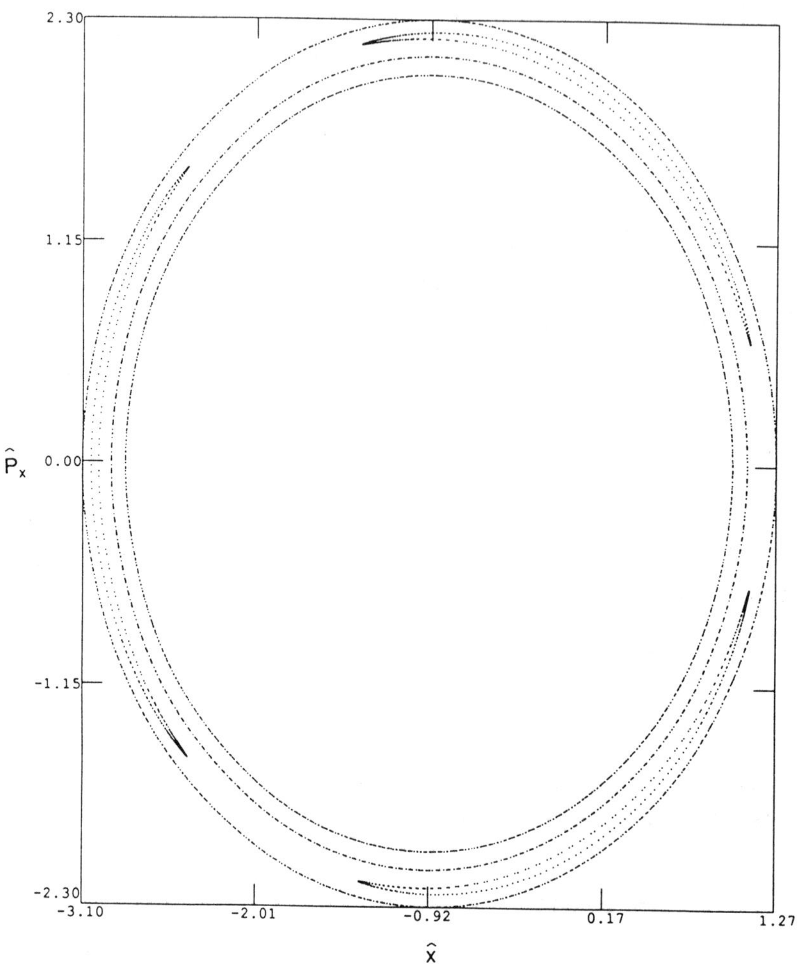


Fig 6 : Surface of section plots showing a period-three island in the linear wiggler case, with $\hat{z} = 0 (\text{mod } 2\pi)$ , for $\widehat{H} = 4$, $B_0 = 1.04$ T, $B_\omega = 0.29$ T and $\widehat{\Omega}_c = 1.05$ .

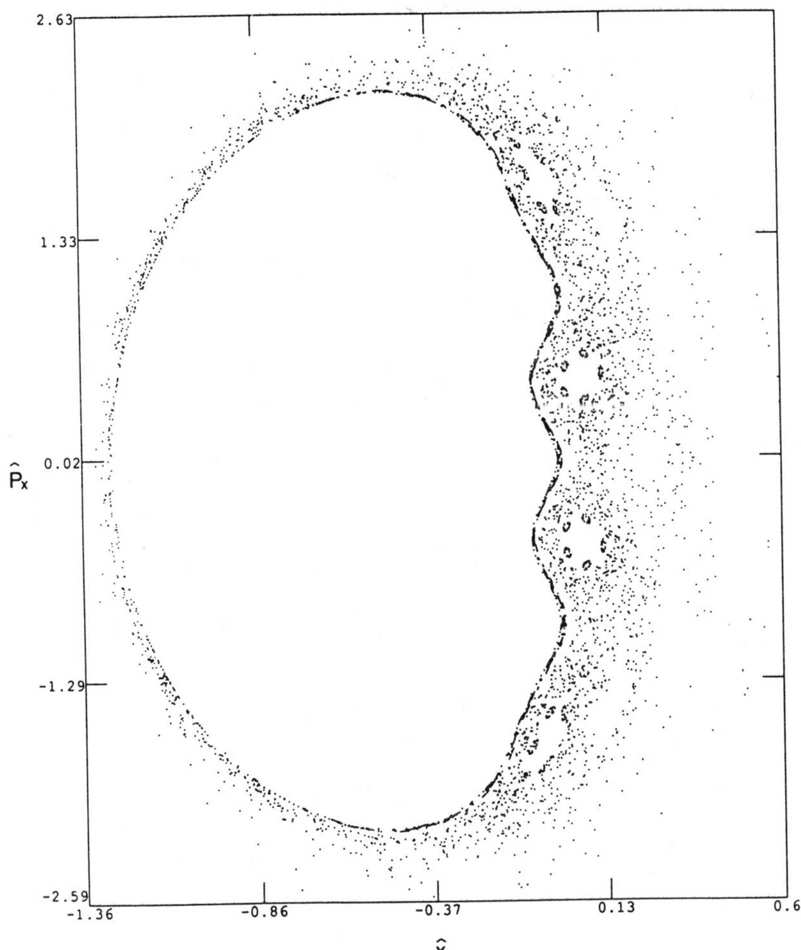


Fig 7 : Surface of section plots for stochastic trajectories in the linear wiggler case with $\hat{z} = 0 \pmod{2\pi}$ , for $\hat{H} = 3$, $B_0 = 2.56$ T, $B_\omega = 0.65$ T and $\hat{\Omega}_c = 2.59$ .

**ACKNOWLEDGMENTS**

The authors wish to thank Professors G. Laval and H. Wilhelmsson, and Dr. H.P. Freund for their useful suggestions.

**REFERENCES**

(1) C. Chen and R.C. Davidson,Phys. Fluids, B2(1), 171 (1990)

(2) M. Hénon, Physica 5d, 412 (1982)

(3) T.C. Marshall, Free-Electron lasers, ed. Macmillan Publishing Company, New-York, London (1985)

(4) K.D. Jacobs, Ph. D. Thesis, M.I.T. (1986)

(5) R.C. Davidson, Theory of nonneutral Plasmas, W.A. Benjamin, Inc., Advanced Book Program, Reading, Massachusetts (1974)

(6) A. Bourdier, J.M. Buzzi, abstracts of the 11th International Free Electron Laser Conference, Naples, Florida (1989)

(7) R.G. Littlejohn, A.N. Kaufman and G.L. Johnson, Phys. Letters 120 A, 291 (1987)

(8) R.D. Jones, Phys. Fluids 24 (3), 564 (1981)

(9) Y.Z. Yin and G. Bekefi, J. Appl. Phys. 55 (1), 33 (1984)

**Research Trends in Physics: Coherent Radiation Generation and Particle Acceleration**
Editorial Board: J.M. Buzzi, A. Prokhorov (Editor-in-Chief), P. Sprangle, and K. Wille
*La Jolla International School of Physics*, The Institute for Advanced Physics Studies, La Jolla, California

# Coherent Scattering - A Possible Origin of Cosmic Rays

**A.G. Morgan*, E.J.N. Wilson, and K. Zioutas****

CERN, Geneva, Switzerland

## 1 Introduction

Direct acceleration in free space by electromagnetic waves remains the "philosophers' stone" of acceleration physics. Attempts to find a practical scheme seem always to founder on the difficulty of keeping wave and particle in step and in directing the transverse field in the direction of particle motion. Compton scattering appears to overcome these problems but it has a very small cross section. We discuss how this cross section may be improved by many orders of magnitude if charged particles are densely packed in a bunch so that they absorb coherently.

## 2 Compton scattering as a means to accelerate

In spite of our reservations about the difficulty of acceleration in-vacuo, a mechanism, *Compton scattering*, by which an electron may acquire a small amount of energy from a photon striking it from behind, seems to do just this.

To accelerate electrons by Compton scattering would be a very laborious and inefficient process for the energy transferred as a photon finds an electron is at most a few eV. Supposing a high power laser could deliver a pulse of 1000 Joules in a mm diameter spot. The cross section an electron presents to this beam of $10^{22}$ photons is $\sigma_c = 0.6 \times 10^{-28} m^2$. Only about 1 of the photons will interact with the electron and the energy each will deposit will be only a small fraction of that of the photon. Although this paper treats the acceleration of electrons, the idea of coherent response could equally be applied to protons or other charged particles. In fact more massive particles may acquire more energy before the $\gamma$ dependent terms in the expressions for acceleration limit the rate of acceleration. Cosmic rays are, of course, mainly these massive particles.

## 3 Acceleration by Coherent Scattering

In the coherent regime the scattered waves can reinforce each other coherently and we can add up the amplitudes of the $N$ scattering centres within a box of one wavelength dimensions *before* squaring them to find the scattered intensity. The scattering probability then rises as $N^2$ where $N$ is the number of centres participating. If the density is high enough for a complete layer, one third to one half of a wavelength thick, to scatter with unit probability,

---

[1]Permanent address: St.Edmund's College Cambridge, England
[2]Permanent address:Physics Department , University of Thessalonika,Greece.

 

the system becomes opaque. The backward scattered wave returning from a smooth layer of such scatterers will be reflected as from a mirror.

Consider $N$ charged particles in a unit cube illuminated with an electromagnetic beam of unit cross section. If the cross section for the incoherent Compton scattering is $\sigma_c$ the probability of an interaction will be $N\sigma_c$ while the probability of an interaction if all the particles interact coherently is $N^2\sigma_c$.

Now take $N$ particles in a cube of edge $\lambda/2$ illuminated by a beam whose dimensions match the face of the cube. The probabilities of incoherent and coherent interaction will be:

$$P_{inc} = \frac{4N\sigma_c}{\lambda^2} \tag{1}$$

and

$$P_{coh} = \frac{4N^2\sigma_c}{\lambda^2} \tag{2}$$

respectively.

The number density of electrons within the box is

$$\rho = \frac{8N}{\lambda^3} \tag{3}$$

so we may substitute $2N/\lambda = \rho\lambda^2/4$ to obtain the density, $\rho'$, at which the probability for the coherent effect is one

$$1 = \left(\frac{\rho'\lambda^2}{4}\right)^2 \sigma_c.$$

The density required becomes smaller as the wavelength increases.

Conversely, the frequency below which light will not penetrate is:

$$\omega' = \frac{2\pi c}{\lambda'} = \pi c \rho^{\frac{1}{2}} \sigma_c^{\frac{1}{4}} = \pi \left(\frac{\sigma_c}{\mu_0^2}\right)^{\frac{1}{4}} \sqrt{\frac{\rho}{\epsilon_0}}.$$

However, the frequency of incident radiation below which the electrons of a plasma are totally reflecting is well known and is called *plasma frequency* , given by the expression,

$$\omega_p = \sqrt{\frac{q^2\rho}{m_0\epsilon_0}} \tag{4}$$

In *MKS* units for electrons ($q = e$) we compare the two expressions,

$$\omega' \approx 83\,\rho^{\frac{1}{2}} \quad \text{and} \quad \omega_p \approx 56\,\rho^{\frac{1}{2}}$$

the difference in the constant is simply due to the fact that we made the rough approximation of using $\lambda/2$ for the side of the effective coherent cube.

## 4 Lorentz Transformation of the Reflecting Layer

In the rest frame of a bunch moving away from source, Expression 4 holds but the wave is seen as red shifted. In the Lab frame the observed plasma frequency is such that,

$$\frac{\omega_{p,L}}{\omega_p} = (1+\beta)\gamma, \tag{5}$$

i.e., $\omega_{p,L}$ is red-shifted to $\omega_p$. Moreover, the Lab observer sees the moving plasma as more dense than a co-moving observer ($\rho_L = \gamma\rho$). Taking both factors into consideration the *observed* plasma frequency of a relativistically moving ($\beta \longrightarrow 1$) plasma is,

$$\omega_{p,L} \approx 2\sqrt{\frac{q^2\gamma\rho_L}{m_0\epsilon_0}}. \tag{6}$$

The depth of the reflecting layer (the coherence length) to an observer in the lab is $\delta \sim \lambda_p/2$ which may be written in terms of observed quantities in the Lab frame,

$$\delta = \frac{\lambda_{p,L}}{2} = \frac{\pi c}{\omega_{p,L}} = \frac{\pi c}{2}\sqrt{\frac{m_0\epsilon_0}{q^2\gamma\rho_L}}. \tag{7}$$

## 5 A linear accelerator

In this section we we treat the bunch as a perfect mirror. There is a problem associated with the rate of transfer of energy to a relativistic particle bunch. The distance that the lab observer sees the bunch has travelled, as it is overtaken by a pulse of electromagnetic radiation, increases rapidly as the bunch becomes relativistic. Consider a laser that produces a pulse of duration $\tau$ measured in the laser lab frame. We take small element ($d\tau$) of this pulse and imagine that its leading edge is just incident on the surface of the bunch. Before the trailing edge has overtaken the surface plane of the bunch the bunch is observed to move a distance, $ds$, in the laser lab frame. The space-time coordinates of the second event are obtained from the intersection of the world line of the bunch, $\beta ct = x$, and that of the trailing edge of the element, $c(t-d\tau) = x$. For a bunch moving away from the laser at speed $\beta c$,

$$ds = \frac{\beta}{1-\beta}c\,d\tau \equiv \frac{(1+\beta)\beta}{1-\beta^2}c\,d\tau \tag{8}$$

which at high speeds ($\beta \longrightarrow 1$) gives

$$ds \approx 2\gamma^2 c\,d\tau. \tag{9}$$

Such a pulse carries with it a small increment of energy from the laser, $dE = \Phi A\,d\tau$ ($\Phi$ is the laser flux, and $A$ the cross sectional area of the beam), which is transferred to the bunch and due to the coherence described above has an accelerating effect.

Now at any velocity the energy of the bunch is given by,

$$E = Nm_0c^2\gamma \tag{10}$$

| Laser of... | today | tomorrow | units |
|---|---|---|---|
| Power Output, $\Phi A$ | $10^{14}$ | $10^{17}$ | $W$ |
| Pulse length, $\tau$ | $10^{-14}$ | $10^{-14}$ | $s$ |
| Initial $\gamma$, | 100 | 1000 | $\gamma_I$ |
| Final energy, $E_F$ | 0.6 | 624 | GeV |
| Linac length, $S$ | 3.8 | $30 \times 10^3$ | $m$ |
| Mean accel.force | 151 | 0.28 | $MeV\, m^{-1}$ |

Table 1: Parameters of Terrestrial Accelerators (for a bunch of $10^{10}$ electrons whose height is $2\,mm$ and whose length $\sim 300\,nm$ - the assumed wavelength of the laser.)

It follows that,

$$\frac{d\gamma}{ds} = \frac{d\gamma}{dE}\frac{dE}{d\tau}\frac{d\tau}{ds} \approx \frac{\Phi A}{2Nm_0c^3\gamma^2}. \tag{11}$$

Thus the distance covered by the bunch in being accelerated from $\gamma_I$ to $\gamma_F$ is,

$$S = \frac{2Nm_0c^3}{3\left(\frac{dE}{d\tau}\right)}\left[\gamma_F^3 - \gamma_I^3\right].$$

We know that the energy gained by the bunch must eventually become that of the whole laser pulse $\Phi A\,\tau$ and so we may substitute for $\gamma_F$ to obtain,

$$S = \frac{2Nm_0c^3}{3\Phi A}\left[\left(\gamma_I + \frac{\Phi A\tau}{Nm_0c^2}\right)^3 - \gamma_I^3\right].$$

Conversely we can consider a bunch in a continuous stream of radiation, here,

$$E_F = m_0c^2\gamma_F = \sqrt[3]{E_I^3 + \frac{3m_0^2c^3\Phi AS}{2N}}.$$

We see that the energy gain per unit length decreases as the particles' energy increases and in particular $E_F \sim S^{\frac{1}{3}}$.

Table 1 contains the parameters for two linac's with lasers at and just beyond the forefront of current technology. (Note: we have chosen $\gamma_I$ to avoid problems with space-charge effects.)

At the present time (or in the near future) we do not envisage the application of this technique for accelerating electrons to high ($\sim TeV$) energies. The acceleration rate one can expect from today's laser is only just comparable to an S-band linac up to 0.6 GeV. Above this energy the S-band linac is better. Tomorrow's laser is comparable with the expected performance of CLIC.

However, the key parameter in this acceleration scheme is $\tau$—how quickly you can transfer the energy: the shorter the pulse the (very much) shorter the Linac.

## 6 A Stellar accelerator

The origin of *cosmic rays* and the mechanisms by which they reach high energies apparently in free space are not understood. Many suggestions have been made so far, starting historically with Fermi's acceleration proposal[1].

The most distinguished feature of the cosmic ray high energy spectrum is the $\sim E^{-2}$ power-law. Any proposed acceleration scenario must explain:

1. the spectrum shape ($E^{-2}$ law)

2. the highest energies observed ($\leq 10^{20}\,[eV]$).

If a coherent mechanism, such as the one presented here, were to exist in the Universe then it would occur in regions of both high charged particle density and high radiation flux.

It is apparent that the sun accelerates particles to high energies during solar flares, which are accompanied by enhanced radio emissions [1].

The radiation energy density in Quasars, and Active Galactic Nuclei is $\sim 10^{18}\,eV\,cm^{-3}$ for a luminosity $\sim 10^{39}\,W$[4]. The corresponding average energy density at near the pulsar surface ($R \sim 10 - 100\,km$) is $\sim 10^{28} eV\,cm^{-3}$; for a luminosity of $\sim 10^{33}\,W$. For comparison consider a laser with 1 J/fs and a spot of 1 mm diameter. The maximum energy density is of the order of $10^{25} eVcm^{-3}$. We find the observed Radio Pulsar luminosities range from $\sim 10^{22}\,W$(for the Crab Nebula) to as low as $\sim 10^{19}\,W$. Whereas X-ray pulsar luminosities reach $\sim 10^{32}\,W$[5, 6] in agreement with the expected Eddington limit (which does not take into account coherent modes of interaction).

We consider a simple model of a neutron star (of luminosity $L$). Over the surface of the star, of radius $r_0$, we find a thin shell of charged particles. If the thickness of the shell is $\delta$ of relation (7) then the star's radiation pressure will continuously expand the spherical envelope—accelerating the particles...The total energy, $E$, of all the particles in this spherical shell is given by relation (10). We can however substitute for $N$; expressing $E$ in terms of the lab-observed particle density at a radius $r$,

$$E = 4\pi m_0 c^2 \gamma r^2 \delta \rho_r.$$

Substituting for $\delta$ we have,

$$E = \frac{2\pi^2 m_0 c^3}{q} \sqrt{m_0 \epsilon_0}\, r^2 \sqrt{\rho_r \gamma}.$$

Now it is not unreasonable to expect continuity of flow which demands that when its radius is $r$ the particle flux represented by the expanding shell must obey the relation,

$$\beta_r r^2 \rho_r = \beta_s r_0^2 \rho_s$$

(the suffix $s$ indicates measured at the surface), and since the particles we are considering are *relativistic* we can write,

$$\rho_r = \rho_s \left(\frac{r_s}{r}\right)^2. \tag{12}$$

Substituting this result into $E$ and differentiating with respect to $\gamma$ we obtain,

$$\frac{dE}{d\gamma} = \frac{\alpha r}{\sqrt{\gamma}}$$

where

$$\alpha \equiv \left(\frac{\pi^2 c^3 r_0}{q} \sqrt{m_0^3 \epsilon_0 \rho_s}\right)$$

Thus in the relativistic limit of relation (9) and following the reasoning of relation (11) we find

$$\frac{d\gamma}{dr} = \frac{L}{2c\alpha}\gamma^{-\frac{3}{2}}r^{-1}$$

with $s$ taken as the distance from the centre of the star ($s \equiv r$) which when integrated gives the energy of a plasma particle at a radius $r$, as,

$$E_F = \left( E_I^{\frac{5}{2}} + \frac{5Lm_0cq}{4\pi^2 r_0\sqrt{\epsilon_0\rho_s}} \ln\left(\frac{r}{r_0}\right)\right)^{\frac{2}{5}}. \tag{13}$$

The implications of equation (13) become apparent when it is *fed* with the parameters of a *typical* pulsar: $r_0 = 10^4\, m$; $L = 6 \times 10^{32}\, W$; $\rho_s = 10^{18}\, m^{-3}$. If we assume that the surface plasma, of electrons and positrons, is energetic (in an outward direction) with $\gamma_I > 1$ and consider the energy of particles reaching a distance $r = 10^5\, m$, we have

$$E_F \approx 4 \times 10^{12}\, eV.$$

We note that the frequency of the radiation that finally accelerates the shell is $\nu \approx 10^{13}\, Hz$— a radio frequency with a wavelength of $\lambda \approx 3 \times 10^{-5}\, m$.

## 7 Discussion

Finding that the acceleration rate to be expected from a man made device is typically only 6 GeV in 250 m and therefore no better than an S-Band Linac is a disappointment. So too is the prediction that acceleration rates get worse at higher energy. The celestial mechanism is even more intriguing. The numerical estimate above suggests a few TeV can be reached before an electron escapes from a pulsars influence. The obvious extension is to repeat the theory for protons to see if they reach higher energies.

## References


[1] J.P.Wefel and W. H. Sorrel, *"Genesis and Propogation of Cosmic rays"*, Mathematical and Phys. Sciences Vol.220 (1987) pp 1 and 227 Edit.M.M.Shapiro and J.P.Wefel D. Reidel Publ. Company, (1987).

[2] G.F.Bignami and W.Hermsen, *"Annual Review of Astronomy and Astrophysics"* , Vol.21,(1983),p.67

[3] , S.L.Shapiro and S.A. Teukolsky, *"Black holes, White Dwarfs and Neutron Stars"*, J.Wiley and Sons, New York (1983) p.289, J.H.Taylor and D.R.Stinebring, *"Annual Review of Astronomy and Astrophysics"* Vol.24 (1986) pp 285,453.

[4] K.R.Lang , *"Astrophysical Formula, 2nd. Edition"* Springer Verlag, Berlin (1980) pp. 52,58

**Research Trends in Physics: Coherent Radiation Generation and Particle Acceleration**
Editorial Board: J.M. Buzzi, A. Prokhorov (Editor-in-Chief), P. Sprangle, and K. Wille
*La Jolla International School of Physics*, The Institute for Advanced Physics Studies, La Jolla, California

# XUV-FELs in Storage Rings

**D. Nölle and K. Wille**

University of Dortmund
4600 Dortmund, Germany

# 1 Introduction

After successful operation in the long–wavelength regime [1], in the visible [2] and near UV[3], in the near future FEL operation will also be obtained in the short–wavelength regime, i.e. the VUV and XUV. Several projects are aiming to produce FEL radiation at these wavelengths [4, 5, 6].

There are two different approaches to reach the short–wavelength regime of the FEL. The first approach is to get short wavelengths by using low–energy electron beams ($\sim 50MeV$) with undulator magnets of very short periods of only some $mm$ length. Moreover, it is planned to radiate at harmonics of the undulator spectrum to further reduce the FEL wavelength [7].

The second approach is to use accelerators of higher energy, i.e storage rings, to drive conventional undulator magnets of considerable length. This approach has the advantage of using rather standard technologies for accelerator and FEL to get to extremly short wavelengths. Due to synchrotron–radiation damping, storage rings provide electron beams of optimum qualtity concerning energy spread, emittance, and high peak currents.

Thus storage rings of moderate energy, low emittance, and strong damping, providing long straight sections for insertion devices, are optimum drivers for FELs in the XUV–wavelength regime, and for FELs at even lower wavelengths in the future.

All these requirements for FEL drivers are nearly perfectly obtained by modern, compact syncrotron–radiation sources. The only argument against is, that FELs are still under development. Thus, much time is needed for testing and optimizing the device, a task that cannot easily performed at user facilities. Therefore, DELTA, the Dortmund ELectron Test Accelerator, the storage ring currently under construction at the University of Dortmund [6], will be an optimum driver for the extreme short–wavelength operation of FELs. It will have all qualities of a compact synchrotron–radiation sources, and will also provide enough beam time necessary for a test facility.

This paper contains the main design criteria of a storage–ring–FEL driver and discusses the requirements by the example of DELTA. Furthermore, the theory for short–wavelength FELs in storage rings will be given and discussed by the example of FELICITA II, a FEL experiment in the $100nm$ regime, which is planned at DELTA[8]. This experiment has the potential to operate also at wavelengths as short as $20nm$.

## 2 Requirements of a FEL-Storage Ring

In order to get high FEL gain the electron beam circulating in a storage ring has to satisfy several important requirements. This has direct consequences on the design of the accelerator. The following chapter presents the design criteria for a dedicated high-gain FEL storage ring in the XUV-range.

### 2.1 Particle density of the electron beam

Generally, the FEL-gain $G$ is directly proportional to the particle density $n_b$ of the electron beam:

$$G \propto n_b = \frac{\hat{I}}{2\pi e c \sigma_x \sigma_z}. \tag{1}$$

$\hat{I}$ is the peak current of one bunch and $\sigma_x$ and $\sigma_z$ are the horizontal and vertical beam dimensions, respectively. They are given by

$$\sigma_{x,z} = \sqrt{\varepsilon_{x,z}\beta_{x,z}} \tag{2}$$

with the beam emittance $\varepsilon$ and the amplitude function $\beta$. Because of the Liouville theorem the uncoupled horizontal emittance is constant around the ring and can be expressed as [9]

$$\varepsilon_0 = \frac{55}{32\sqrt{3}}\frac{\hbar}{mc}\gamma^2\frac{\langle\frac{1}{R^3}H(s)\rangle}{J_x\langle\frac{1}{R^2}\rangle} \tag{3}$$

with

$$H(s) = \gamma(s)D^2(s) + 2\alpha(s)D(s)D'(s) + \beta(s)D'^2(s) \tag{4}$$

and the optical functions $\beta(s)$, $\alpha(s) = -\beta'(s)/2$ and $\gamma(s) = (1+\alpha^2(s))/\beta(s)$ and the Dispersion $D(s)$ and its slope $D'(s)$. In most of the storage rings we have $J_x \approx 1$ and the bending radius $R$ is the same in all magnets, i.e. $R = \text{const.}$ With these simplifications we get

$$\varepsilon_0 \propto \frac{E^2}{R}\int_{bends} H(s)\,ds. \tag{5}$$

$E$ is the beam energy in [GeV]. With the emittance coupling $k$ the horizontal and vertical emittance becomes

$$\varepsilon_x = \frac{\varepsilon_0}{1+k} \quad \text{and} \quad \varepsilon_z = \frac{k\,\varepsilon_0}{1+k} \tag{6}$$

It is obvious from (4) and (5) that a small emittance requires small betafunctions and a small dispersion inside the bending magnets. Actually, the minimum of the emittance is provided if these optical functions have a waist in the bending magnets. This important constraint has to be fullfilled by the beam optics of a FEL-storage ring.

Very high particle densities in a bunch causes a reduction of the beam lifetime due to the *Touschek-effect* [10, 11]. For a given lifetime $\tau_b$, the maximum peak current is

$$\hat{I}_{max} = 3.8\cdot 10^{17}\frac{E^2\varepsilon_x\sqrt{\langle\beta_x\rangle\langle\beta_z\rangle}}{D(E)\,\tau}A_E^3. \tag{7}$$

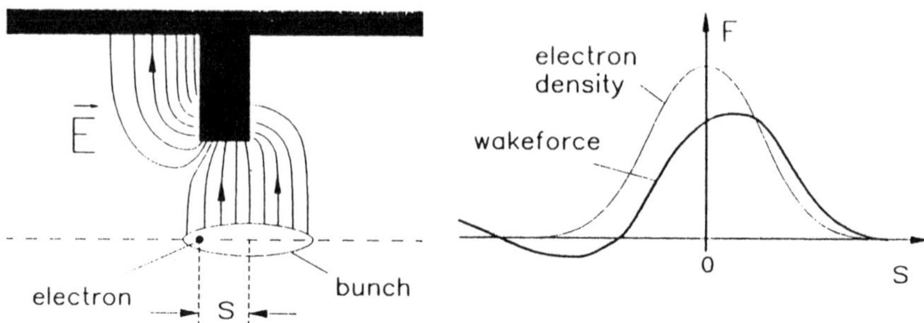


Figure 1: Wake fields exited by a bunch passing an obstacle in the vacuum chamber. The right curve shows the wake force on an electron with a distance $s$ from the center of the bunch

Since the betafunctions $\beta_x$ and $\beta_z$ are rather small because of the emittance requirements, high bunch currents are only possible with a large energy acceptance $A_E$ of the storage ring.

The peak current $\hat{I}$ in (1) also depends on the bunchlength $\sigma_s$ as

$$\hat{I} = I_b \frac{L}{\sqrt{2\pi}\sigma_s}. \tag{8}$$

$I_b$ is the average bunch current and $L$ the circumference of the ring. The bunchlength is given by

$$\sigma_s\ [m] = 4.1 \cdot 10^4 \frac{\alpha\ E}{f_s \sqrt{R}} \tag{9}$$

with the momentum compaction factor $\alpha$ and the synchrotron frequency $f_s$, which is mainly determined by the accelerating voltage $U_{rf}$ according to $f_s \propto \sqrt{U_{rf}}$. Therefore, a high cavity voltage is required, in order to obtain large peak currents. Unfortunately, a very intense bunch is normally significantly longer than calculated from (9). The reason comes from the wake fields [12] excited by the bunch at any obstacle in the vacuum chamber as shown in figure 1. These wake fields generate forces on the electrons causing a lengthening of the bunch. This effect has been measured the first time at SPEAR [13]. Additional investigations have been made at all electron storage rings in the meantime. Qualitatively the results are the same. Above a specific threshold $I_{th}$ the bunchlength increases linearly with the bunch current as shown as an example in figure 2. The slope of the curve depends on the impedance of the vacuum chamber. The consequence of the bunchlengthening due to wake fields is a careful design of all components of the vacuum system. It is necessary to avoid any sudden step in the cross section of the chambers. A smooth tapering is highly required.

## 2.2 Radiation Damping

As shown by Renieri [14] the power of the FEL-beam is proportional to the power of the synchrotron radiation emitted by the circulating beam

$$P_{FEL} \approx \frac{3}{2} A_E \frac{E}{\tau_s} N_b \tag{10}$$

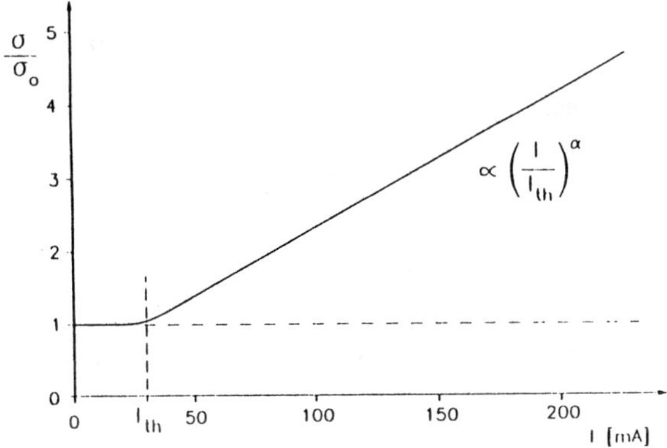


Figure 2: Bunchlength as a function of the bunch current

where $N_b$ is the number of electrons in one bunch. The strong synchrotron radiation causes short damping times of the synchrotron oscillation namely

$$\tau_s = \frac{E\ t_r}{\Delta E_{syn}} \tag{11}$$

where $t_r$ is the revolution time of the ring and

$$\Delta E_{syn} = 8.85 \cdot 10^{-5} \frac{E^4}{R} \qquad \text{with} \qquad E, \Delta E_{syn} \text{ in [GeV]} \tag{12}$$

the energy radiated during one revolution. Because of $R \propto E/B$ we can write

$$\tau_s \propto \frac{1}{B\ E^2}. \tag{13}$$

Therefore, the product $B\ E^2$ has to be as large as possible in order to get high FEL output power, or if the beam energy is given by the experimental requirements, the magnetic field $B$ must be high. On the other hand, a high field leads to a small bending radius $R$ and according to (5) the beam emittance becomes large.

Fortunately, the emittance of a periodic magnet structure also depends on the bending angle $\Theta$ per magnet cell

$$\varepsilon_0 \propto \Theta^3 \tag{14}$$

With this relation we get both a strong radiation damping and a small emittance simultaneously by designing the magnet lattice with a relatively large number of cells containing short bending magnets with rather strong fields.

## 2.3 Additional design considerations

In particular in the XUV-range the FEL-gain is rather small and has to be compensated by a sufficient length $l$ of the undulator. For small signals, i.e. at the entrance of the undulator, the gain goes like $G \propto l^3$, whereas at very high signal intensities the gain grows exponentially with the length. Under all circumstances a long undulator gives a significant contribution to the FEL-gain. Therefore, a free straight section of a length of at least 15 m is necessary.

Because of the coherence condition the divergence of the particle trajectories in the beam has to be as small as possible, or with other words, the betafunctions have to be almost constant along the undulator. Assuming the straight section as a drift space, the betafunction is given by

$$\beta(s) = \beta^* + \frac{s^2}{\beta^*} \tag{15}$$

where $s$ is the distance from the center of the section and $\beta^*$ the betafunction at this point. It is obvious that $\beta^*$ must not be too small. Values between 10 and 20 m are recommended.

Since orbit distortions have a very sensitive influence on the emittance and the coupling, a precise beam–position measurement and an effective orbit–correction scheme is important. The absolute accuracy of the monitors is $\Delta x < 100\ \mu$m and the resolution should be one order of magnitude smaller. In order to avoid beam blow–up due to scattering of the electrons on the atoms of the restgas, the vacuum pressure is supposed to be less than $p < 10^{-7}$ Pa. This also will cause a long beam lifetime.

Besides the wake fields high beam currents may generate various modes of instabilities, limiting the peak current. Therefore, a FEL storage ring needs an effective feedback system for stabilisation of the transverse and longitudinal beam oscillations. In addition, the accelerating cavities need damping of the dangerous modes. Finally, a fast and reliable injection is required, to keep the injection time as short as possible.

## 3 The Dortmund FEL-Storage Ring DELTA

Based on the criteria presented in chapter 2 the FEL storage ring DELTA has been designed at the University of Dortmund [15, 6]. During the design phase of the machine two different possible magnet lattices have been studied: the FODO-structure as used for the SLC-damping ring [16] and an arrangement of three quadrupoles between the bending magnets (triplet structure). As a result of these studies it was shown that the triplet structure is the better solution. The optics of one triplet cell of DELTA is to be seen in figure 3. With the triplet structure one can easily provide the required waists of the horizontal betafunction and the dispersion in the bending magnets. In addition, depending on the quadrupole strength this structure results in a wide range of the beam emittance, which is shown in figure 4. The entire ring is arranged as a racetrack and consists basicly of two half rings containing the triplet cells separated by two about 20 m long straight sections (figure 5). These straight sections are foreseen in particular for the installation of the required long FEL undulators. The flexibility of the beam optics allows very different experiments in these straights. At the very beginning, a fourfold symmetry with quadrupoles in the center of the straight sections is chosen because of simplicity. The corresponding beam optics of one quadrant of this machine is shown in figure 6.

For later experiments with very long FEL-undulators one can remove the quadrupoles in the middle of the straight section resulting in an approximately 15 m long free

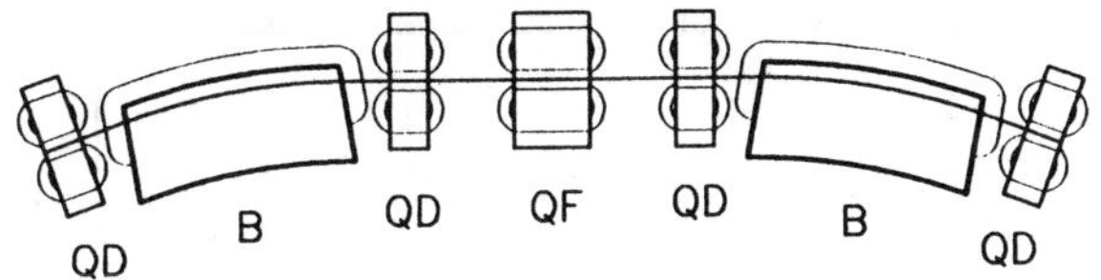


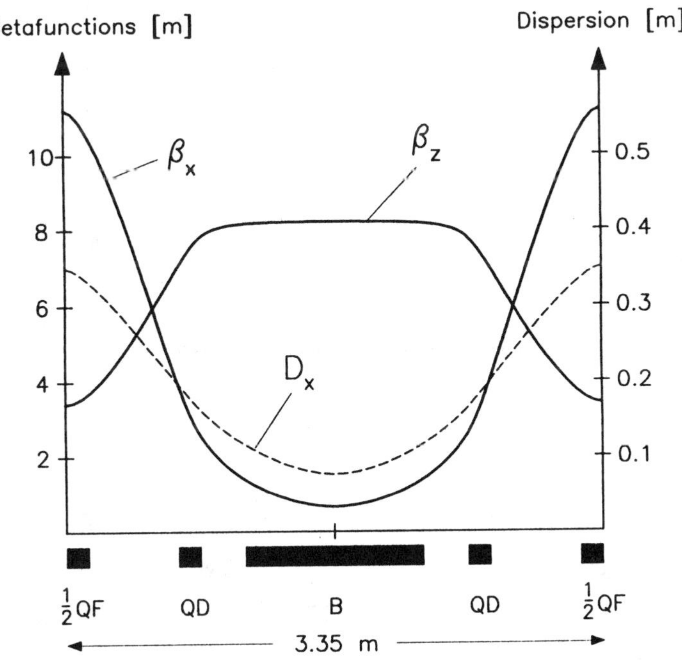


Figure 3: Betafunctions and dispersion of one triplet cell of the DELTA lattice

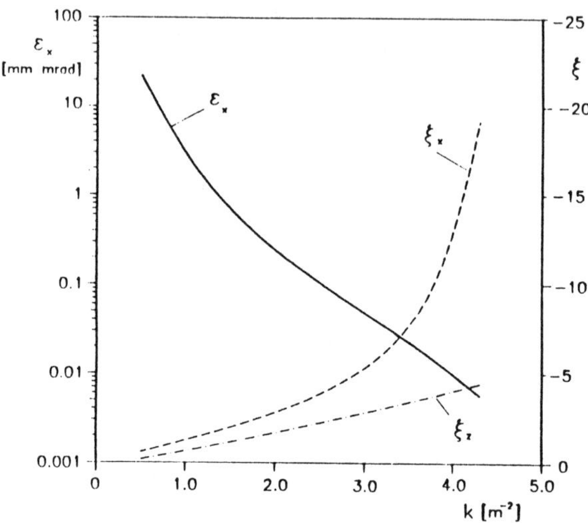


Figure 4: Beam emittance as a function of the quadrupole strength of the DELTA triplet cell. The resulting chromaticities in both planes are also drawn in the diagram.

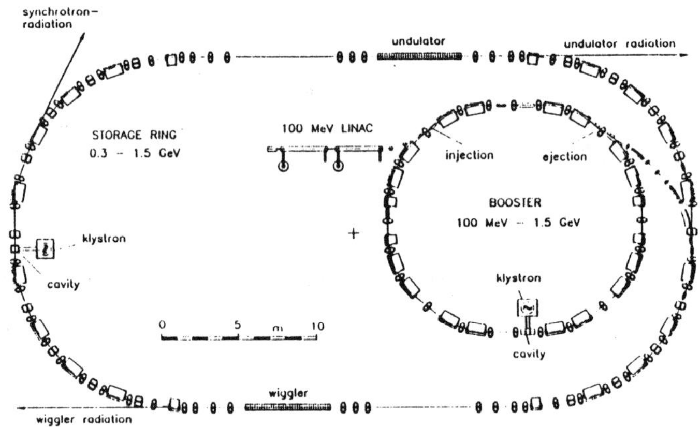


Figure 5: The storage ring complex of the Dortmund FEL-ring DELTA. The figure shows the most simple forefold symmetry with four insertions for wigglers and undulators. The injection scheme consists of a 100 MeV LINAC and a 1.5 GeV booster synchrotron.

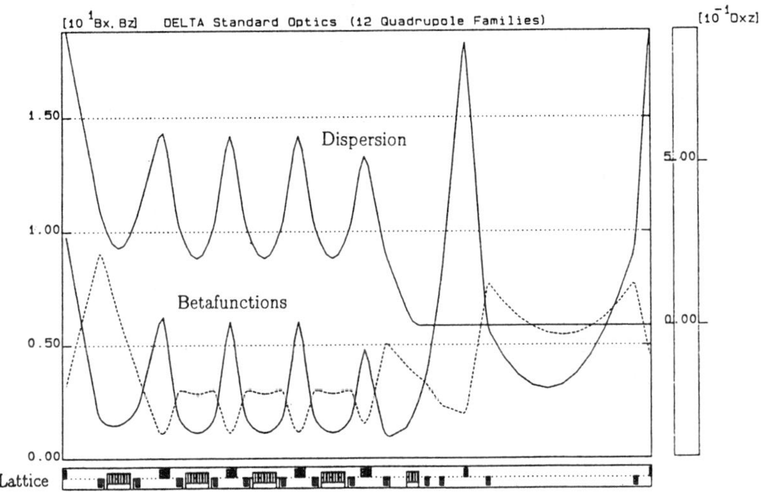


Figure 6: Optical functions of one quadrant of the DELTA storage ring with fourfold symmetry

space. The quadrupoles at both ends provide a powerfull means of matching of the optics between the insertion and the arcs. In the middle of the arcs the injection and the rf-cavities are installed. The most important parameters of the machine are listed in table 1.

The vacuum system of DELTA has to provide low vacuum pressure and low microwave impedance. The best solution seems to be an arrangement consisting of two chambers connected by an 8 mm high, 27 mm wide pumping slit, as shown in figure 7. The right chamber is used by the beam and has a water cooled absorber at the outside against the heat load form synchrotron radiation. The whole system is mainly pumped by integrated ion–getter pumps (DIP) and non-evaporable getter pumps

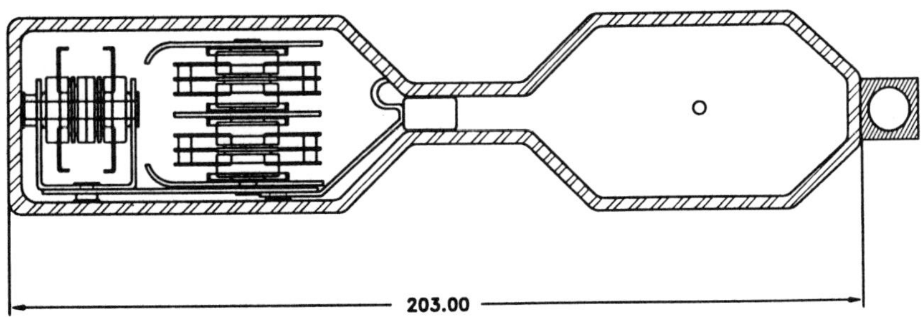


Figure 7: Cross section of the DELTA vacuum chamber. The right channel is used by the electron beam and the left one containes the compact pump module with ion getter pumps and NEG pumps.

Table 1: The most important parameters of the Dortmund FEL-storage ring DELTA. All values are at $E = 1$ GeV

| | | | |
|---|---|---|---|
| Tunes | $Q_x$ | | 9.711 |
| | $Q_z$ | | 2.835 |
| Natural emittances: | | | |
| 0 % coupling | $\varepsilon_x$ | [m·rad] | $4.84 \cdot 10^{-9}$ |
| 10 % coupling | $\varepsilon_x$ | [m·rad] | $4.42 \cdot 10^{-9}$ |
| | $\varepsilon_x$ | [m·rad] | $4.42 \cdot 10^{-10}$ |
| Momentum compaction factor | $\alpha$ | [%] | 0.42 |
| Natural chromaticity | $\xi_x$ | | -22.68 |
| | $\xi_z$ | | -5.89 |
| Energy spread | $\frac{\Delta E}{E}$ | | $4.7 \cdot 10^{-4}$ |
| Damping times | $\tau_x$ | [ms] | 29.3 |
| | $\tau_z$ | [ms] | 28.6 |
| | $\tau_s$ | [ms] | 14.1 |
| Damping partition numbers | $J_x$ | | 0.977 |
| | $J_z$ | | 1.000 |
| | $J_x$ | | 2.023 |
| Revolution frequency | $f_r$ | [kHz] | 2602.37 |
| Energy loss / turn | $\Delta E$ | [keV] | 26.9 |
| Accelerating frequency | $f_{RF}$ | [MHz] | 500 |
| Harmonic number | $q$ | | 192 |
| Maximum average current: | | | |
| (1 bunch) | $I_{max}$ | [mA] | 100 |
| ( $\geq$ 10 bunches) | $I_{max}$ | [mA] | 500 |
| Number of cavities | $n_c$ | | 1 |
| Shunt impedance / cavity | $R_s$ | [MΩ] | 3.0 |
| Maximum RF-power | $P_{RF}$ | [kW] | 60 |

(NEG) simultaneously. These pumps are mounted together in the pumping channel at the inner side of the vacuum chamber. Both types will provide a pressure low enough to guarantee a beam life time of at least 10 h [17].

In order to get the required low orbit distortions, a system of 40 position monitors will be installed around the ring. Each monitor consists of four symmetrically arrangeded pick-up electrodes and a narrow-band rf-electronic for the 500 MHz signals. This monitor system has been tested in the laboratory, and an absolute accuracy of $\Delta x < 50\ \mu$m and a resolution of better than $5\mu$m was obtained [18]. This is sufficient under all circumstances.

The storage ring project DELTA has been approved by the end of 1989 and is now in the construction phase. The building is expected to be finished in mid 1992, and, based on the present schedule, all technical components of the machine are ready by the end of 1993. After commissioning the first FEL experiments will start in 1994.

This will be the investigation of an optical klystron, called FELICITA I [42]. This experiment will be a FEL–device, using an electromagetic undulator magnet of 16 periods. As each period can be powered separately, it is possible to realize both the optical klystron and conventional FEL configuration with a single undulator. Furthermore, it is possible to use such a flexible device for first tests of more exotic FEL devices, as a distributed optical klystron [19].

Further goals of this experiment are the test of the FEL properties of DELTA and the optimisation of the machine for the XUV-FEL experiments. It also provides tools for beam diagnostics during FEL operation[20].

# 4 Theoretical Modeling of High–Gain FEL in Storage Rings

This chapter gives a short overview of the theory and the numerical modeling of high–gain FELs and their operation in storage rings, as implemented in the FEL–Simulation code FELS [22].

## 4.1 Theory of High–Gain FELs and their Interaction with Storage Rings

The interaction mechanism of the FEL can be described very simply. A highly relativistic electron beam with its main velocity component in longitudinal direction passes through an undulator magnet, to produce an oscillating transverse velocity component. Synchronously, a radiation field is transversing the undulator, spatially overlapping with the electron bunch. Due to the transverse motion of the electrons, governed by the Lorentz force equation, the radiation field can couple to the electron beam and exchange energy with the electrons. Of course, amount and sign of the energy exchange depend on the relative phase between the velocity of the single electron and the electrical field vector of this wave front. This phase $\eta$ has to be a slowly variing function over the whole undulator to initiate effective energy exchange. For the individual electron the evolution of the energy ($\nu_i$) and the FEL–phase ($\eta_i$) reads in dimensionless units [21] as defined in table2 on page 27:

$$\begin{aligned} \frac{d}{d\tau}\nu_i &= -\Re\left[i\sum_{f=1}^{\infty}\left\{\tilde{a}_o^f(\tilde{z}_i,\tilde{r}_\perp^s,\tau) - \frac{\partial}{\partial x}\tilde{a}_o^f(\tilde{z}_i,\tilde{r}_\perp^s,\tau)\right\}\exp\{if\eta_i\}\right] \\ \frac{d}{d\tau}\eta_i &= \nu_i - \tilde{Z}_r\beta_\perp^{s\,2} \end{aligned} \tag{16}$$

According to the resonant condition of the FEL

$$\lambda_r^f = \frac{l_o}{2f\gamma_r^2}(1+\frac{k^{W2}}{2}) \tag{17}$$

an energy exchange between the radiation field and electron beam is possible for each wavelength, corresponding to the spontaneous line spectrum of the used undulator magnet.

The macroscopic path of a single electron through the undulator magnet is described by:

$$\begin{aligned} \tilde{x}_i(\tau) &= \tilde{X}_o^\beta\cos(\tilde{K}_x^\beta\tau + \phi_x^\beta) \\ \tilde{y}_i(\tau) &= \tilde{Y}_o^\beta\cos(\tilde{K}_y^\beta\tau + \phi_y^\beta) \\ \text{with} \quad \tilde{K}_l^\beta &= \frac{k^W\kappa_l}{\sqrt{2}\gamma} \quad k^W = \frac{eB^oL_o}{2\pi mc} \end{aligned} \tag{18}$$

$$\kappa_x^2 + \kappa_y^2 = \frac{2\pi}{l_o}$$

The parameters $\kappa_l$ in eq.(18) characterize the focussing properties of the magnetical field of the given undulator magnet. For instance $\kappa_x$ is zero for the common undulator type with plane pole faces [23].

The evolution of the elm. field due to the FEL interaction is given by the wave equation, describing the collective effect of all electrons interacting with the elm. field. As the change of the field amplitude is slow compared to the change of the phase of the elm. field, the evolution of the laser field can be described by means of the paraxial wave equation [24].

$$\begin{aligned} \{\nabla_\perp^2 - i4f\tilde{Z}_r\partial_\tau\}\tilde{a}_o^f(\tilde{z},\tilde{r}_\perp,\tau) &= 4f\tilde{Z}_r\tilde{S}_f(\tilde{z},\tilde{r}_\perp,\tau) \\ \tilde{S}_f(\tilde{z},\tilde{r}_\perp,\tau) &= -j_f\langle\exp(if\eta)\rangle_{\tilde{z}+\tau}, \quad f \quad \text{odd} \\ \tilde{S}_f(\tilde{z},\tilde{r}_\perp,\tau) &= -\frac{\xi}{2}\frac{\partial}{\partial\tilde{x}}\langle\exp(if\eta)\rangle_{\tilde{z}+\tau}, \quad f \quad \text{even} \end{aligned} \tag{19}$$

This equation describes the evolution of the elm. field averaged over time intervals short compared to the time an electron needs to pass over one undulator period, and averaged over spatial distances of the order of the wiggle amplitude of a single electron ($\xi$).

Due to the symmetry of the single–electron sources [25], over which one averages to form the source functions given by eq.(19), the even harmonics average almost to zero, and contribute only if the wiggle amplitude becomes comparable to the width of the driving current. In some special cases this could become important in the XUV–regime, for FELs operating with electron and laser beams of extremely small diameters. But in almost all cases the source function is dominated by the contributions of the odd harmonics. As these contributions are dominated by the fundamental, in most cases a code following only the evolution of the fundamental is sufficient [22].

In a storage ring the electrons normally circulate over many hours, corresponding to roundtrip numbers of the order of $10^{10}$ or more. Even the damping time of the order of some $\mu s$ corresponds to $10^4 - 10^5$ roundtrips. The interaction time of the FEL with the storage ring can be characterized by the rise time $\tau_r$ defined by eq.(20).

$$\tau_r = \frac{\tau_U}{G - P} \tag{20}$$

$\tau_U$ is the roundtrip time of the storage ring, $G$ and $P$ denote the gain of the FEL and the losses of the optical cavity, respectively.

The rise time $\tau_r$ can vary over a wide range, but is typically very small compared to the damping time and, in the high–gain case, also to the period of the synchrotron oscillation of the storage ring.

Due to the different time scales, long–time effects as radiation damping or synchrotron oscillation, have no effect on the evolution of the FEL pulse, that is totally decoupled from the longitudinal phase space of the storage ring.

A storage ring FEL operates in a naturally pulsed manner [26]. Assuming the laser is switched on at a certain time. Due to the gain process the FEL pulse evolves. The higher the amplitude rises, the stronger the electron beam is disturbed by the interaction. Especially, the energy spread strongly increases during the interaction. This leads to gain degradation. Accordingly, the rising of the pulse is stopped before the FEL gain saturates due to the high intensity, as in case of single–pass devices. Due to the enlarged energy spread the gain of the FEL drops to levels insufficient to compensate for the cavity losses. As a consequence, the FEL pulse decays corresponding to the resonator losses and the laser is switched off for times of the order of the damping time of the storage ring.

This is not the whole truth, since it describes only the evolution of the very first FEL pulse after starting operation. Without external influence, the second FEL pulse starts with an electron beam, not completely damped down to its initial high equibibrium quality. Since it starts with an electron beam of minor quality, it does not reach the same power level as the previous pulse. The following laser pulses start with even worse electron–beam quality, and the power level drops to an equilibrium level. The FEL operates in cw mode and the heating of the laser is balanced by the damping of the storage ring. The power level of the laser is related to the synchotron radiation power of the storage–ring as described by the so called "Renieri limit" [14].

In order to obtain laser pulses of high peak power, the laser is operated in a Q–switched pulsed mode. In this mode the laser is switched off, until the electron beam is damped to the initial high quality. Then the laser is switched on and the high power is reached with every pulse. The repetition rate is determined by the damping time of the storage ring. Although, the peak power is very high, the average power is at the low level determined by the Renieri limit [26].

Due to the different time scales, the FEL simulation code has to deal with the transverse phase space of the storage ring only. The influence of synchrotron–radiation damping and synchrotron oscillation can be totally neglected in the simulation runs. Furthermore, due to the relatively large bunch length compared with the slippage length [27], short–pulse effects, as they can be seen at LINAC based FEL–devices in the long–wavelength regime [28], do not appear.

Therefore, the recirculation of the electron beam, back to the entrance of the undulator, can be simply calculated using the R–matrix, describing the transformation of the transverse coordinates and velocities of a single electron through the optics of the storage ring.

## 4.2 'FELS', short Description of a 3D–FEL–Simulation Code

The theoretical model presented before, is implemented in the 3D–FEL code 'FELS' developed at the University of Dortmund [22]. This code describes the FEL in the long–pulse regime, i.e. the regime with the electron bunch length much bigger than the slippage length. Furthermore, the evolution of the fundamental and also the gain process for the higher harmonics of the spontanous spectrum of the undulator can be included.

The electron beam is described in terms of 'macroparticles' i.e. the content of a small volume of the electron beam, that contains a couple of electrons ('microparticles'), all of the same energy, and with the same macroscopic coordinates and velocities. The propagation of these macroparticles is described by eq.(18). Their initial distribution in space and energy is modeled by random generators, producing a Gaussian energy distribution of the macroparticles, and a two-dimensional correlated Gaussian distribution in each transverse direction[29], providing the distribution of initial position and velocity. The transverse distributions have to be correlated in order to model the shape of the phase-space volume, determined by the optics of the storage ring, or any other type of accelerator used.

The microparticles confined to a macroparticle are treated as a statistical ensemble in the FEL-phase $\eta$. $\eta$ is distributed homogeneously over the range from $[-\pi, \pi[$, corresponding to a 'length' of the macroparticle of one fundamental wavelength. The microparticles are initially distributed with equal spacing homogeneously over the given range, in order to avoid large numbers of microparticles, needed, if a random generator is used to form uniform distribution [30].

In the undulator magnet the microscopic "evolution" of each macroparticle is modeled by solving the system of differential equations (16) for every microparticle. As the evaluation of trigonometric functions always consumes much CPU time, the code does not calculate the evolution of the FEL phase $\eta$ directly, but follows the sine and cosine of it. This has to be paid by some more computer memory, but reduces the run time by the order of 5.

To calculate the evolution of the elm. field, eq.(19) is split into real and imaginary part, and is rewritten in form of finite differences using ADI methods [31] on a Cartesian mesh. The system of linear equations obtained by this discretisation is then solved by an implicit algorithm, taking care of the sparseness of the system [31].

For each time step the source functions (eq.(19)) are calculated, projecting the distribution of macroparticles onto the field mesh and using the microscopic distribution of the FEL phase to calculate the source contribution of each macroparticle. The contribution assigned to each meshpoint is calculated by an area-weighted algorithm [32].

Outside the undulator, the paraxial wave equation has to be solved without sources. Instead of solving the partial differental equation numerically, it is possible to calculate the evolution analytically in the frequency domain for a given input field [33]. This is used in combination with a two-dimensional FFT algorithm to model the recirculation of the laser pulse through the optical cavity.

As two different types of optical resonators are common to FELs in the short-wavelength regime, both are included in the simulation code. The first type of resonator is the simple confocal resonator with two spherical mirrors on axis. The second one is the ring resonator providing high-Q resonators in the short-wavelength regime due to the rather high reflectivity at grazing incidence [36].

Although the ring resonator is constructed of a system of rather complicated mirrors — paraboloids and multifacet mirrors, forming the arcs — it can be treated as a periodic sequence of optical elements on the axis defined by the center of 'mass' of the laser pulse traveling through the resonator [34]. All mirrors are treated as simple elements, producing a phase shift of the elm. field according to the displacement of the corresponding mesh point from the optical axis [35].

# 5 FELICITA II: The Propsed High–Gain SRFEL in the 100 nm Regime

The storage ring DELTA provides two straight sections of 20 m length each. 14 m can be used for insertion devices. FELICITA II will be the second FEL experiment at DELTA, operating in the wavelength regime from 100 to $20nm$. This chapter will discuss the design of of this device, and will give first simulation results on its operation. The data presented in the following concern the operation in the long–wavelength regime around $100nm$.

## 5.1 Design Paramters of the Undulator Magnet and the Optical Cavity of FELICITA II

In order to design a FEL device with maximum potential for short wavelength operation, one has to use the maximum available length and to construct the undulator with the maximum number of periods possible under the given constraints. The constraints for the FELICITA II undulator have been defined as follows:

- The wavelength fixed to 100 nm at the lowest energy
- The electron–beam energy has to be higher than 500 MeV. DELTA is designed for a maximum energy of 1.5 GeV, so that operation below 500 MeV might result in a nonstable beam and / or low peak current.
- The maximum length of a straight section, available for insertion devices at DELTA is 14 m.
- A magnetic field of the order of 1 T peak is estimated to be the upper limit, that can be obtaind using standard technology for the undulator magnet.

In this wavelength regime the optical cavity has to be designed as a ring resonator Fig.(8), using the highest possible reflectivity, which can be obtained only at grazing incidence. Furthermore, the highest peak current in DELTA will be reached in single bunch–operation. Therefore, the the roundtrip time of the laser pulse in the optical cavity has to be equal to the roundtrip time of the electrons in DELTA, corresponding to a path length of $115.2m$. Four optical elements will be used in this ring resonator. After separation of the laser beam from the electron beam at the beginning of the arcs of the storage ring (Fig.8), the laser beam will be deflected by a parbolical mirror under 20° from the undulator axis. The defocussing properties of this mirror reduce the energy load of the laser beam on the second optical element, a multifacet mirror, as proposed for the Los Alamos short–wavelength–FEL facility [36]. It represents the arc of the 'optical storage ring', which provides a 160° bend and focussing of the laser beam into the long recirculation. Furthermore, the widening of the beam to 10 times of its minumum diameter provides a simple method for output coupling. For instance a flat mirror could be used as a scraper in the recirculation section to deflect radiation of the edge of the laser field. The second arc, also a multifacet mirror, focuses the laser beam onto the second parabolical mirror, that is used to deflect the laser beam back onto the undulator axis, and to match the beam to the inital waist in the center of

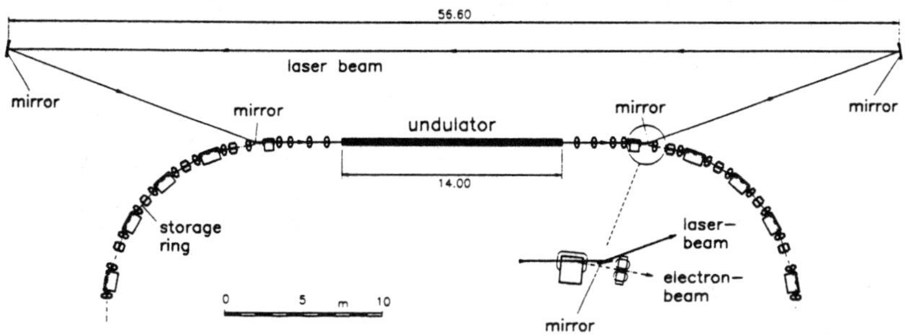


Figure 8: Layout of DELTA with the undulator magnet of FELICITA II installed in the upper straight section. The laser beam and the electron beam are separated after the first bending magnet of the arc, providing a deflection of 10° for the electron beam. The laser beam is deflected and defocussed by a first parabolical mirror, mounted only a few cm from the orbit of the electron beam. On its path to the mutifacet mirror the beam widens, giving a consideably reduction of the power load on this mirrors. The mirrors of the first multifacet arc compensate this divergency of the beam. After passing the long straight of the recirculation the laser beam is focussed back by the second arc and matched to the initial spot size by the second parabolical mirror.

the undulator. As the parabolical mirrors and the arcs are designed to have identical parameters the optics of the resonator has twofold symmetry (Fig.9).

The parameters of the undulator magnet and the optical cavity following from these considerations are listed in table 3.

## 5.2 Numerical Simulation of FELICITA II

The main question of FEL operation is the attainable amount of gain. The device has to provide enough gain to overcome mirror losses for the whole range of wavelengths, the device is designed operate in. When calculating the gain one has to take care of all effects, that can result in gain degradation. Since such a FEL has to operate in the exponential–gain regime, analytical calculation of the FEL gain as in the small–signal small–gain regime is no longer possible [38]. All effects have to be included in the calculations in a selfconsistent manner [39].

For storage–ring–FEL devices the situation is even worse, as they normally opperate with a flat electron beam, that does not have the axial symetry known from LINAC devices. Therefore, only 3D calculations including all degrading effects due to emittance and energy spread can give realistic results.

Figure (10) shows the gain of FELICITA II versus detuning $\nu$ – the normalised energy deviation from the resonance energy – for different input parameter from the FEL code 'FELS'. The curve with maximum gain (10.1) shows the results with almost no degrading effects included. Only the influence of the emittance on the overlapp

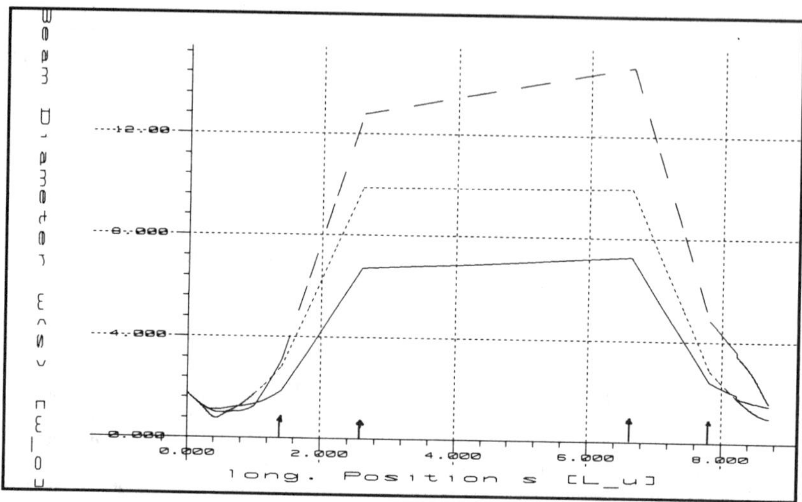


Figure 9: Beam diameter of the laser beam during one roundtrip through the optical cavity, starting at the entrance of the undulator. The arrows within the figure label the positions of the optical elements within the ring resonator. The solid and the dashed line represent the evolution of the beam diameter in horizontal and vertical direction for the loaded cavity, whereas the dotted line corresponds to the empty cavity. In the long straight of the recirculation the beam has 10 times of its minimum diameter, to reduce the power load on the arc mirrors.

of elm. field and electron beam has been taken into acount. The next step is to include energy spread. An energy spread of $0.7 \cdot 10^{-3}$ assumed for DELTA gives a gain degration factor of the order of 10 (Fig10.2). Including all emittance effects, i.e. the influence of the transverse betatron motion on the energy exchange, results in additional degradation of the gain Fig(10.3). The emittance has been adjusted to $1.5 \cdot 10^{-8} mrad$ in horizontal and $1.5 \cdot 10^{-9} mrad$ in vertical direction.

A more detailed study of the influence of energy spread and emittance is shown in figures(11,12) with logarithmic plots of FEL gain versus emittance and energy spread, respectively.

Now let us have a more detailed look on one single pass. In Fig.(14) the evolution of the elm.–field intensity is shown with the y–axis with logarithmic scale. One can see that FELICITA II enters the exponential gain regime. First the elm. field is only amplified proportionally to the third power of the distance the electrons have passed already in the undulator. In the second half of the undulator this evolution changes into an exponential rise. In the exponential–gain regime only little changes of the driving current due to energy spread or small variations of the emittance, result in big changes of the gain, as shown in Fig.(12) and Fig.(11).

In such a high–gain FEL optical guiding has to be considered. As DELTA is usually operated with electron beams of quite different emittances in vertical and horizontal direction, the guiding will lead to different focussing properties of the electron beam for both directions. As the vertical emittance is usually much smaller – in our case it is one tenth of the horizontal emitance – the gradient of the electron density in vertical direction is much bigger than in horizontally. Thus, the electron beam is focussed stronger in horizontal direction than in vertical direction (Fig.(15)). This, of

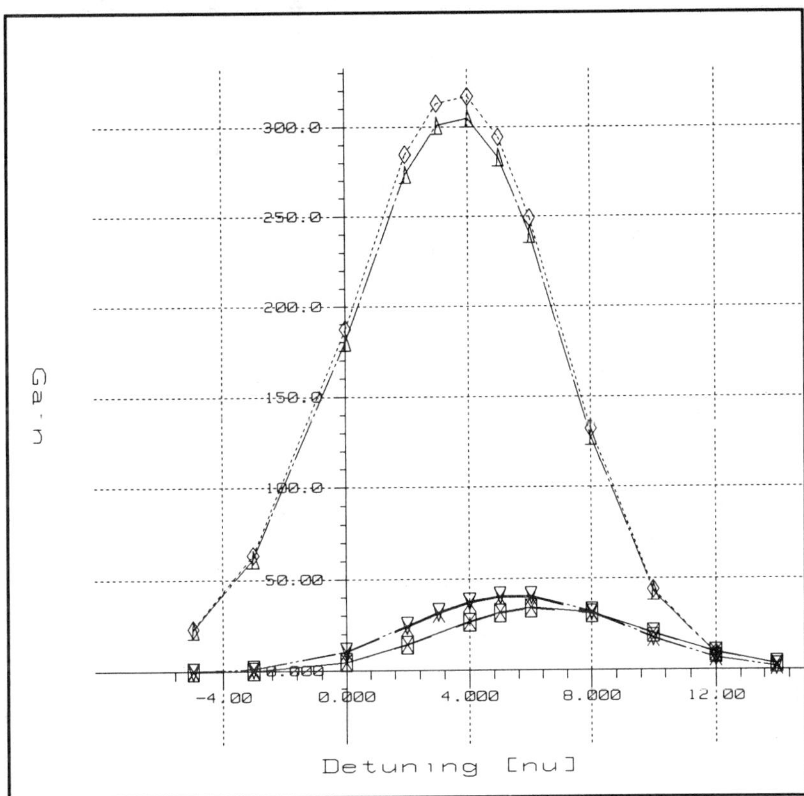


Figure 10: Gain curve of FELICITA II calculated with 'FELS'. The gain is shown versus detuning for different kinds of electron beams and with different flags set in the 'FELS' runs. All curves seem to be almost doubled. However, one of the two curves shows the gain calculated from the rise of the elm. field energy and the second is calculated from the energy loss of the electron beam. This gives a a check of 'numerical' energy conservation.

1. Gain of FELICITA II for an electron beam without energy spread and influence of the emittance.
2. Gain calculated with energy spread ($\frac{\Delta\gamma}{\gamma} = 0.7 \cdot 10^{-3}$), but no emittance effects included.
3. Energy spread and the effects of the finite emittance included for ( $\epsilon_x = 1.5 \cdot 10^{-8} m \cdot rad$, $\epsilon_y = 1.5 \cdot 10^{-9} m \cdot rad$)

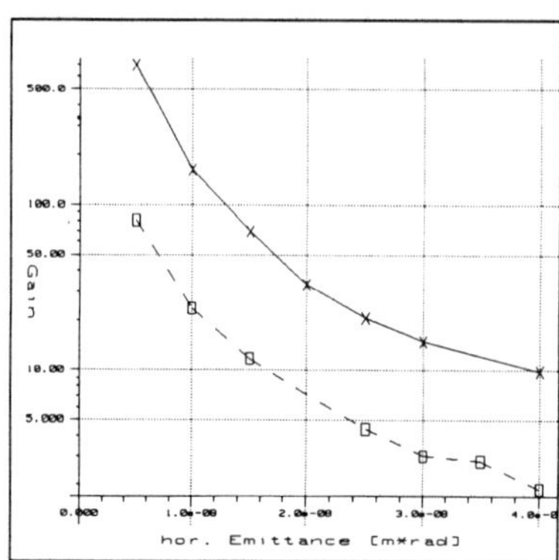


Figure 11: Logarithmic plot of the gain of FELICITA II versus emittance for a given detuning and energy spread. Om the x–axis the value for the horizontal emittance is given, the vertical emittance was choosen to be 10 times less. The energy spread was ajusted to $0.5 \cdot 10^{-3}$ for the upper and to $1.0 \cdot 10^{-3}$ for the lower curve.

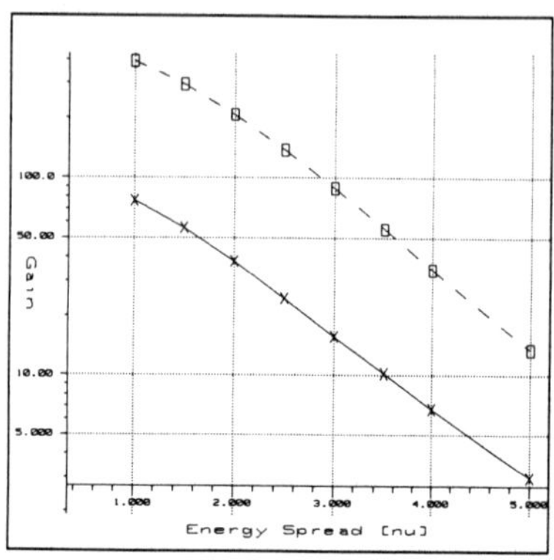


Figure 12: Logarithmic plot of the gain of FELICITA II versus energy spread for a given emittance and energy ($\nu = 6$). The emittances have been choosen as $\epsilon_x = 1 \cdot 10^{-8} mrad$ (horizontal) and $\epsilon_y = 1 \cdot 10^{-9} mrad$ (vertical) for the upper curve, and $\epsilon_x = 2 \cdot 10^{-8} mrad$ (horizontal) and $\epsilon_y = 2 \cdot 10^{-8} mrad$ (vertical) for the lower curve.

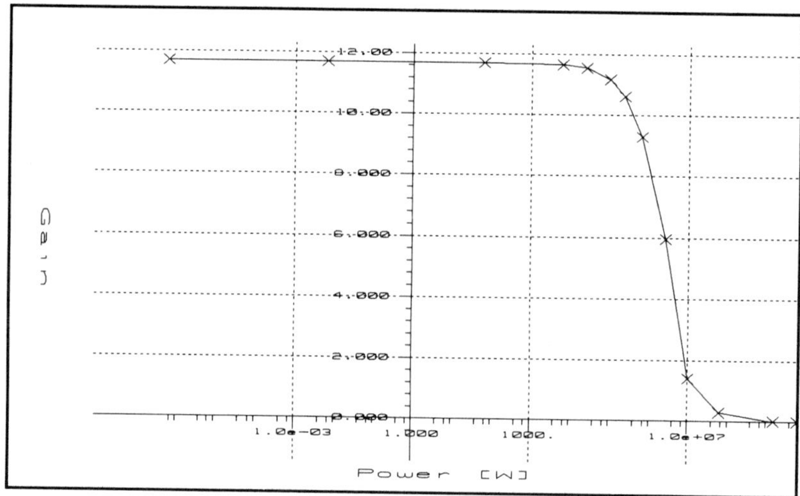


Figure 13: Dependence of the FEL gain on the power of the elm. field in the otical cavity. With increasing field amplitude the synchotron–oscillation frequency of the electrons in the FEL–bucket gets bigger, recently reaching the frequency corresponding to one revolution in phase–space. There run were made with an energy spread of $1.0 \cdot 10^{-3}$, vertical emittance of $1.5 \cdot 10^{-8} mrad$ and horizontal emittance of $1.5 \cdot 10^{-9} mrad$.

course, will cause problems with the optical cavity, as mode matching for the loaded optical resonator is more complicated [41]. Such a mode mismatch is the reason for the gain fluctuations, which can be seen in Fig.(17) and Fig.(21), respectively. But, although it is possible to operate storage rings with circular beams also, it is more effective to operate the FEL with a flat beam, because the gain is much higher due to the extremly small vertical phase-space volume of the electron beam. Fig.(14) exhibits the evolution of the field intensity, operating with an electron beam with 100% coupling of the optics of the storage ring. A nearly circular beam can be obtained, with the horizontal emittance reduced to the half of the uncoupled beam, but now with the same emittance in vertical direction.

In the following the results of computer simulations will be presented, that describe the evolution of a FEL-pulse. Both cases, single-pass operation and storage-ring operation of the FEL, will be discussed.

In a storage-ring device the electron beam is recirculated every time during the laser pulse. Therefore, changes of the electron-beam quality due to the interaction are influencing the gain process during the next pass. As the electron-beam quality is the stronger disturbed the higher the amplitude rises in the optical cavity, the effect of gain degradation will increase during the pulse. Damping effects due to synchrotron radiation do not influence this process, as they act on a different time scale. Due to the gain degradation the laser saturates, and does not reach the saturation power level of single pass devices (Fig.(20)). If the gain has dropped below the mirror losses, it takes a long time to get the electron beam damped down to a beam quality, sufficient to overcome the oscillator threshold again. To obtain the high power level of the first pulse in each of the following FEL pulses, the laser has to switched off by a Q-switch mechanism, in order to let the beam damp down to its initial equilibrium quality.

After reaching the maximum intensity the pulse decays due to the losses of the optical cavity, typically a very quick process in the VUV and XUV regime. Figure(16) shows the evolution of elm. field and electron beam during the building up of the pulse from the spontanous radiation level up to its peak amplitude.

For single-pass FEL oscillators, like LINAC-driven FELs, the temporal structure is completely different. As the electron-beam quality remains constant within the range determined by the stability of the accelerator, the FEL gain saturates due to the field intensity (Fig(13)).

Rewriting eq.(16) in form of a second-order differential equation and keeping only the fundamental, one gets

$$\frac{d^2}{d\tau^2}\eta_i = |\tilde{a}_o| \sin(\eta_i + \phi). \tag{21}$$

This is the pendulum equation, often used to describe the FEL in the small-signal small-gain regime [21]. One can easily see, that the revolution frequency of the single electron in the longitudinal phase space of the FEL is identical with the elm. field amplitude. Therefore, the gain of the field or the energy loss of the electrons will drop, if this frequency gets so high, that the electrons circulate by about one half period in phase-space during the interaction. This effect is the synchotron-oscillation, well known from storage rings. At this intensity the FEL power is reachin an equilibrium level, where gain and losses are balanced by the field intensity.

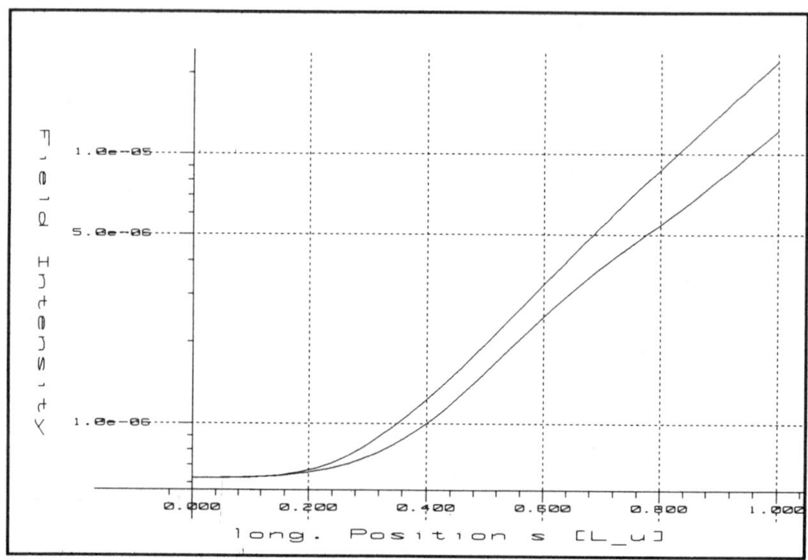


Figure 14: Logarithmic plot of the evolution of the elm.–field amplitude during one pass through the undulator for a flat electron beam a) and a more circular beam b), i.e. an electrom beam with the same parameters, but 100% coupling. First the intensity rises only with the third power of the distance, the electrons have passed in the undulator, changing into an exponential rise in the second half.

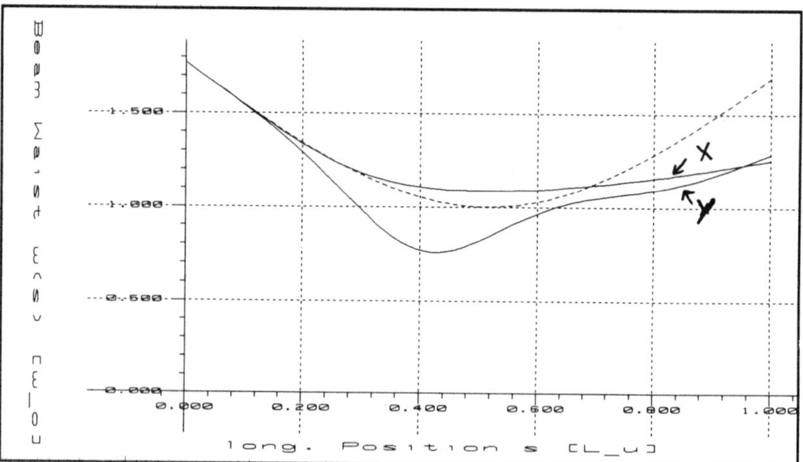


Figure 15: This figure shows the evolution of the diameter of the laser field, when passing through the FEL. The dotted curve corresponds to an empty cavity. The solid and the dashed cure give the diameter in case of the loaded cavity in horizontal and vertical direction, respectively. Since the index of diffraction in the two directions is different for a the device operated with a flat beam, common for storage rings, optical guiding acts with different strength on the laser beam in each transverse direction.

These considerations do not include effects such as sidebands or synchro–betatron resoances, whicg produce a more complicated form of the signal [43]. All these effects are not included by the 'FELS' code, since the additional modeling of longitudinal effects would require an extremly high amount of computer–storage and CPU time.

Calculating the FEL peak power from the absolute value of the dimensionless amplitdude $\tilde{a}_o$ gives:

$$\begin{aligned} P &= \frac{\epsilon_o c}{2} \int dx dy |E_o|^2 \\ &= \frac{m^2 c^5 \epsilon_o}{4\pi^2 e^2} \frac{\gamma_r^2 w_{o1}^2}{N_u L_u k^W} JJ_f^{-1} \left( \frac{f k^{W^2}}{(1 + \frac{k^{W^2}}{2})} \right) \int d\tilde{x} d\tilde{y} |\tilde{a}_o^f(r_\perp, \tilde{z}, \tau)|^2 \end{aligned} \quad (22)$$

According to the previous discussion less FEL power expected from the storage–ring–FEL device than from a single–pass FEL. The simulation runs for FELICITA II predict peak powers of the order of 20 MW (Fig.(16)) for the storage–ring device and of 55 MW in case of single–pass device (Fig.(20)). Furthermore, one can see that one obtains a pulsed operation for the storage–ring FEL. The duration time corresponds to 10 – 20 roundtrips in the optical cavity or DELTA, respectively. With a roundtrip time of $360ns$ of DELTA, a pulse duration from start–up to saturation of the order of $10\mu s$ is expected.

The duration of each micropulse, i.e. the wave–packet circulating in the resonator cannot be determined by this simulation. A rough estimate of the micropulse duration results in $1 - 10ps$.

Due to the shape of the electron beam gain and guiding are different for both transverse directions. Thus, the shape of the initial axisymmetric laser beam will change, according to the shape of the electron beam. Fig.(22) and Fig.(23) show the spatial energy distribution of the laser pulse at highest intensity. In both cases the shape of the laser beam gets flat in horizontal direction. Furthermore, one can see, that a higher–order cavity mode is excited in vertical direction. In case of the simulation of single–pass–device the effect is even stronger, as the field intensity is much higher.

As the FEL–pulses for storage–ring operation are rather short, one could think of pulsed undulator magnets, with a pulse length of 10 – 100 $\mu s$. Such operation would minimize the influence of the FEL on the storage ring, as it is operating only for times much shorter than the time, instabilties or even resonances need to rise or the ring needs to 'find' a new orbit as a consequence of the influence of the FEL. Thus, there would be no need for by–pass operation of the undulator.

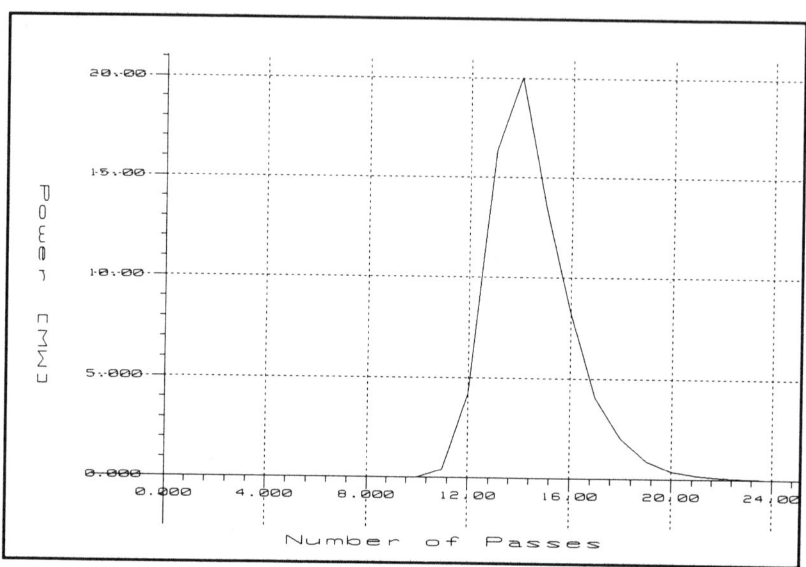


Figure 16: FEL–power in Watts versus number of passes through the undulator. During start–up the pulse grows exponentially, the growth stops, as gain degradation rises (Fig.(17,18)), and the pulse decays with the cavity decay time.

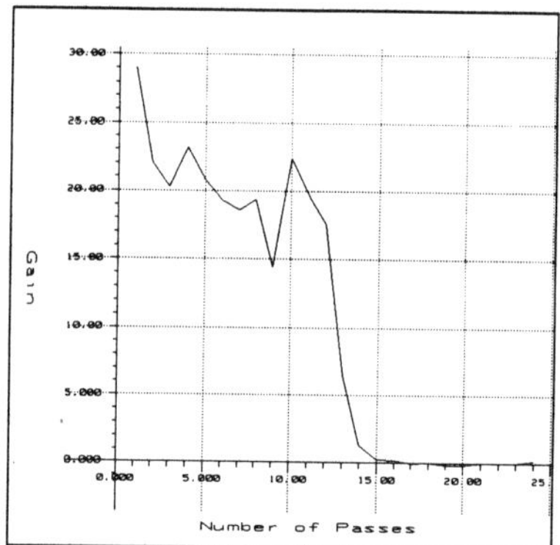


Figure 17: Evolution of the gain during the FEL pulse. The gain stays constant, except of fluctuations due to nonideal mode matching in the resonator[41]. After the first 10 pulses the gain shows a sudden break down, when the amplidude reaches the theshold, where gain degradation gets severe, as can be seen by the rise of the enery spread shown in figure(18).

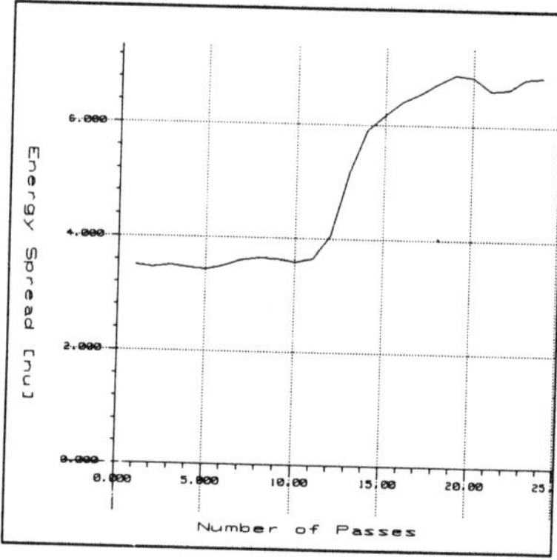


Figure 18: The evolution of the energy spread of the electron beam. The rise of the energy spread strongly depends on the intensity of the laser field within the optical resonator. Only at high intensities, there is a large increase, giving strong gain degradation.

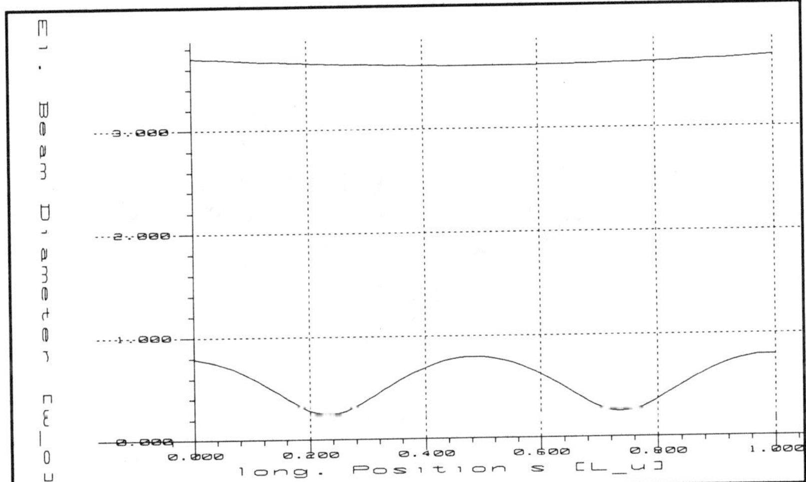


Figure 19: This picture shows the focussing properties of the FEL undulator on the electron beam [40]. In horizontal direction the undulator provides no focussing and acts just like a drift space whereas, in vertical direction there is a kind of quadrupole focussing, leading to two betatron oscillations over the whole undulator length. Thus, this undulator produces a tune shift of 2 for the storage-ring optics, so that there is no need for special matching of the undulator to the optics of the storage ring.

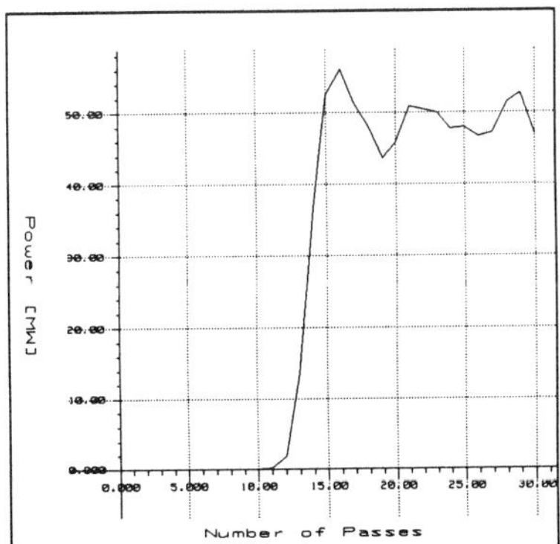


Figure 20: Evolution of the field power in a single-pass device versus the number of passes in the optical resonator. In this case the length of the laser pulse is determined by the length of the macropulse comming from the accelerator. After building-up of the pulse, the laser saturates due to the high intensity (Fig.(21)).

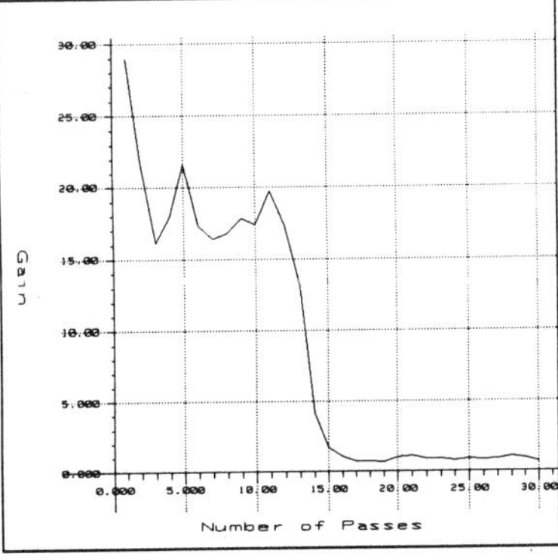


Figure 21: Evolution of the gain during the rise of the FEL pulse. After reaching a certain level, the intensity saturates and the gain drops to the value of the resonator losses. The fluctuations of the gain value at the beginning of the pulse are due to mode mismatch of the laser field caused by optical guiding [41].

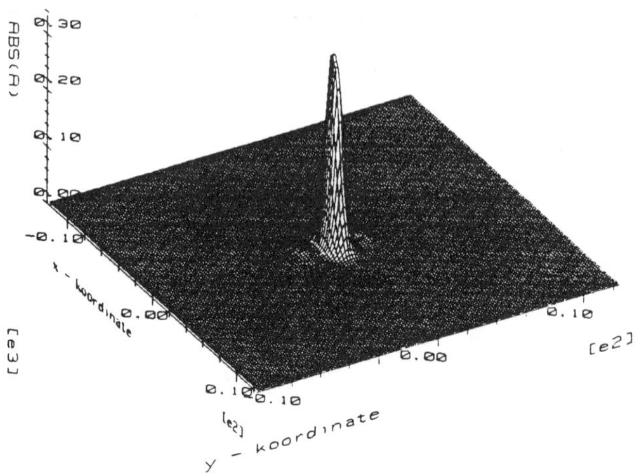


Figure 22: 3D–plot of the elm. field at highest intensity for the SR–FEL simulation. The intensity of the FEL pulse is shown in dimensionless units, versus the transverse position of the field mesh point, measured in units of the minimum beam waist. The intensity corresponds to a peak power of $20MW$.

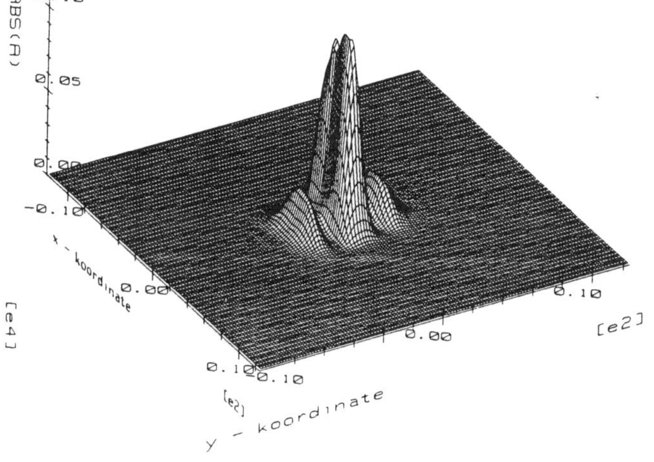


Figure 23: 3D plot of the elm. field at highest intensity for the simulation of a single–pass device. The intensity of the elm. field in this picture corresponds to a peak power of $55MW$.

Table 2: Dimensionless set of quantities relevant for describing the FEL appropriately. All internal parameters of a FEL device are combined to a few essential quantities, facilating a comparison of different FEL devices and a formulation of design criteria.

|  |  |
| --- | --- |
| $\tau = \frac{c}{L_u}, \quad 0 \leq \tau \leq 1$ | dimensionless interaction time |
| $\tilde{z} = \frac{1}{N_u \lambda_1}(z - ct)$ | distance from the center of the elm. pulse in the comoving frame, normalised to the slippage length |
| $\tilde{z}_i = \frac{1}{N_u \lambda_1}(z_i - \beta_o ct) = \tilde{z}_i(\tau = 0) - \tau$ | postion of the i$^{th}$ electron relative to the center of the radiation pulse. $\tilde{z}_i(\tau = 0)$ corresponds to the initial distance between the electron and the pulse center |
| $\tilde{x} = \frac{x}{w_{o1}}$ | horizontal position x normalized to the beam waist of the fundamental empty–cavity mode [24]. |
| $\tilde{y} = \frac{y}{w_{o1}}$ | vertical coordinate y normalized to the beam waist of the fundamental. |
| $\tilde{\beta}_x = \frac{L_u}{w_{o1}} \beta_x$ | normalized horizontal velocity |
| $\tilde{\beta}_y = \frac{L_u}{w_{o1}} \beta_y$ | normalized vertical velocity |
| $\frac{d}{d\tau} = \frac{L}{c} \frac{d}{dt}$ | total 'time' derivative |
| $\frac{\partial}{\partial \tau} = L_u \left(\frac{\partial}{\partial z} - \frac{\partial}{c \partial t}\right)$<br>$\frac{\partial}{\partial \tilde{z}} = N_u \lambda \frac{\partial}{\partial z}$<br>$\frac{\partial}{\partial \tilde{x}} = w_{o1} \frac{\partial}{\partial x}$<br>$\frac{\partial}{\partial \tilde{y}} = w_{o1} \frac{\partial}{\partial y}$ | partial derivatives |
| $\tilde{Z}_r = \frac{\pi w_{o1}^2}{\lambda_1 L_u}$ | Raileigh length of the fundamental normalized to the undulator length. |
| $f$ | harmonic number |
| $JJ_f(x) = \begin{cases} (JJ_{\frac{f-1}{2}}(x) - JJ_{\frac{f+1}{2}}(x))^2, & f \in 2n+1 \\ (JJ_{f-1}(x) - JJ_{f+1}(x))^2, & f \in 2n \end{cases}$ | JJ - factor, determining the coupling to the harmonics |
| $\tilde{a}_o^f(\tilde{r}_\perp, \tilde{z}, \tau) = \frac{2\pi e}{mc^2} \frac{N_u L_u k^W}{\gamma_r^2} \sqrt{JJ_f\left(\frac{f k^{W2}}{4(1+\frac{k^{W2}}{2})}\right)} \cdot E_o(\tilde{r}_\perp, \tilde{z}, \tau)$ | dimensionl. field amplitude |
| $j_f = \frac{\pi e^2}{\epsilon_o mc^2} \frac{N_u l_u^2 k^{W2}}{\gamma_r^3} JJ_f\left(\frac{f k^{W2}}{4(1+\frac{k^{W2}}{2})}\right) \rho(\tilde{\vec{r}})$<br>$= \frac{\pi e}{\epsilon_o mc^3} \frac{N_u l_u^2 k^{W2}}{\gamma_r^3} JJ_f\left(\frac{f k^{W2}}{4(1+\frac{k^{W2}}{2})}\right) \frac{\hat{I}}{w_{o1}^2}$<br>$\cdot \frac{1}{2\pi \tilde{\sigma}_x \tilde{\sigma}_y} \exp\{(\frac{\tilde{x}-\tilde{x}_o)^2}{2\tilde{\sigma}_x^2}\} \exp\{(\frac{\tilde{y}-\tilde{y}_o)^2}{2\tilde{\sigma}_y^2}\}$ | dimensionless current density |

Table 3: Preliminary parameters of FELICITA II, the second FEL experiment at the storage ring DELTA.

| Parameter | Symbol | Value |
|---|---|---|
| **Parameters of the Undulator Magnet** | | |
| Period Length | $l_o$ | 3.4 cm |
| Number of Periods | $N_u$ | 411 |
| K Value | $K^w$ | 3.4 |
| Peak Magnetic Field | $B_o$ | 1.07 T |
| Length of the Undulator | $L_u$ | 13.94 m |
| **Parameters of the Ring Resonator** | | |
| Focus of the Defocussing Elements | $F_d$ | $-12.26$ m |
| Effective Focus of the Arcs | $F_f$ | 22.92 m |
| Length of the Undulator Straight Section | $s_u$ | 25.70 m |
| Distance of the Defoc. Elements and the Arcs | $s_d$ | 16.45 m |
| Length of the Recirculation Section | $s_r$ | 56.61 m |
| Beam Waist of the Undulator Section | $w_{ou}$ | $3.9 \cdot 10^{-4}$m |
| Beam Waist of the Recirculation Section | $w_{or}$ | $3.9 \cdot 10^{-3}$m |
| Raileigh Length | $z_r$ | 4.778 m |
| Total Reflectivity | $R_{tot}$ | 0.52 |
| **Estimated Parameters of Electron Beam** | | |
| Peak Current | $\hat{I}$ | 150 A |
| Electron Energy | $E$ | 500 MeV |
| Energy Spread | $\frac{\Delta E}{E}$ | $0.7 \cdot 10^{-3}$ |
| Vertical Emittance | $\epsilon_x$ | $1.5 \cdot 10^{-8}$ m·rad |
| Horizontal Emittance | $\epsilon_y$ | $1.5 \cdot 10^{-9}$ m·rad |
| **Resulting Dimensionless FEL Parameters** | | |
| 'Colson' Current | $j$ | 1185 |
| Detuning ( max. Gain ) | $\nu_{max}$ | 6.0 |
| Energy Spread [$\nu$] | $\sigma_\nu$ | 3.5 |
| **FEL Output Power** | | |
| Average Power ( SR Operation ) | $\bar{P}_{SR}$ | 0.23 W |
| Spontanous Power | $\bar{P}_{Spont}$ | 110 W |
| Peak Power ( SR Operation ) | $\hat{P}_{SR}$ | 20 MW |
| Peak Power ( LINAC Operation ) | $\hat{P}_{SP}$ | 55 MW |

## References

[1] G. Ramian, *Nucl. Instr. and Meth.*, to be publ.

[2] M.E. Couprie et al., *Nucl. Instr. and Meth.*, **A296**, 13, 1990

[3] G.N. Kulipanov et al., *Nucl. Instr. and Meth.*, **A296**, 1, 1990

[4] J.C. Goldstein et al., *Nucl. Instr. and Meth.*, **A296**, 282, 1990

[5] S. Benson, private comunication

[6] DELTA Group, *DELTA, a Status Report*, University of Dortmund, Institute of Physics, August 1990

[7] R. Warren, *Nucl. Instr. and Meth.*, to be publ.

[8] D.Nölle et al., *FELICITA II, a Feasibility Study for a XUV–FEL at DELTA*, DELTA Int. Rep., to be publ.

[9] M. Sands, *The Physics of Electron Storage Rings. An Introduction.* Proceedings of the International School of Physics "Enrico Fermi", ed. B. Touschek, 1971

[10] C. Bernadini et al, *Nuovo Cimento*, **34**, 1473, 1964

[11] H. Bruck, *Accélérateurs Circulaires de Particules*, Presses Universitaires de France, 1966

[12] T. Weiland, *On the Qualitative Prediction of Bunch Lengthening in High Energy Electron Storage Rings* DESY 81/88, 1988

[13] A.W. Chao, J. Gareyte, Stanford University, PEP Note 224, 1976

[14] A. Renieri, *Nuovo Cimento* **B53**, 160, 1979

[15] K. Wille, *DELTA, Vorschlag für einen Testspeicherring an der Universität Dortmund*, University of Dortmund, Institute of Physics, Mai 1986

[16] G.E. Fischer et al., *A 1.2 GeV Damping Ring Complex for the SLC*, SLAC-PUB-3170, July 1983

[17] M. Michel, *Überlegungen zum Ultrahoch-Vakuumsystem für DELTA*, Diploma Thesis, University of Dortmund, 1988

[18] S. Brinkmann, R. Heisterhagen, K. Wille, *Report on the Electronics for the Beam Position Monitor*, DELTA Int. Rep. 90-003, 1990

[19] V.N. Litvinenko, *Nucl. Instr. and Meth.*, to be publ.

[20] K.J. Kim, *LBL-23007* , 1976

[21] W.B. Colson, *Proc. SPIE*, **Vol 738**, 2, 1980, ed. B.E. Newnam

[22] D. Nölle, *FELS User guide*, DELTA Int. Rep., to be publ.

[23] E.T. Scharlemann, *J. Appl. Phys.*, **58**, 2154, 1985

[24] H. Kogelnik, T. Li, *Appl. Opt.*, **5**, 1550, 1966

[25] M.J. Schmidt, C.J. Elliot, *Phys. Rev. A* **41**, 3853, 1990

[26] P. Elleaume, *J. de Phys.*, **45**, 997, 1984

[27] R. Bonifacio, B.W.J. NcNeil, *Nucl. Instr. and Meth.*, **A272**, 280, 1988

[28] S. Benson, Ph. D. Thesis, Stanford 1985

[29] F. James, *CERN DD/80/6*, 1980

[30] D. Nölle, *Optische Klystrons in Speicherringen*, Diploma Thesis, University of Dotmund, 1988

[31] W.H. Dress, b.P. Flannery, S.A. Teukolsky, W.B. Vetterling *Nummerical Recipes in C*, Cambridge University Press, 1988

[32] M. Berz, H. Wollnik, Nucl. *Instr. and Meth.*, **A267**, 15, 1988

[33] E.A. Sziklas, A.E. Siegman, *Appl. Opt.*, **14**, 1874, 1975

[34] R.L. Tokar et al. *IEEE*, **QE-25**, 73, 1989

[35] B. Faatz et al., *Nucl. Instr. and Meth*, **A296**, 654, 1990

[36] B.E. Newnam, *Multiface, Metal Mirror Design for Extreme–Ultraviolet and Soft X–Ray Free–Electron Laser Resonators*, 1987

[37] D.H. Dowel et al., *Nucl. Instr. and Meth*, to be publ.

[38] U. Bizzarri et al., *Rivista Nuovo Cim.*, **10**, 1, 1987

[39] D. Nölle, *Nucl. Instr. and Meth*, to be publ.

[40] N.M. Kroll et al., *IEEE*, QE-17, 1436, 1984

[41] K.C. Sun et al., *Nucl. Instr. and Meth.*, **A296**, 659, 1990

[42] D. Nölle et al., *Nucl. Instr. and Meth.*, **A296**, 263, 1990

[43] J.E. Solid et al., *Nucl. Instr. and Meth.*, **A285**, 153, 1989

**Research Trends in Physics: Coherent Radiation Generation and Particle Acceleration**

*La Jolla International School of Physics*, The Institute for Advanced Physics Studies, La Jolla, California

# Tunable Radiation Generation Using Underdense Ionization Fronts

**R.L. Savage, Jr., W.B. Mori, and C. Joshi**
Department of Electrical Engineering
University of California, Los Angeles
Los Angeles, CA 90024

**T.W. Johnston**
INRS-Énergie
Varennes, Québec J3X-1S2, Canada

**G. Shvets**
Plasma Fusion Center
Massachusetts Institute of Technology
Cambridge, MA 02139

## ABSTRACT

Recently, underdense, relativistically-propagating, laser-produced ionization fronts have been used to upshift the frequency of impinging electromagnetic radiation from 35 GHz to more than 138 GHz in a continuously tunable fashion [1]. These experiments utilized a resonant microwave cavity operating near its cutoff frequency through which an ionizing laser pulse propagated. We present a derivation of the expected upshifted frequencies, carried out in the laboratory frame using a space-time picture of the upshifting process, and a detailed derivation of the reflection and transmission coefficients within the waveguide

## INTRODUCTION

Almost 25 years ago [2], it was proposed that the frequency of electromagnetic radiation directed toward, and reflected from, a rapidly moving boundary between a region of plasma and one of neutral gas would be altered by the relativistic Doppler effect [3, 4]. Thus, if such a boundary, or ionization front, were to propagate at a velocity close to that of light, frequency upshifts of many orders of magnitude would be obtained. However, in order for the incident radiation to reflect from the front, the density of the plasma would have to be extremely high. Advances in laser technology have enabled the creation of relativistically propagating ionization fronts via photo-ionization [5]. These have led to a renewed interest in reflection front ionization fronts [6, 7, 8]. As an ionizing laser pulse passes through a region of neutral gas, a front which propagates at the velocity of the laser light is created. While such a front may be highly

 

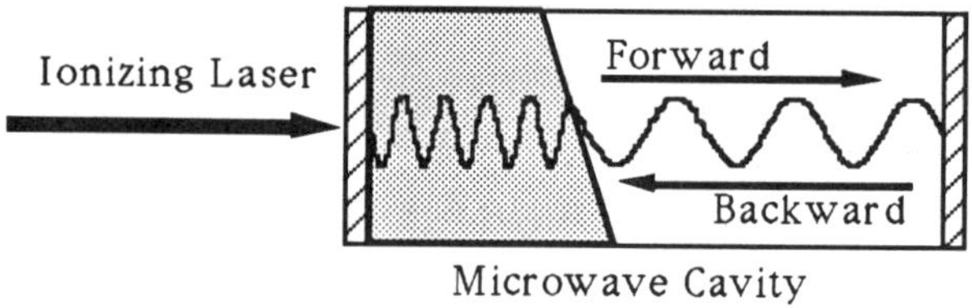


Figure 1: Schematic of the experimental set-up.

relativistic, in most cases it is not possible to make it dense enough to reflect an impinging electromagnetic wave. In order to reflect the impinging radiation, the front must be overdense; that is, the density of the plasma in the front must be above the critical density for electromagnetic radiation *when viewed from the rest frame of the front.* The critical density is defined as that density for which the characteristic plasma frequency, $\omega_p$, equals the radiation frequency. The plasma frequency is defined by $\omega_p^2 = 4\pi n_0 e^2/m$ where $n_0$ is the plasma density, $m$ is the electron's mass, and $e$ is its charge. Because the frequency of the incident radiation is much higher when viewed from this frame and the plasma frequency is Lorentz invariant, creation of an overdense front requires that the plasma be more dense than for a stationary boundary.

The early theoretical work in this area concentrated on reflection from overdense ionization fronts, but more recently it has been shown that significant upshifts are possible even for underdense fronts, where the incident wave is transmitted into the plasma [6]. Furthermore, under certain conditions the upshifted radiation may propagate in the plasma in the same direction as the ionization front. This type of "reflection" from an underdense front yields frequency shifts that are much different from those given by the usual Doppler effect.

We have experimentally confirmed most of the predictions of the underdense ionization front theory in a recent set of experiments. These experiments utilize a resonant microwave cavity that is filled with neutral gas (see Fig. 1). As an ionizing laser pulse passes through the cavity, the frequency of the microwave radiation is upshifted. The front encounters a standing microwave field. It is composed of radiation which propagates in the same direction as the ionizing laser pulse, which we label the *forward* wave, and radiation which counterpropagates with respect to the laser pulse, which we call the *backward* wave. The plasma density in the ionization front is controlled by adjusting the neutral gas density in the cavity. Source radiation at 35 GHz has been upshifted to more than 138 GHz in a continuously tunable fashion.

## SPACE-TIME ANALYSIS

The derivation of the upshifted frequencies can be carried out either in the laboratory frame or in the rest frame of the ionization front. By making a Lorentz transformation to the front's rest frame, the derivation is simplified to that of a stationary boundary with neutral gas and a higher frequency electromagnetic wave impinging upon it from one side and plasma and the upshifted radiation streaming away on the other [1]. However, in certain geometries the front may propagate faster than the speed of light. Because one cannot transform to a frame moving faster than $c$, the analysis for superluminous fronts must be undertaken in the laboratory frame. Space-time plots are particularly useful because they help one to visualize the upshifting mechanism over the full range of front velocities, from stationary to infinitely fast.

Fig. 2 is a space-time plot for an ionization front interacting with radiation confined within a microwave cavity operating near cutoff. The ionizing laser pulse, which passes through the cavity from right to left, is propagating at approximately $c$, and is represented by the heavy black line at 45 degrees to the space and time axes. The parallel vertical lines in the figure represent the stationary cavity windows from which the circulating microwave radiation reflects before the arrival of the laser pulse. The group velocity of the circulating radiation inside the cavity, $v_g$, is $.5c$. The parallel red lines in Fig. 2 show the propagation along the cavity axis of points of equal phase in the source radiation. Because the product of the group veloctiy and the phase velocity, $v_p$, is $c^2$, the phase velocity of the source radiation, given by the inverse of the slope of the red lines, is greater than $c$.

As the source radiation is transmitted into the underdense front, its frequency is upshifted. The degree of upshift can easily be derived by taking a close look at the interaction of the radiation with the moving boundary. Fig. 3 shows a magnified detail of Fig. 2 along the front's space-time line. The parallel red lines track points of constant phase for the

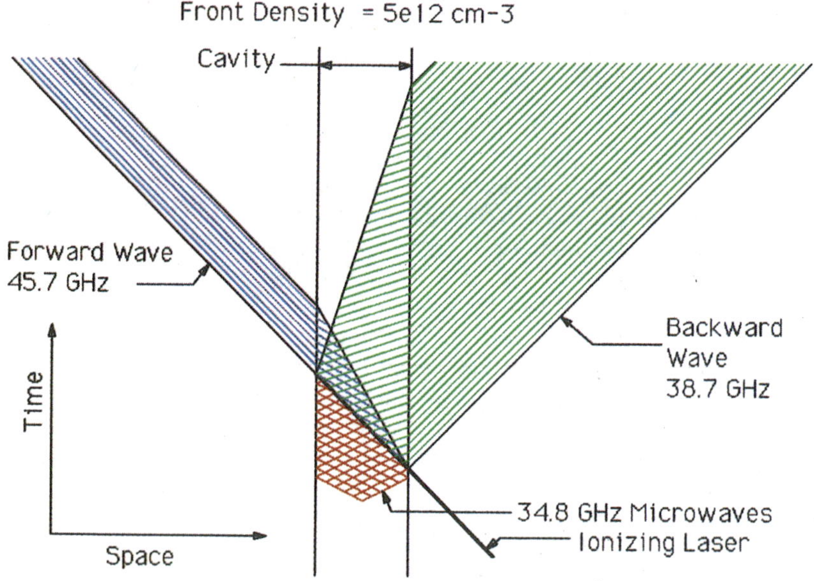


Figure 2: Space-time diagram for an ionizing laser pulse passing through a resonant microwave cavity.

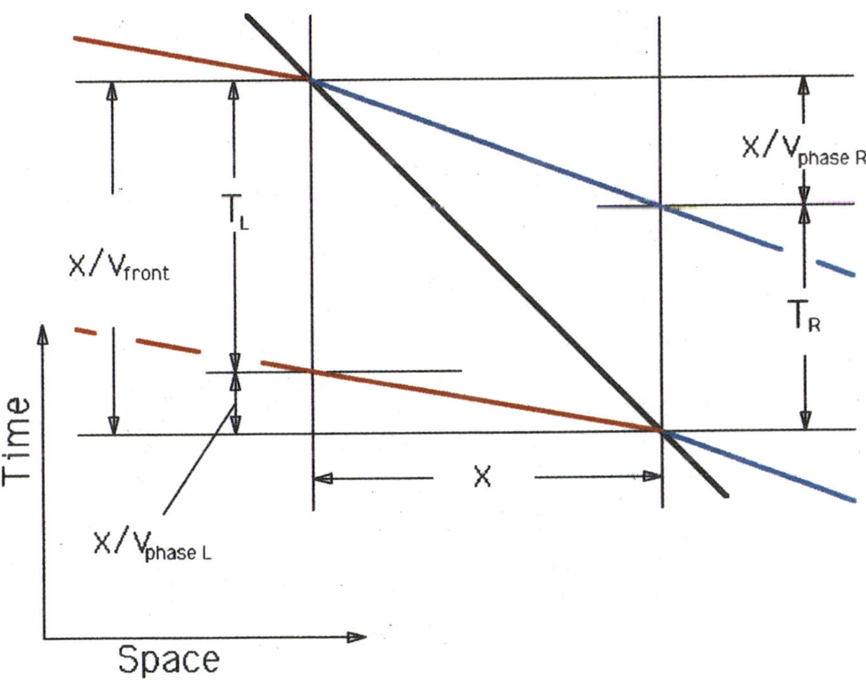


Figure 3: Detail of Fig. 2 along the ionization front's space-time line for the *forward* radiation.

radiation inside the cavity that is propagating in the same direction as, and being overtaken by, the ionization front, the forward wave. The parallel blue lines track points of constant phase for the upshifted, transmitted radiation. The derivation proceeds from three simple requirements: 1) the source radiation must obey the dispersion relation (Lorentz-invariant in our geometry) for an electromagnetic wave in a conducting waveguide, $\omega_0^2 = \omega_c^2 + c^2k^2$, where $\omega_c$ is the cutoff frequency for the particular operating mode in the waveguide; 2) the number of cycles of the source radiation impinging on the front must equal the number of cycles of upshifted radiation leaving the front; and 3) the upshifted radiation must obey the dispersion relation for an electromagnetic wave in a plasma filled waveguide (also Lorentz-invariant), $\omega_{up}^2 = \omega_c^2 + \omega_p^2 + c^2k^2$, where $\omega_p$ is the plasma frequency. The first and third requirements thus dictate the slope of the red and blue lines in Fig. 2, and the second requirement is simply that the spacing between the lines, measured along the front's space-time line, must be equal on both sides.

The full range of front velocities can be considered by simply changing the slope of the front's space-time line. The two limiting cases are a stationary boundary and one that propgates at an infinitely high veloctiy. For the stationary boundary, the front's space-time line is vertical and requirement 2 is simply that the frequency is fixed. As is the usual case for stationary interfaces, the wavenumber must adjust itself in order to obey requirement 3. For the infinitely fast front, also known as "flash ionization" [9], the boundary is horizontal in the space-time plot. Thus, the wavenumber is fixed by requirement 2 and the frequency adjusts in order to satisfy requirement 3.

It can easily be seen in Fig. 3 that the temporal periods of the radiation on the left and right sides of the front are given by:

$$T_L = \frac{2\pi}{\omega_L} = X\left(\frac{1}{v_{front}} - \frac{1}{v_{phase_L}}\right) \text{ and } T_R = \frac{2\pi}{\omega_L} = X\left(\frac{1}{v_{front}} - \frac{1}{v_{phase_R}}\right).$$

Using these expressions, we can write the upshifted frequency as a function of the source frequency as:

$$\omega_R = \omega_L \frac{(1/v_{front} - 1/v_{phase_L})}{(1/v_{front} - 1/v_{phase_R})}.$$

Viewed another way, this relationship follows from the requirement that the phase be continuous across the boundary, which is a consequence of the requirement that E and B be continuous at the front. The dispersion relations to the left of (before) and to the right of (after) the front are

$$\omega_L^2 = \omega_c^2 + c^2 k_L^2 \text{ and } \omega_R^2 = \omega_c^2 + \omega_p^2 + c^2 k_R^2.$$

By eliminating $k_R$ and $k_L$ from the above expressions, we obtain a quadratic equation in $\omega_R$ which can be solved to give

$$\omega_R = \omega_L \gamma^2 (1 - \beta v_g/c) \left\{ 1 - \beta \left[ 1 - \frac{\omega_c^2 + \omega_p^2}{\omega_L^2 \gamma^2 (1 - \beta v_g/c)^2} \right]^{\frac{1}{2}} \right\}. \quad (1)$$

Here $\beta = v_{front}/c$ and $\gamma = (1 - \beta^2)^{-1/2}$ is the relativistic Lorentz factor associated with the front. $v_g$ is the group velocity of the source radiation in the cavity, and is related to $v_{phase_L}$ by $v_g v_{phase_L} = c^2$. A similar derivation for the radiation inside the cavity which is propagating in the opposite direction to the ionization front (left to right in Fig. 2), the backward wave, yields an upshifted frequency given by

$$\omega_R = \omega_L \gamma^2 (1 + \beta v_g/c) \left\{ 1 - \beta \left[ 1 - \frac{\omega_c^2 + \omega_p^2}{\omega_L^2 \gamma^2 (1 + \beta v_g/c)^2} \right]^{\frac{1}{2}} \right\}. \quad (2)$$

The upshifted forward wave is shown by the blue lines in Fig. 2, and the backward wave is shown by the green lines. These parallel lines represent the propagation of points of equal phase, and are bounded by black lines which represent the group velocity of the radiation. Eq. 1 and Eq. 2 both predict upshifted frequencies that are proportional to the front density, with the forward wave always upshifted to a higher frequency than that of the backward wave for a given plasma density. We note that $\beta$ is not limited to being less than unity, in fact, we recover the flash-ionization results for $\beta \to \infty$ [9].

The group velocity of the upshifted radiation inside the waveguide is plotted as a function of the front density in Fig. 4. The group velocity of the forward wave, which is always greater than that of the source radiation, increases as the plasma density in the front increases. On the other hand, the group velocity of the backward wave initially decreases as the front density increases. It goes to zero when the front density is such that $\omega_p^2 = \omega_L^2(1+\beta v_g/c)^2 - \omega_c^2$. At this point the group velocity of the backward wave changes sign, then increases in the forward direction as the plasma density continues to increase. For our experimental parameters, this occurs at a plasma density of $2\times10^{13}$ cm$^{-3}$.

Fig. 5 is a space-time plot for a front density of $1\times10^{13}$ cm$^{-3}$, where the group velocity of the backward wave is small, but it is still propagating

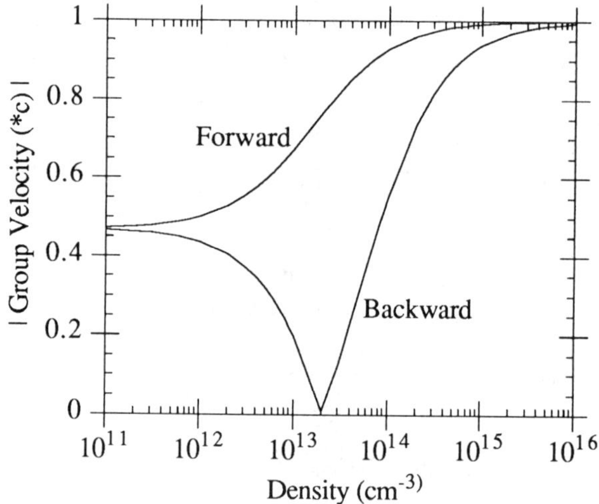


Figure 4: Group velocity of the *forward* and *backward* upshifted radiation vs. the plasma density in the front.

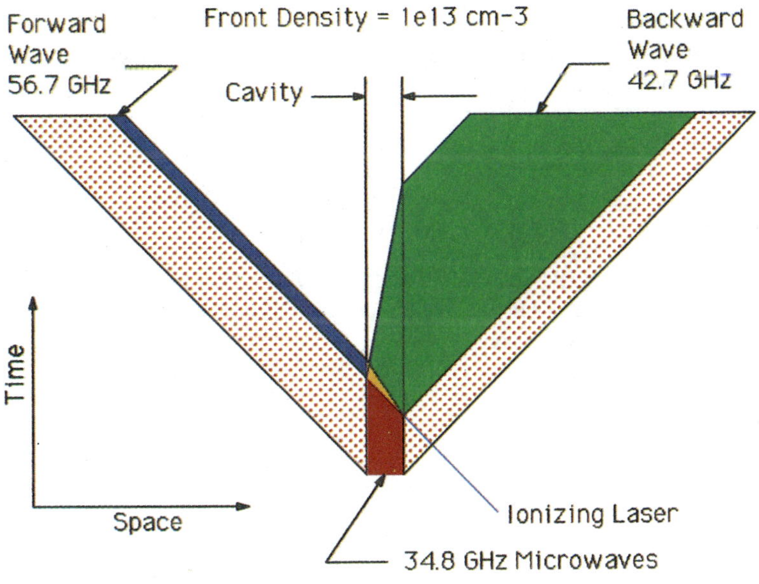


Figure 5: Space-time diagram for a front density of $1\times10^{13}$ cm$^{-3}$.

in the backward direction. The red stippled area indicates source radiation that leaks out of the cavity before the arrival of the laser pulse. For a front density of $4\times10^{13}$ cm$^{-3}$, shown in Fig. 4, the backward wave has turned around and is propagating in the same direction as the ionizing laser. At this density the forward wave has already been significantly compressed in duration. Also, because the backward wave is now following along behind the front, its duration decreases significantly as its group velocity approaches that of the front at higher plasma densities.

## REFLECTION AND TRANSMISSION COEFFICIENTS

The reflection and transmission coefficients for a TE wave polarized perpendicular to the plane of incidence (s-polarization) can easily be calculated in the front's frame (recall that in this frame the problem is reduced to that of a stationary boundary). We first derive the modes that can exist in the streaming plasma. The momentum equation for the electrons in the front's frame is

$$\frac{\partial \vec{v}}{\partial t} + \vec{U}\frac{\partial \vec{v}}{\partial z} = -\frac{e}{m_0 \gamma}(\vec{E} + \frac{1}{c}\vec{U} \times \vec{B}). \tag{3}$$

$\vec{U}$ is the velocity of the neutral gas, which in this frame is equal to the velocity of the front, and $m_0$ is the rest mass of the electrons. We use two of Maxwell's equations,

$$\nabla \times \vec{E} = -\frac{1}{c}\frac{\partial \vec{B}}{\partial t} \tag{4}$$

and

$$\nabla \times \vec{B} = \frac{1}{c}\frac{\partial \vec{E}}{\partial t} - \frac{4\pi}{c}\vec{J}, \tag{5}$$

where $\vec{J}$, the free current, is $ne\vec{v}$. By taking the curl of Eq. 4, considering only TE waves ($\nabla \cdot \vec{E} = 0$), and substituting the right side of Eq. 5 for $\nabla \times \vec{B}$, we can write

$$\nabla^2 \vec{E} = \frac{1}{c^2}\frac{\partial^2 E}{\partial t^2} - \frac{4\pi n e}{c^2}\frac{\partial v}{\partial t}. \tag{6}$$

We fourier analyze by assuming solutions of the form $e^{i(\vec{k}\cdot\vec{r}-\omega t)}$. Differentiating Eq. 6 gives

$$\left(\frac{\omega^2}{c^2} - k^2\right)\vec{E} = \frac{-i4\pi n e \omega}{c^2}\vec{v}. \tag{7}$$

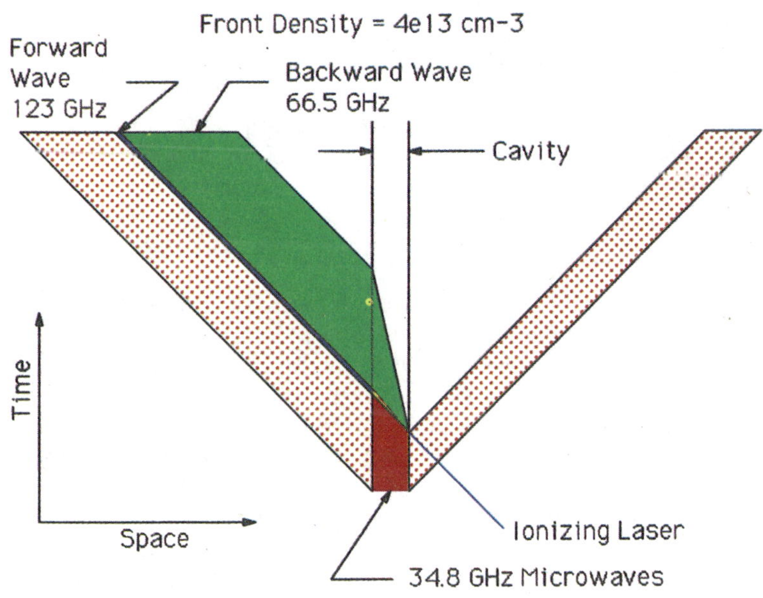


Figure 6: Space-time diagram for a front density of $4\times10^{13}$ cm$^{-3}$. At this density the upshifted *backward* wave is "reflected" into the *forward* direction.

We can now differentiate Eq. 3 and eliminate $\vec{E}$ and $\vec{B}$ using Eqs. 4 and 7 to write, after a bit of algebra,

$$(\omega - Uk_{\|})\left(1 - \frac{\omega_p^2}{\omega^2 - c^2k^2}\right) = 0, \tag{8}$$

where the subscript $\|$ denotes along the direction of $U$, which is opposite to the direction of propagation of the front. This equation has three roots,

$$k_{1\|} = \frac{\omega}{U}, \tag{9}$$

$$k_2 = \frac{\omega}{c}\left(1 - \frac{\omega_p^2}{\omega^2}\right)^{\frac{1}{2}}, \tag{10}$$

and

$$k_3 = -\frac{\omega}{c}\left(1 - \frac{\omega_p^2}{\omega^2}\right)^{\frac{1}{2}}. \tag{11}$$

We neglect the third root because we assume a positive value for the wavenumber of the source wave.

Now that we know that there are two possible modes in the streaming plasma, we can apply the boundary conditions to solve for the reflection and transmission coefficients. We assume a discontinuous boundary between the neutral gas and the plasma, which lies in the x-y plane. $U$ is along the z axis and an electromagnetic wave with its electric field along the x axis is incident upon the boundary with its wave vector in the y-z plane (s-polarization). there are three conditions that must be satisfied at the plasma-neutral boundary: 1) continuity of the tangential component of $\vec{E}$, 2) continuity of the normal component of $\vec{B}$, and 3) the total current must be zero because the electrons and ions have identical initial velocities when they are created at the boundary. Considering an incident mode (subscript $i$), a reflected mode (subscript $r$), and the two transmitted modes derived earlier (subscripts 1 and 2) these boundary conditions can be written as:

$$\left(\vec{E}_i + \vec{E}_r\right)_{\perp} = \left(\vec{E}_1 + \vec{E}_2\right)_{\perp} \tag{12}$$

$$\left(\vec{B}_i + \vec{B}_r\right)_{\|} = \left(\vec{B}_1 + \vec{B}_2\right)_{\|} \tag{13}$$

and

$$\nabla \times (\vec{B}_1 + \vec{B}_2) = \frac{1}{c}\frac{\partial}{\partial t}(\vec{E}_1 + \vec{E}_2). \tag{14}$$

In our geometry, Eqs. 12 and 13 become

$$E_i + E_r = E_1 + E_2 \tag{15}$$

and

$$B_i \frac{k_{i\|}}{k_i} - B_r \frac{k_{r\|}}{k_r} = B_1 \frac{k_{1\|}}{k_1} - B_2 \frac{k_{2\|}}{k_2}. \tag{16}$$

Realizing that $B = (ck/\omega)E$ for each mode and that $k_{i\|} = k_{r\|}$, we can eliminate the reflected mode from the two equations above and write

$$E_1 = \frac{2E_i - E_2\left(1 + \frac{k_{2\|}}{k_{i\|}}\right)}{\left(1 + \frac{k_{1\|}}{k_{i\|}}\right)} \tag{17}$$

Eq. 14 can be differentiated and rearranged to give

$$E_1 = E_2 \left( \frac{\omega^2 - c^2 k_2^2}{c^2 k_1^2 - \omega^2} \right). \tag{18}$$

We can now equate Eqs. 17 and 18 and reduce them to

$$\frac{E_2}{E_i} = \frac{2}{\left(1 + \frac{k_{2\|}}{k_{i\|}}\right) + \left(1 + \frac{k_{1\|}}{k_{i\|}}\right)\left(\frac{\omega^2 - c^2 k_2^2}{c^2 k_1^2 - \omega^2}\right)}. \tag{19}$$

Using this and Eq. 18, we obtain

$$\frac{E_1}{E_i} = \frac{2}{\left(1 + \frac{k_{1\|}}{k_{i\|}}\right) + \left(1 + \frac{k_{2\|}}{k_{i\|}}\right)\left(\frac{c^2 k_1^2 - \omega^2}{\omega^2 - c^2 k_2^2}\right)}. \tag{20}$$

And finally, Eq. 15 simply gives us

$$\frac{E_r}{E_i} = \frac{E_1}{E_i} + \frac{E_2}{E_i} - 1. \tag{21}$$

Thus Eqs. 19, 20, and 21 give us the reflection and transmission coefficients in the front's frame. In order to obtain the laboratory frame coefficients, we must transform back to the laboratory frame.

We make the Lorentz transformation using the relations

$$\vec{E}_{lab} = \gamma(\vec{E} + \vec{\beta} \times \vec{B}) - \frac{\gamma^2}{\gamma + 1}\vec{\beta}(\vec{\beta} \cdot \vec{E}) \tag{22}$$

and

$$\vec{B}_{lab} = \gamma(\vec{B} + \vec{\beta} \times \vec{E}) - \frac{\gamma^2}{\gamma + 1}\vec{\beta}(\vec{\beta} \cdot \vec{B}). \tag{23}$$

Using these relations, we can write the electric fields in the laboratory frame as

$$\vec{E}_{ilab} = \vec{E}_i\gamma\left(1 - \beta\frac{k_{i\|}c}{\omega}\right), \tag{24}$$

$$\vec{E}_{rlab} = \vec{E}_r\gamma\left(1 + \beta\frac{k_{i\|}c}{\omega}\right), \tag{25}$$

$$\vec{E}_{1lab} = 0, \tag{26}$$

and

$$\vec{E}_{2lab} = \vec{E}_2\gamma\left(1 - \beta\frac{k_{2\|}c}{\omega}\right). \tag{27}$$

Because $\vec{E}_1$ transforms to zero in the laboratory frame, we see that this mode is simply a static magnetic field. We can use the relation $B_1 = (ck_1/\omega)E_1$ to obtain $B_1$, then transform back to the laboratory frame to give

$$\vec{B}_{1lab} = E_1\gamma\left\{\left(\frac{ck_{1\|}}{\omega} - \beta\right)\hat{\perp} + \frac{ck_{1\perp}}{\gamma\omega}\hat{\|}\right\} \tag{28}$$

or

$$B_{1lab} = E_1\gamma\left\{\left(\frac{ck_{1\|}}{\omega} - \beta\right)^2 + \left(\frac{ck_{1\perp}}{\gamma\omega}\right)^2\right\}^{\frac{1}{2}}. \tag{29}$$

We can now write, after a considerable amount of algebraic manipulation, the laboratory frame reflection and transmission coefficients as

$$r \equiv \frac{E_{rlab}}{E_{ilab}} = \frac{\left(\sqrt{\epsilon_i^{'}} - \sqrt{\epsilon_t^{'}}\right)}{\left(\sqrt{\epsilon_i^{'}} + \sqrt{\epsilon_t^{'}}\right)}, \tag{30}$$

$$t_2 \equiv \frac{E_{2lab}}{E_{ilab}} = \frac{2\sqrt{\epsilon_i^{'}}}{\left(\sqrt{\epsilon_i^{'}} + \sqrt{\epsilon_t^{'}}\right)}, \tag{31}$$

and

$$t_1 \equiv \frac{B_{1lab}}{E_{ilab}} = \left\{\frac{1-\beta^2}{1-\beta^2\epsilon_i^{'}}\right\}^{\frac{1}{2}} 2\beta\sqrt{\epsilon_i^{'}}\frac{\left(\sqrt{\epsilon_i^{'}} - \sqrt{\epsilon_t^{'}}\right)}{1-\beta\sqrt{\epsilon_t^{'}}}. \tag{32}$$

Here $\epsilon_i^{'} \equiv 1 - \omega_c^2/\omega'^2$ and $\epsilon_t^{'} \equiv 1 - (\omega_c^2 + \omega_p^2)/\omega'^2$ with primed quantities specified in the front's frame.

For the range of plasma densities and $\sqrt{\epsilon_i'}$ values used in our experiments, $r \approx 0$ and $t \approx 1$. Note that Eqs. 30, 31, and 32 reduce to the vacuum expressions in the limit $\epsilon_i' \to 1$ ($\omega_c = 0$). Although the ratio of the transmitted to incident powers at the front is always expected to be nearly unity, for large upshifts most of the incident energy is left in the free-streaming mode's static $B$ field as the transmitted wave's duration shortens. The above model assumes that the ionization front is discontinuous. Similar results are obtained for finite length underdense fronts, except that the energy in the free-streaming mode is instead converted to thermal energy in the plasma, as is the case for finite length overdense fronts [3].

ACKNOWLEDGEMENTS

We gratefully acknowledge many useful discussions with T. Katsouleas, A. M. Sessler, C. E. Clayton, J. M. Dawson, and R. P. Brogle.

This work was supported by the U.S. Department of Energy under contract number DE-AS03-83-ER40120, grant number DE-FG03-91-ER12114, and LLNL subcontract number 1828703.

## References

[1] R. L. Savage Jr., C. Joshi, and W. B. Mori, Phys. Rev. Lett. **68**, 946 (1992).

[2] V. I. Semenova, Sov. Radiophys. **10**, 599 (1967).

[3] M. Lampe, E. Ott, J. H. Walker, Phys. Fluids **21**, 42 (1978).

[4] The degree of upshift is equal to what would be expected from a moving mirror, but because the moving boundary, or ionization front, carries no energy, the reflected electromagnetic energy is reduced for reflection from an ionization front.

[5] Wm. M. Wood, C. W. Siders, and M. C. Downer, Phys. Rev. Lett. **67**, 3523 (1991) .

[6] W.B. Mori, Phys. Rev. A **44**, 5118 (1991).

[7] P. Sprangle, E. Esarey, and A. Ting, Phys. Rev. A **41**, 4463 (1990).

[8] H. C. Kapteyn and M. M. Murnane, J. Opt. Soc. Am. B **8**, 1657 (1991).

[9] S. C. Wilks, J. M. Dawson, and W. B. Mori, Phys. Rev. Lett. **61**, 337 (1988) and C. J. Joshi *et al.*, IEEE Trans. Plasma Sci. **18**, 814 (1990).

# Part II

# PARTICLE ACCELERATION

**Research Trends in Physics: Coherent Radiation Generation and Particle Acceleration**
Editorial Board: J.M. Buzzi, A. Prokhorov (Editor-in-Chief), P. Sprangle, and K. Wille
*La Jolla International School of Physics*, The Institute for Advanced Physics Studies, La Jolla, California

# Dynamics of Wave Beams Self-Action in Media With Strictive Nonlinearity

**N.E. Andreev[(1)], L.M. Gorbunov[(2)], A.I. Zykov[(3)], and S.V. Tarakanov[(4)]**

[(1)]Institute of High Temperature, Russian Academy of Sciences, Moscow
[(2)]Lebedev Physics Institute, Russian Academy of Sciences, Moscow

[(3)]Institute of Informatics and Automatics, Russian Academy of Sciences, St. Petersburg
[(4)]P. Lumumba Peoples Friendship University, Moscow

## Introduction

The quasistationary theory of wave beams self-action is based on assumption that the nonlinear response establishment time is much less than measurement time or the time of intensity variation of incident radiation (Ref.[1]). In reality this approximation is correct for fast enough nonlinear responses ( Kerr' nonlinearity in liquids, relativistic nonlinearity in plasma). However for many other nonlinearity types the medium response is connected with relatively slow processes ( thermal, ionization, striction nonlinearities ). Their characteristic times are considerably longer both the time of measurement and the time of intensity variation. Under these conditions the dynamics of self-action is mainly determined by the nonlinear response formation.

For the case of strictive nonlinearity the dynamics of cylindrical wave beam self-action were discussed in Ref.[2]. Using paraxial approximation it was shown that created initially by means of external lenses focal region ( beam width minimum ) moves with time upstream from the medium depth to its boundary. New focal regions appearing during self-action process also move upstream. Analogous result was obtained in Ref.[3], where self-similar solutions of coupled quasioptic and sound equations were considered. Similar set of equations but for flat wave beam with plane wavefront at the boundary was investigated in Ref.[4] by means of perturbation theory in time. Contrary to Refs.[2,3] the results of Ref.[4] say that during self-action process intensity maximum appears near the boundary of the medium and moves inside it. But perturbation theory used in Ref.[4] allowed to investigate only an initial stage of self-action process.

In the given report we present the results of numerical solution of coupled quasioptic and sound equations for cylindrical and flat wave beams with various boundary conditions.The main result is that during self-action process on the beam axis appear virtual intensity maxima moving inside the medium with speed considerably higher than sound velocity. By motion each intensity maximum first increases and then decreases and disappears. The next maximum appears farther inside the medium and follows the evolution of the preceding one. The stationary state is set between the boundary and the appearance point of the following maximum. When the first ( counting from the boundary ) stationary intensity maximum is established the "birth" process of virtual maxima continues deeper in the medium for flat beams. The obtained results allow to retrace the wave beam field and medium density space-time evolution. The physical reason for the above described wave beam self-action dynamics is discussed.

## Basic equations

To describe self-action dynamics of electromagnetic wave beam in nonlinear medium we use the equation for slowly varying in time ( for the period $1/\omega_0$) and along the beam axis OZ ( by the wave length ) complex amplitude $E(\vec{r}_\perp,z,t)$ of the wave electric field strength $\vec{E}(\vec{r},t)=\mathrm{Re}\{\vec{e}E(\vec{r}_\perp,z,t)*\exp(-i\omega_0 t + ikz)\}$ :

$$2i\left(\frac{\omega_0}{c^2}\frac{\partial E}{\partial t} + k\frac{\partial E}{\partial z}\right) + \Delta_\perp E + \frac{\omega_0^2}{c^2}\left[\varepsilon_{nl} + i\varepsilon''\right]E = 0 , \qquad (1)$$

where $k=\frac{\omega_0}{c}\sqrt{\varepsilon_0}$ is wave number defined by the real part of the linear dielectric constant of the medium $\varepsilon_0$, $\varepsilon_{nl}$ is real part of nonlinear addition to dielectric constant due to electric field action, $\varepsilon''$ is imaginary part of linear dielectric constant governing wave dissipation. For a plasma ($\varepsilon_0= 1-n_0/n_c$) the value $\varepsilon_{nl}=-\delta n/n_c=-(n_e-n_0)/n_c$ is defined by the deviation of the electron concentration $n_e$ from the initial value $n_0$ ($n_c= m\omega_0^2/4\pi e^2$ - critical density ). For an arbitrary medium it may be assumed that $\varepsilon_{nl}\simeq \left[\partial\varepsilon_0/\partial\rho\right]_T\delta\rho$, where $\delta\rho$ - density perturbation.

For strictive nonlinearity the dynamics of small density perturbations are governed by the sound wave equation with ponderomotive driving force (Ref.[1]) :

$$\left\{\frac{\partial^2}{\partial t^2} + 2\hat{\gamma}_s\frac{\partial}{\partial t} - V_s^2\left[\Delta_\perp + \frac{\partial^2}{\partial z^2}\right]\right\}\delta\rho = -\left[\rho_0/16\pi\right]\left(\frac{\partial\varepsilon_0}{\partial\rho}\right)_T\left[\Delta_\perp + \frac{\partial^2}{\partial z^2}\right]|E|^2 \qquad (2)$$

where $V_s$ is the sound speed, $\hat{\gamma}_s$ is the sound wave dissipation operator, $\rho_0$ is initial undisturbed density. For a plasma $V_s= \left[(\bar{z}T_e+\gamma_i T_i)/M_i\right]^{1/2}$, where $\gamma_i$ is ion adiabatic index, $\left[\partial\varepsilon_0/\partial\rho\right]_T= -(M_i n_{ci})^{-1}$, where $M_i$ and $n_{ci}=n_c/\bar{z}$ are the ion mass and the critical concentration respectively, $\bar{z}$ is the ion charge.

For further consideration of the problem it's convenient to introduce dimensionless variables defined by characteristic transversal beam size $a_0$ at the medium boundary :

$$\tilde{r}_\perp = r_\perp / a_0 \;;\quad \tilde{z} = z / (2ka_0^2) \;;\quad \tilde{t} = V_s t / a_0 \qquad (3)$$

Thus Eqs.(1),(2) for dimensionless amplitude $\tilde{E}=E/E_0$, normalized by maximum amplitude value at the boundary $E_0$, and for dimensionless density perturbation $A = \left(\frac{\omega_0 a_0}{c}\right)\delta\rho\left(\frac{\partial \varepsilon_0}{\partial p}\right)_T$ take the form

$$\left\{ i\left( \beta \frac{\partial}{\partial t} + \frac{\partial}{\partial z} + \Gamma_E \right) + \Delta_\perp \right\} E = A\, E \qquad (4)$$

$$\left\{ \frac{\partial^2}{\partial t^2} + 2\,\hat{\Gamma}_s \frac{\partial}{\partial t} - \Delta_\perp \right\} A = \alpha\, \Delta_\perp |E|^2 \qquad (5)$$

where tilde sign is omitted, $\beta=2\,\frac{V_s}{C}\,\frac{\omega_0 a_0}{C}$, $\Gamma_E=\left(\frac{\omega_0 a_0}{C}\right)^2 \varepsilon''$, $\hat{\Gamma}_S = \frac{a_0}{V_s}\,\hat{\gamma}_S$. In Eq.(5) we neglect small terms, which are proportional to the squared ratio of transversal scales to longitudinal ones. Parameter $\alpha$ characterizes the influence of ponderomotive force on the density perturbations and is given by $\alpha = \frac{1}{V_s^2}\left(\frac{\omega_0 a_0}{C}\right)^2\left(\frac{\partial \varepsilon_0}{\partial \rho}\right)^2 \frac{\rho_0}{16\pi} E_0^2$ . In particular, for a plasma $\alpha = \frac{1}{4}\,\frac{n_0}{n_c}\left(\frac{\omega_0 a_0}{C}\right)^2 \frac{2E_0^2}{E_s^2}$, where $E_s = \left[4\pi n_c (T_e + \gamma_i T_i)/\bar{z}\right]^{1/2}$ is striction nonlinearity characteristic field.

We shall investigate both flat beam with $\Delta_\perp = \partial^2/\partial y^2$ and cylindrically symmetric beam when $\Delta_\perp = \frac{1}{r}\frac{\partial}{\partial r}\left[r\frac{\partial}{\partial r}\right]$. Density perturbation damping operator $\hat{\Gamma}_S$ is assumed to be in the next form :

$$\hat{\Gamma}_S = \Gamma_{SO} + \Gamma_{SV}\,\Delta_\perp \qquad (6)$$

where $\Gamma_{SO}, \Gamma_{SV}$ are constants. In a plasma under the condition that ion-ion collision frequency $\nu_{ii}$ is greater than $\omega = 2\pi V_S/a_o$ the second term in Eq.(6) describes viscosity and $\Gamma_{SV}=0.8\,\frac{V_S}{a_o\nu_{ii}}\,\frac{T_i}{\bar{z}\,T_e}$. In the opposite case when $\nu_{ii}<\omega_s$, dumping decrement doesn't depend on wavelength of perturbation and is defined by $\Gamma_{SO}= 0.8\,\frac{a_o\nu_{ii}}{V_S}\,\frac{T_i}{\bar{z}\,T_e}$.

The set of Eqs.(4),(5) should be accomplished by initial and boundary conditions. Let us consider that a medium occupies half-space $z\geqslant 0$, and a beam with fixed field amplitude falls on the medium boundary from the region $z<0$. The main results will be obtained for a Gaussian beam with the unity dimension transversal scale :

$$E(r_\perp, z=0, t) = E^{(0)}(t)\,\exp(-r_\perp^2 + i\varphi(r_\perp)), \quad t>0 \qquad (7)$$

where $E^{(0)}(t)$ determines radiation amplitude time dependence, $\varphi(r_\perp)$ is phase front curvature. For plane phase front $\varphi(r_\perp)=0$. For time-independent incident beam intensity $E^{(0)}(t)=1$.

Boundary conditions on transversal coordinate correspond to the usual require of symmetry in relation to the beam axis and field asymptotic decrease at the boundary of the region under consideration $r_\perp=R_{max}\gg 1$ :

$$\left.\frac{\partial E}{\partial r_\perp}\right|_{r_\perp=0} = 0 \;;\quad E(r_\perp=R_{max}, z, t)=0 \;;\quad \left.\frac{\partial E}{\partial r_\perp}\right|_{r_\perp=R_{max}} = 0 \qquad (8)$$

The boundary conditions for density perturbations $A$ have the same form.

Initial conditions for Eq.(5) correspond to the absence of density perturbation at $t=0$ :

$$A(r_\perp,z,t=0) = 0 \quad ; \quad \left.\frac{\partial A}{\partial t}\right|_{t=0} = 0 \ ,$$

and for Eq.(4) it correspond to the solution of linear stationary field equation :

$$\left( i\frac{\partial}{\partial z} + \Delta_\perp \right) E(r_\perp,z,t=0) = 0 \qquad (9)$$

The formulated problem (4)-(9) was solved numerically. In order to prevent density perturbation reflection from the boundary $r_\perp=R_{max}$ the "soft absorbing wall" model was used. To obtain that the damping decrement $\Gamma_{SO}$ (6) was made to increase exponentially near the boundary $r_\perp=R_{max}$ of the region under consideration.

To control the solution accuracy two conservation laws were used.

The electromagnetic field conservation law is derived from Eq.(4). Taking into account boundary conditions (8) in the region under consideration $Z \in [0,D]$, $r_\perp \in [0,R_{max}]$, this law takes the next form :

$$Q + T + \beta\frac{\partial W}{\partial t} = 1 \qquad (10)$$

where the term $Q = 2\int_0^D dz \int_0^{R_{max}} \Gamma_E \ |E|^2 ds \Big/ \bar{S}_{ZO}$ characterizes absorbing of radiation, the term $T = \int_0^{R_{max}} |E(r_\perp,z=D,t)|^2 ds \Big/ \bar{S}_{ZO}$

stands for transmission of radiation, the term

$W = \int_0^D dz \int_0^{R_{max}} |E|^2 ds \Big/ \bar{S}_{zo}$ determines the electromagnetic field energy, normalized by the incident radiation energy flow

$\bar{S}_{zo} = \int_0^{R_{max}} |E(r_\perp, z=0, t)|^2 ds$. The integrals are taken over the element **ds** , which is equal **dy** for plane geometry and **ds**$=2\pi r$**dr** for cylindrical beam.

The matter conservation law is derived from Eq.(5) and for any fixed **z** with $\Gamma_{SO}, \Gamma_{SV}$ being constants takes the form :

$$\int_0^{R_{max}} A(r_\perp, z, t) ds = 0 \qquad (11)$$

## Flat wave beam self-action

We begin the investigation of self-action dynamics by the same simplifications in Eqs.(4),(5), as in Ref.[4]. We neglect the electromagnetic field dissipation ($\Gamma_E=0$) and consider the beams with transverse dimension satisfying the condition $a_0 \ll \frac{c^2}{\omega_0 V_S}$. This assumption allows us to omit the time derivative in Eq.(4). Besides we'll consider the beam with time-independent amplitude ($E^{(0)}=1$).

Let's summarize shortly the analytic results obtained in Ref.[4].

Under the assumption of zero sound wave damping ($\hat{\Gamma}_S=0$) the set of Eqs.(4),(5) takes the following form :

$$\left( i \frac{\partial}{\partial z} + \frac{\partial^2}{\partial y^2} \right) E = AE \qquad (12)$$

$$\left( \frac{\partial}{\partial t^2} - \frac{\partial^2}{\partial y^2} \right) A = \alpha \frac{\partial^2 |E|^2}{\partial y^2} \qquad (13)$$

The incident at t=0 on the boundary z=0 wave beam penetrates instantaneously ($\beta=0$) in a medium without density perturbations (A=0). Thus beam field at the initial moment is defined by linear Eq.(9) and takes the form

$$E_0(y,z,t=0) = \int_{-\infty}^{\infty} dy' f(y') \; G(y-y',z) \qquad (14)$$

where $G(y-y',z) = (4i\pi z)^{-1/2} \exp\left[ i(y-y')^2/4z \right]$, $f(y)=E_0(y,z=0)$ is given beam field on the boundary.

It's natural to assume that during some time interval density perturbations dynamics is governed by the field (14). Substituting it into the right-hand side of Eq.(13) and using zero initial conditions for density perturbations one will find:

$$A(y,z,t) = -\frac{\alpha}{2} \left[ 2\,|E_0(y,z)|^2 - |E_0(y+t,z)|^2 - |E_0(y-t,z)|^2 \right] \qquad (15)$$

The first term in right-hand side of Eq.(15) is the stationary solution of Eq.(13). Other two terms describe density "splashes" leaving beam localization region along OY axis with the sound speed.

Once appeared density disturbance changes the beam field which during a sufficiently short time interval may be expressed as $E=E_0+E_1$, where $|E_1| \ll |E_0|$. In the linear relative to $E_1$ approximation with the boundary condition $E_1(y,z=0,t)=0$ the solution of Eq.(12) is :

$$E_1(y,z,t) = -i\int_0^z dz'\int_{-\infty}^{\infty} dy' \; A(y',z',t)\; E_0(y',z')\; G(y-y',z-z'), \tag{16}$$

where the expression for G function is written above.

We'll be interested in the absolute value of dimensionless electric field amplitude :

$$\mathcal{E} = \left[\left(\mathrm{Re}\ E_0 + \mathrm{Re}\ E_1\right)^2 + \left(\mathrm{Im}\ E_0 + \mathrm{Im}\ E_1\right)^2\right]^{1/2} \tag{17}$$

The expressions (14)-(16) give the analytic solution to the considered problem concerning the initial stage of flat beam self-action dynamics. It should be noted that an attempt to get similar solution for an axially symmetric wave beam was unsuccessful.

It should be pointed out that the described above method is valid only for the time interval during which the difference between $\mathcal{E}$ and an undisturbed field amplitude $\mathcal{E}_0 = \left[\left(\mathrm{Re}\ E_0\right)^2 + \left(\mathrm{Im}\ E_0\right)^2\right]^{1/2}$ is sufficiently small.

As an example let's consider a Gaussian beam with plane phase front ($\varphi=0$) when $f(y)=\exp(-y^2)$. Introducing notation $æ=4z$ from Eqs.(14)-(16) we get:

$$E_0(y,æ)=(1+æ^2)^{-1/4}\exp\left(-y^2/(1+æ^2)\right)\exp\left[-i\left(0.5\mathrm{arctg}(æ)-\frac{æ\, y^2}{1+æ^2}\right)\right] \tag{18}$$

$$A(y,æ,t)= -\alpha\ (1+æ^2)^{-1/2}\exp\left(-2y^2/(1+æ^2)\right)\left[1-\exp\left(-2t^2/(1+æ^2)\right)* \right.$$
$$\left.\mathrm{ch}\left(4yt/(1+æ^2)\right)\right] \tag{19}$$

$$E_1(y,æ,t) = \alpha\ \exp(i\pi/4)\int_0^{æ} dq\ (PQ)^{-1/2} \exp\left[y^2(3i+q)/Q\right] *$$
$$\left\{1-\cos\left[4yt/Q\right]\exp\left[2it^2(1+iæ)/(PQ)\right]\right\} \qquad (20)$$

where $P=1+iq$, $Q=3(æ-q)-i(1+æq)$.

It's clear from Eq.(18) that at the initial moment the beam diffract and its on-axis amplitude decreases as $(1+æ^2)^{-1/4}$ and transversal size grows by increasing of the distance from the medium boundary.

It follows from Eq.(19) that during the time $t<t_c=\sqrt{(1+æ^2)/2}$ an on-axis density hole grows proportionally to the squared time. It means that in this time we can neglect the thermal pressure (the second term at the left-hand side of Eq.(13)). By $t>t_c$ the thermal pressure becomes important and the quasistationary state is settled near the medium boundary.This state is achieved the sooner the closer the considered point is to the medium boundary.

In an arbitrary case it's impossible to calculate the integral in Eq.(20). Therefore we'll consider the on-axis (y=0) field perturbations during a sufficiently short time interval ($t<t_c$). Then in the linear with respect to $E_1$ approximation one may find the on-axis beam amplitude from Eqs.(17),(18) and (20). Calculating the derivative with respect to the z-coordinate for the mentioned amplitude and setting it to zero one can get an equation for the position of amplitude on-axis extreme :

$$\sqrt{1 + æ_0^2} = 1/16\ \alpha t^2\ \Phi(æ_0^2) \qquad (21)$$

where $\Phi(\ae^2)\simeq 76-600\ae^2$ for $\ae^2<1$. The Eq.(21) has a solution only for $t>t_0=4/(19\alpha)$, and $\ae_0=0$ for $t=t_0$. The value $\ae_0$ increases with time and follows the equation

$$\ae_0=\frac{\left[19\alpha t^2/2-2\right]^{1/2}}{\left(1+75\alpha t^2\right)^{1/2}} \tag{22}$$

The presented below numerical results show that obtained extreme is the maximum. Thus it may be concluded that the medium expelled out of the beam localization region produces the intensity maximum on the beam axis. At first this maximum appears at the medium boundary through some time after beginning of the self-action process. The maximum increases and moves downstream with time.

As it was pointed out the formulated analytical method is valid only for the limited time intervals. In particular it's not applicable for investigation of the stationary state achievement in the regions far from the boundary, where the huge electric field amplitude modifications appeared during rather long time interval.

In order to represent the final beam state we begin our consideration with stationary solution of Eqs.(12),(13). In this state the pressure of high frequency field is equal to the pressure of medium and the field is obtained from the non-linear Schrödinger equation. The solution of this equation shows that on the beam axis the sequence of intensity maxima and minima exists. The field maxima values decrease by increase of distance from the boundary (Fig.1).

The maxima positions and their height depend on parameter $\alpha$ in Eq.(12). It should be noted that the solution of nonlinear Schrödinger equation is in qualitative agreement with the results of paraxial approximation (Ref.[5]), but they have significant quantitative difference at least by $\alpha=6$.

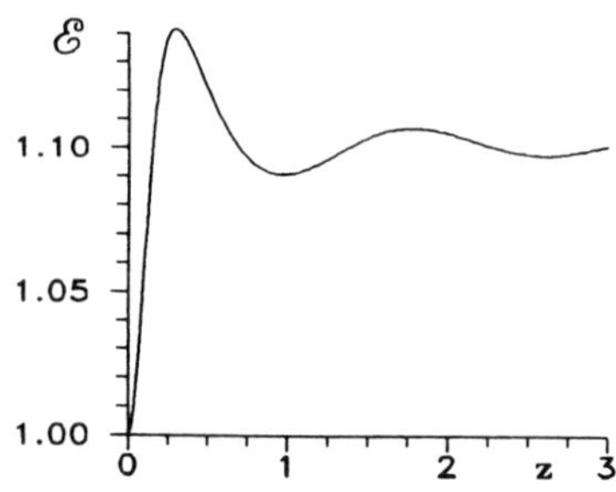


Fig.1. Stationary state distance dependence of on-axis electric field amplitude ($\alpha$=6).

In order to investigate the self-action dynamics for long time intervals the set of Eqs.(12),(13) was solved numerically. The sound wave damping was taken into account ($\Gamma_S \neq 0$). The simulations were carried out for the Gaussian beam with width $a_0$= 30μm and frequency $\omega = 2 \cdot 10^{15} s^{-1}$ for plane and curved phase front on the medium boundary. Other numerical parameters correspond to hydrogen plasma with electron temperature 100 eV and $n_0/n_c = 5 \cdot 10^{-2}$. Average laser intensity was taken to be $4.7 \cdot 10^{12}$Wt/cm$^2$ ( $\alpha$=6 ). In the Eq.(6) the sound damping coefficients are $\Gamma_{SO}$=0.05 and $\Gamma_{SV}$=0.03.

We begin the investigation of self-action dynamics from the beam with plane phase front on the boundary. Fig.2 shows the time evolution of the on-axis electric field amplitude maxima $E_m$ and of their $z_m$ coordinates. It may be seen that at a non-zero time moment the first intensity maximum appears near the medium boundary. The electric field amplitude $E_m$ of the maximum increases with time and the maximum moves away from the boundary. The amplitude $E_m$ reaches the greatest value and after that decreases, and instantaneously the second maximum appears on some distance from the boundary. In small time interval after

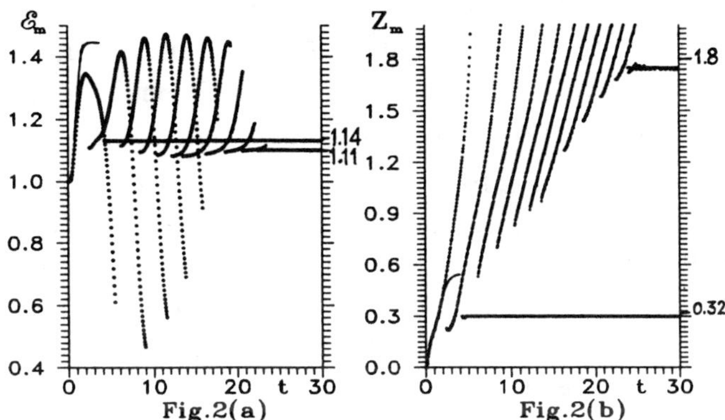


Fig.2. Time evolution of the on-axis electric field amplitude maxima $E_m$(a) and of their $Z_m$ coordinates(b). Solid lines represent the results of Ref.[4]. The values at the right-hand side stand for the stationary problem solution.

appearance this maximum approaches to the boundary and then moves the same way the first maximum does. Some time later the next (third) amplitude maximum appears and its amplitude and longitudinal coordinate are practically constant. Just the very field maximum is remained in the stationary state.

Along with that at longer distances from the boundary other intensity maxima continue to appear. Their dynamics is similar to the dynamics of the first two maxima but the point of appearance shifts deeper inside the medium.

As it follows from Fig.2(b) the speed of all transient maxima is approximately constant. This speed has the value $V_m \simeq 2ka_0V_s$ and it is much higher than the speed of sound.

The solid lines at the Fig.2 show the analytical results from Ref.[4] under the same range of parameters. The agreement with numerical simulations is rather good for the time interval $t<1$, where the theory developed in Ref.[4] is valid.

The Fig.1 shows the maximum amplitude and the position of the maxima in the stationary state.

The field space-time evolution is connected with the density evolution. The Fig.3 shows constant density and field contours at three time moments. For $t=1.5$ there are the only one on-axis density hole and only one field maximum situated a bit far away from the boundary. At the time $t=3.0$ there are two density holes and accordingly two field maxima. It should be noted that the farther from the boundary density holes and field maximum have the greater values. For $t=5.0$ the previous two holes being separated by the density peak remain. The density minimum nearest to the boundary also remains and it is saved in stationary state.

The Fig.4 shows the 3-dimensional density plot at the time moment $t=4.0$ . It may be seen that the sound waves appearing near the medium boundary are far away from the region occupied by the beam. In the farther from the boundary regions the density perturbations behave in an another way.In particular there are the density splash on axis and holes in both z- and transversal (y-) directions.

The self-action process for the beams with non-constant phase front on the medium boundary has analogous form. We investigated the beams with initially parabolic phase front : $\varphi = -y^2\delta$, where $\delta = ka_0^2/2R$, R being the front curvature radius. Fig.5 shows time dependence of the on-axis field maxima values and their positions for the initially defocused beam ($\delta=-0.1$). It may be seen that the first transient maximum appears at some distance from the boundary with greater time delay in comparison with the case of the plane wave front beam. Later the self-action process of the defocused beam develops the same manner as for the plane wave front beam (Fig.2).

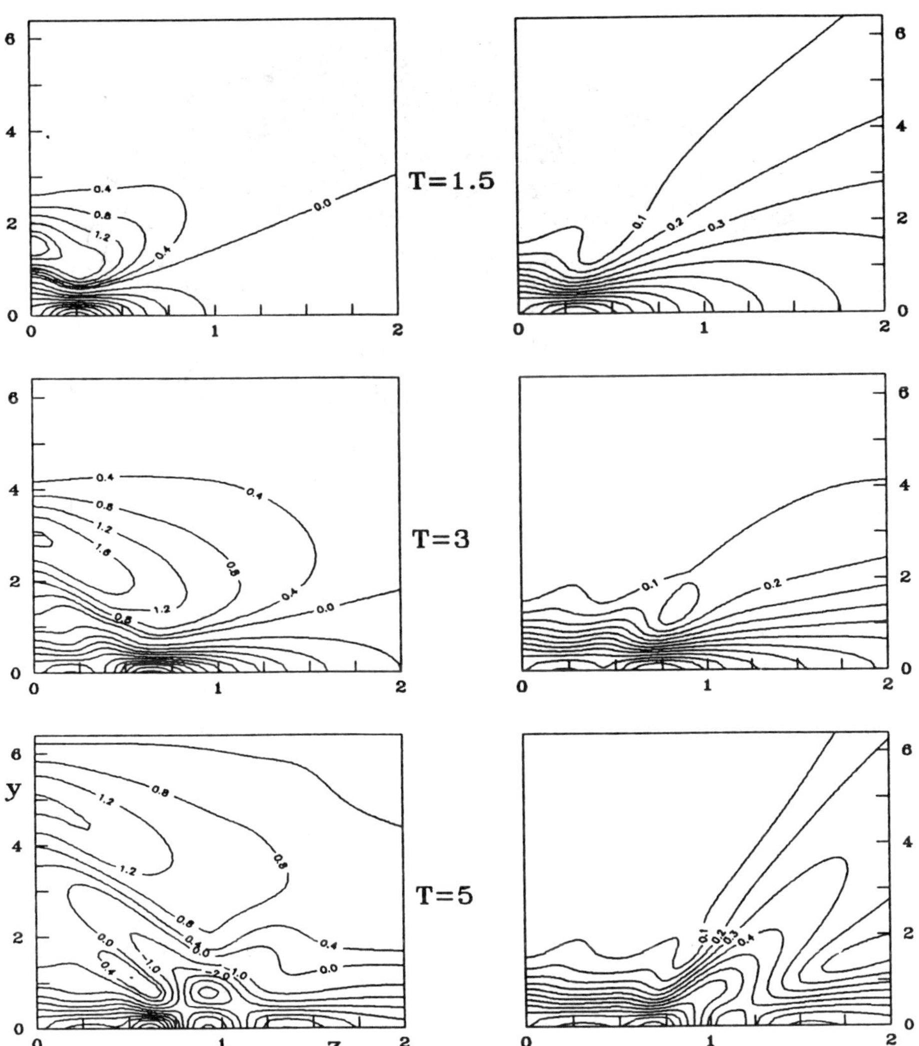


Fig.3. Constant density and field amplitude contours at time moments t=1.5, 3.0, 5.0.

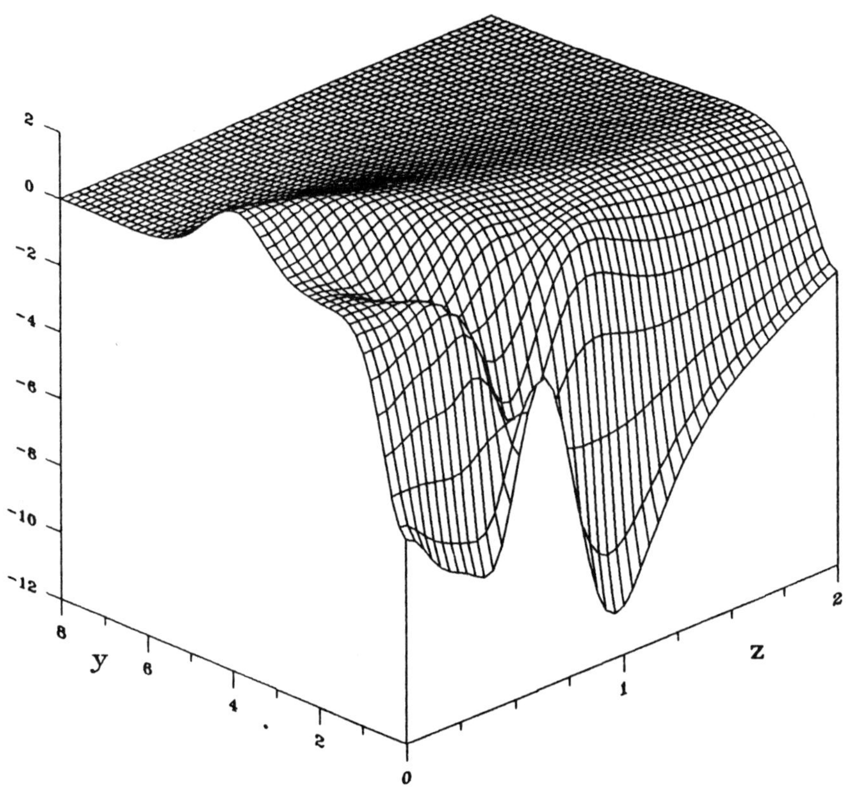


Fig.4. Three-dimensional density plot at t=4.0.

Fig.6 shows the maxima values $E_m$ and corresponding coordinates $Z_m$ for the beam focused at the boundary ($\delta$=+0.1). It may be seen that field maximum arising at the initial moment moves at first inside the medium with increasing amplitude. On the contrary to the plane wave front beam in this case the first stationary maximum appears earlier and closer to the boundary with greater amplitude. The subsequent evolution is analogous to the one for the plane phase-front beam.

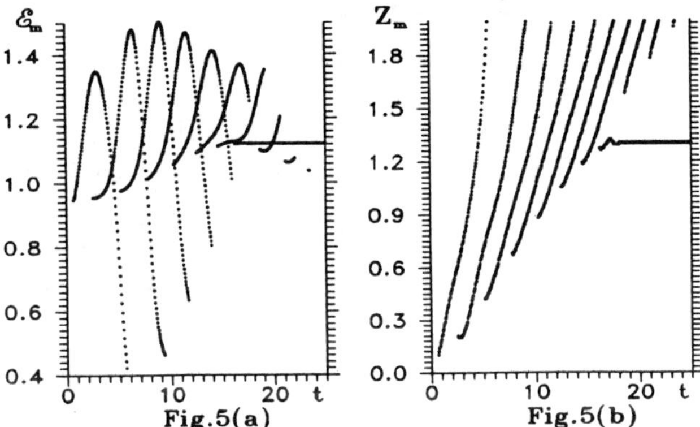


Fig.5. Time dependence of the on-axis field maxima values(a) and their positions(b) for the initially defocused beam ($\delta$=-0.1).

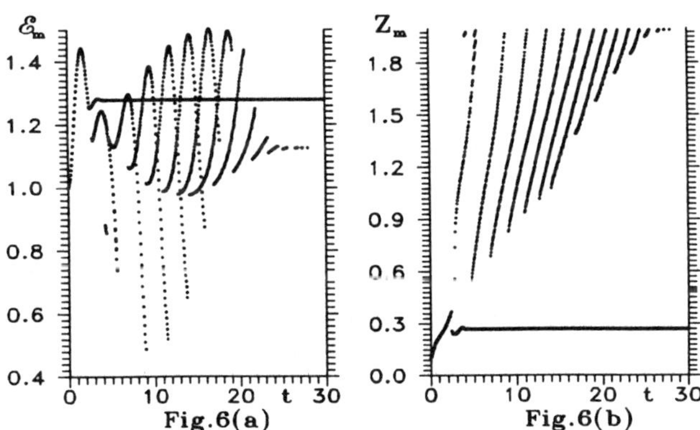


Fig.6. Time dependence of the on-axis field maxima values(a) and their positions(b) for the initially focused beam ($\delta$=+0.1).

## Self-action of cylindrical beam

We investigate also the dynamics of nonlinear propagation of azimutally-symmetric cylindrical beam not only in the case $\beta=0$ in Eq.(4) but taking into account the term with time derivative ($\beta\neq0$). It's well known (Ref.[5]) that in this geometry without nonlinear dissipation in the field equation ($\Gamma_E=0$) by exceeding threshold power the singularity (focus) is formed on axis in the stationary state. For the Gaussian incident beam the corresponding to threshold nonlinear parameter is equal $\alpha_{th}= 7.54$.

The calculations were carried out for the value $\alpha=8.4$, which is higher the threshold of stationary self-focusing. In order to eliminate the unlimited field growth in the focus we introduced the nonlinear dissipation in the form $\Gamma_E=\gamma_E|E|^4$, modeling, for example, two-photon absorption. The value of parameter $\gamma_E$ was chosen in such a way that the ratio of the maximum radiation intensity to the boundary one was restricted to $10^2$.

In Figs.7-9 the results are presented for the beam with constant amplitude on the nonlinear medium boundary at $t>0$. Fig.7 shows contour lines $|E(r=0,z,t)|^2$=const for $\beta=1$, $\Gamma_{SO}=0.5$, $\Gamma_{SV}=0.02$, $\gamma_E=0.02$. This figure illustrates time evolution of the on-axis intensity distribution. An arrow shows the stationary focus position derived by means of solution the nonlinear Schrödinger equation, which can be obtained from Eqs.(4),(5) by neglecting of the terms with time derivatives ($A=-|E|^2$). The self-action process is characterized the appearance of the transient field maxima. The first of them appears just near the nonlinear medium boundary and the subsequent ones arise at longer distances from the boundary. All of the maxima move inside the medium towards the stationary focus position. When the density perturbations and the field pressure balance each

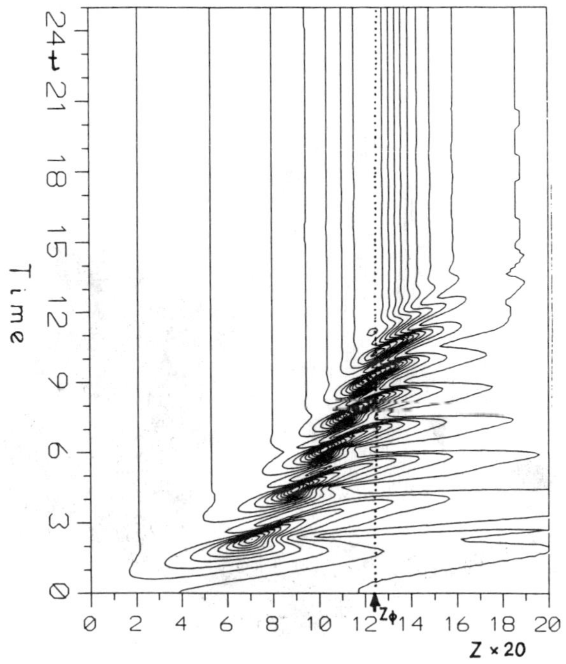


Fig.7. Time evolution of the on-axis electric field amplitude contour levels. An arrow points out the stationary focus position ( β=1, $\Delta E$=0.4 ).

other in the region close to the boundary, the transient maxima appear farther from the boundary. As it may be seen from Fig.7 each maximum after appearance increases and moves with the speed $V_m \simeq 4V_s ka$. In the time when the first maximum reaches the greatest value beside it closer to the boundary another transient maximum arises. With time the first maximum decreases and the second increases. The transient intensity maxima continue appear till the moment of time their production point reaches the stationary focus position. After that the stationary state is settled.

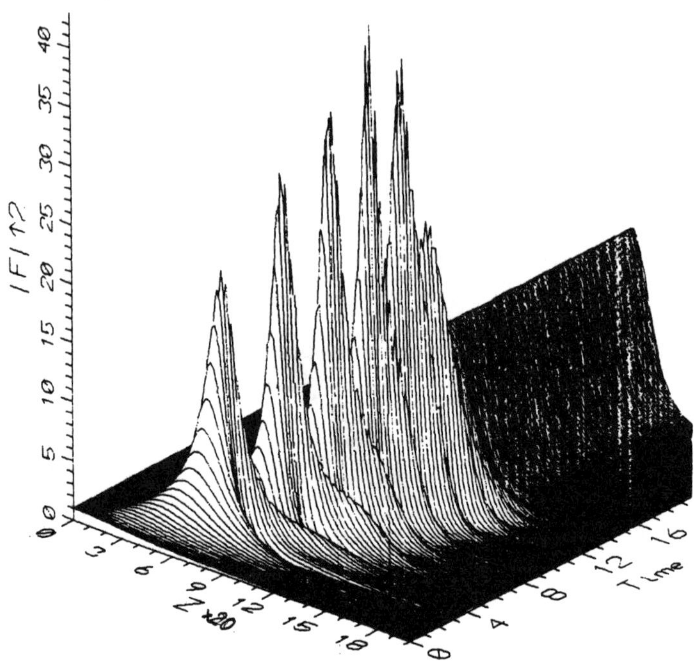


Fig.8. Orthographic projection of the on-axis radiation intensity time evolution.

Dynamics of intensity $|E(r=0,z,t)|^2$ on the beam axis in orthographic projection is shown in Fig.8.

Fig.9 represents the results of solution of Eq.(4) without the time derivative. The comparison of Fig.7 and Fig.9 shows that taking into account the finite velocity of radiation propagation at least for $\beta \leqslant 1$ doesn't change qualitatively the nonlinear self-action beam dynamics.

Fig.10 shows the trajectories of the on-axis field maxima when the boundary amplitude is non-constant and changes with time as $E^{(0)}(t)=\exp\left[(t-t_1)^2/(2t_2^2)\right]$, where $t_1=2$, $t_2=\sqrt{2}$.

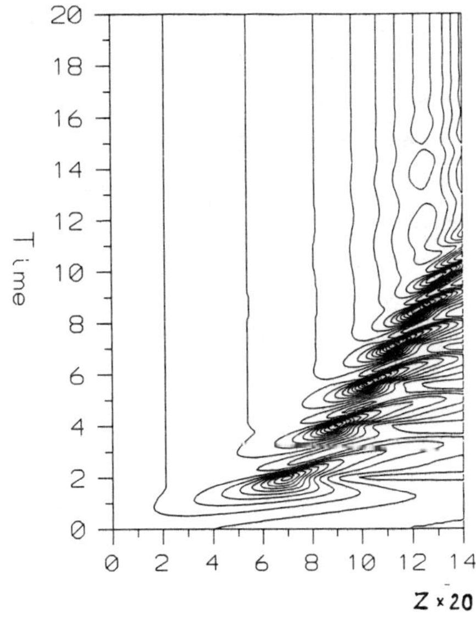


Fig.9. Time evolution of the on-axis electric field amplitude contour levels. ( $\beta$=0 in Eq.(4), $\Delta E$=0.4 ).

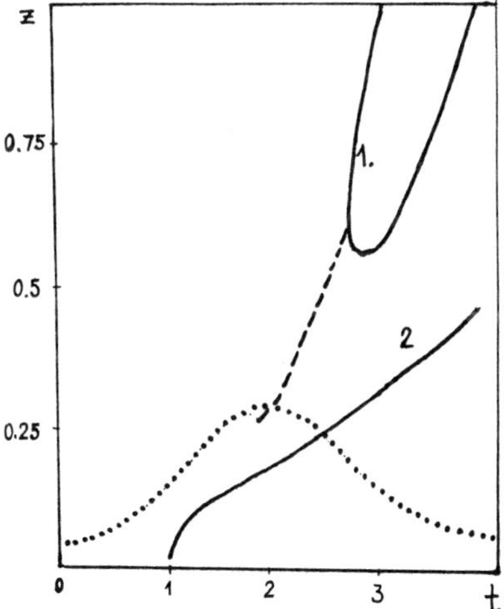


Fig.10. Focus(1), quasistationary maximum(2) and dynamic maximum(3) trajectories. Dotted line stands the amplitude time dependence on the boundary.

Other parameters were the same as later. The curve 1 defines the focus trajectory ( the singularity in solution ), obtained in the quasistatic approximation from the nonlinear Schrödinger equation (Ref.[5]). The dashed line which shows the intensity maximum movement in quasistatic approximation approaches the curve 1. The curve 2 shows maximum intensity trajectory obtained from our dynamic calculations. It's clear that the velocity of this maximum is considerably lower than the light speed, though it is much higher than the sound speed.

## Conclusions

As it was shown in our simulations the self-action dynamics of the wave beams in media with strictive nonlinearity is rather complicated and may be approximately split into a number of stages. At first stage nonlinear response is still small and the beam field differs little from the field in homogeneous linear medium . Strictive (ponderomotive) forces are strongest at the medium boundary where the beam is very narrow. Just in this place the matter is forced out most quickly and the region of depressed density appears. This region acts as focusing lens for the radiation. When the focusing action of the region suppresses the defocusing action of diffraction, the rays will approach to the beam axis. This effect first appears close to the boundary for a very narrow region near the beam axis where the diffraction is most week and the matter is forced out most quick. With time the density hole deepens and expands. The greater number of rays change their directions and collimate to the beam axis, though this takes place farther from the boundary. The intensity maximum moves away from the medium boundary and increases. Just that very initial stage is investigated in Ref.[4] for for the case of flat beam.

The second stage begins after the intensity in the maximum becomes large enough and the strictive force considerably exceeds pressure force in this region. It's natural that the region of maximum intensity produces the effective matter ejection. As the result the focusing region of lower density moves into the area where the intensity maximum was situated. In its turn intensity maximum moves farther forward. Thus moving intensity maximum created by moving behind region of depressed density creates in turn the intensity maximum. The propagation speed of such a formation may be estimated by dividing the distance between the intensity maximum and density minimum by characteristic time during which the matter is expelled by strictive forces. This speed is approximately equal $V_m \simeq 4ka_0V_s$ and is much greater the velocity of sound.

The third stage of beam self-action is connected with the dynamics of density perturbations created by the field maximum moving with speed $V_m$. The density perturbations being not supported by strictive forces begin to move away normal to the beam axis as the sound waves and after the time interval of order $a_0/V_s$ the density at the beam axis increases. This density splash deflects the rays from the beam axis and prevents the achievement of the first intensity maximum. At the same time the new field maximum formation process begins closer to the medium boundary at the edge of the region where an equilibrium between field and medium pressures takes place. As a number of rays collected to the beam axis inside the area of the second maximum increases this number decreases in the region of the first maximum, and the latter disappears.

Thus the transient intensity maxima leave out the region where the electromagnetic field pressure equals medium pressure. The velocity of these maxima are greater than speed of the stationary region boundary, and their intensity exceeds the stationary maximum intensity. It may be thought that the specialized experiments would be able to justify the results of our considerations.

## References

[1] Y.R.Shen. "The principles of nonlinear optics" John Wiley & Sons,Inc. 1984

[2] E.L.Kerr, Phys.Rev.A.,1971,4,1195

[3] N.Lontano,A.M.Sergeev,A.Cardinali, Phys.Fluids,1989,B1,(4),901

[4] L.M.Gorbunov,S.V.Tarakanov,JETP,1991,91,58 (in Russian)

[5] M.B.Vinogradova,O.V.Rudenko,A.P.Suchorukov. "The theory of waves",Nauka Publishers,Moscou,1979,p.28 (in Russian)

**Research Trends in Physics: Coherent Radiation Generation and Particle Acceleration**
Editorial Board: J.M. Buzzi, A. Prokhorov (Editor-in-Chief), P. Sprangle, and K. Wille
*La Jolla International School of Physics*, The Institute for Advanced Physics Studies, La Jolla, California

# Electron Acceleration in a Strong Laser Field and a Static Transverse Magnetic Field

**V.V. Apollonov, A.I. Artemyev, Yu.L. Kalachev, A.M. Prokhorov, and M.V. Fedorov**

General Physics Institute
Russian Academy of Sciences
117942 Moscow, Russia

## 1 INTRODUCTION

During some last years the problem of laser acceleration of electrons (LAE) attracts a rather large attention [1-3]. In all the previously proposed LAE methods the laser intensity I is assumed to be not too large ( $I < 10^{14} W/cm^2$ ). Besides almost in all these methods some resonance conditions are assumed to be fulfilled, and this is not a simple problem to adjust these schemes for a wide energy range. In the present paper we analyze a new approach to the LAE problem different both from the inverse free-electron laser methods and from the plasma methods.

In fact the method proposed does not imply the presence of a plasma. The laser intensity I is not so strongly restricted as in other LAE methods ( our schemes works up to $I \sim 10^{17}$ $W/cm^2$ ). An efficiency of this method has not a sharp resonance dependence on the initial electron energy. Acceleration occurs without any assumptions about electron-light phase matching. In principle these features would make possible a cyclic acceleration of electrons in a single focal volume in a system using a train of short laser pulses.

## 2 THE FORMULATION OF THE PROBLEM

Let us examine the motion of an electron in the (X,Z) coordinate plane in a uniform static magnetic field Ho directed along the Y axis ( Fig. 1). Let us assume, that the time t=0 is a time when in the absence of a laser field electron trajectory crosses the X=Y=Z=0 point.

 

So in the absence the laser field the equations of the electron's motion have a solution

$$Z=R\sin(\Omega t) \qquad v_z=v_o*\cos(\Omega t)$$
$$X=R[\cos(\Omega t)-1] \qquad v_x=v_o*(-\sin(\Omega t)) \tag{1}$$

where c=1, $\Omega=eH_o/\gamma_o$ is the Larmour frequency, $R=V_o/\Omega$ is the Larmour radius, $V_o$ and $\gamma_o$ are the electron velocity and energy.

We assume that the laser field is linearly polarized along the X axis and it is focused along the Z axis so that, the diameter d and length L of the focal volume obey inequalities

$$d \ll L \ll R \tag{2}$$

Under these conditions the longitudinal part of laser field strength E is negligible and E can be taken in the form

$$E=H=E_o(z)\cos[\omega(t-z)+\phi_o] \tag{3}$$

where $E_o(z)$ is an laser field amplitude which varies slowly over distances on the order of the wavelength, $\phi_o$ is the initial phase, $\omega$ is the laser field frequency.

In fact, $E_o$ depends not on z only, but on the transverse coordinates x,y also. Nevertheless, in accordance with inequalities (2), we will ignore the transversal non uniformity of E assuming that the electron enters to the laser focus at very large |z| where E is negligible (look at the fig.1).

To solve the classic relativistic equations of the electron's motion in the fields E (3) and $H_o$ we use the perturbation theory, with respect to the field E. An alternative approach is to use the computer simulation.

### 3 QUALITATIVE DESCRIPTION OF THE METHOD PROPOSED

The main idea of the method proposed is the following. When the electron passes the focal volume it i subjected to the oscillating electromagnetic field. The period of these oscillations is determined by the phase of the wave in electron trajectory. The letter turns due to the action of the magnetic field. As a result, the projection of the electron speed on the direction of propagation of the wave axes is small at the beginning and at the end of the focal volume. So the force acting on the electron oscillates very fast at the

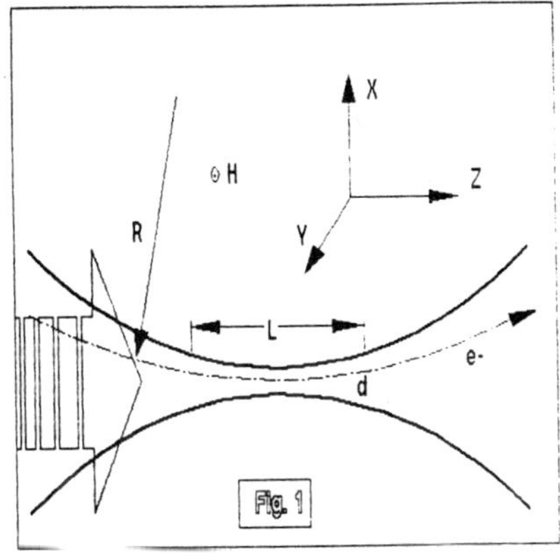


Fig.1 Geometry of this method for accelerating electrons.

beginning and at the end of the focal volume. And there exists a region inside the focal volume where the velocity of the electron is approximately parallel to the direction of the light wave. The force acting on the electron changes very slowly within this region (2-a). Substituting the unperturbed solution (Eq.(1)) to the laser field phase (Eq.(3)) we find a characteristic time $\delta t$ or distance $\delta l$ at which the laser field (3) at the electron orbit $E(z(t),t)$ changes essentially (reverses its sign)

$$\delta l=\delta t=\min \{ \lambda*\gamma^2, (6\pi/\omega\ \Omega^2)^{1/3} \} \tag{4}$$

where $\lambda=2\pi/\omega$ is the laser wavelength, $\gamma$ is the relativistic factor, $\gamma=\varepsilon_0/m \gg 1$.

The electron gains the main part of increase of it's energy just on passing this region. At other places the force acting on electron oscillates too fast to provide a sufficient change in electron's energy (see fig. 2-a ).

We assume that $\delta l \ll L$ . (5)

If the magnetic field $H_0$ is zero or small then on passing the

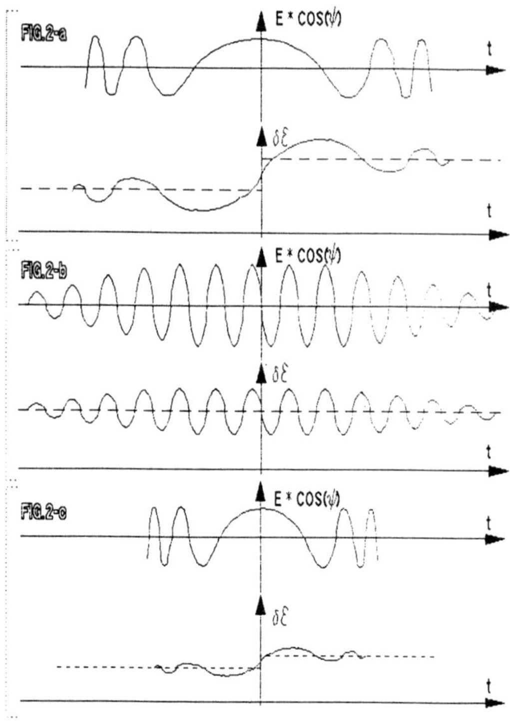


Fig.2 The dependence of the electromagnetic force acting on the electron on it's trajectory for different values of it's initial relativistic factor $\gamma_o$

focal volume the electrons feel that the oscillations of the light force are uniform . As a result the energy the electron gains through one semi period of an oscillation is compensated by the decrease in energy through the next semi period ( fig 2-b ). So if the magnetic field is small or zero the electron's energy at the end of the focal volume is practically equal to that at the beginning of the focal volume ( see fig. 2-b ). If the magnetic field is turned on,it violates the uniformity of oscillations of the force by which the light wave acts on electron. As a result of this violation the increase in energy is possible ( see fig. 2-a ).

If the magnetic field is extremely high (if $eH_0 \gg mc\omega/\gamma^2$ ) then the further increase of the magnetic field leads to the decrease of the length of the interval along the z-axes where the electron's speed V is parallel to z-axes and the phase of the light wave that is is practically constant. So the length of the interval along the z-axes where the light field oscillates slowly on electron's trajectory decreases because the trajectory of electron turns too rapid ( see fig. 2-a ) .

So there exists an optimal value of the magnetic field

$$eH_0 = mc\omega/\gamma^2 \tag{6}$$

In other terms for a given value of the magnetic field $H_0$ there exists an optimal value of the initial energy of electrons $\gamma_0$ for which the interaction of electrons with light provides the greatest change in energy . If the electron moves slowly

$$\gamma \ll \gamma_0 = \sqrt{mc\omega / eH_0} \tag{7}$$

the oscillations are too fast ( fig.2-a ). We want to note here that the optimum conditions are not very sharp ( fig. 3 ) .

## 4 THE ANALYTIC SOLUTION OF THE ELECTRON MOTION EQUATIONS AND THE CALCULATION OF THE INCREASE IN IT'S ENERGY

We describe the motion of electrons the relativistic analogous of Newton equations . Using the perturbation theory we can find the change of electron energy in the first $\delta\gamma^{(1)}$ and second $\delta\gamma^{(2)}$ order of laser field E. The first order effect is interesting if we want to accelerate ore decelerate a single electron [4]. But if we want to accelerate a beam of electrons it is necessary to perform the averaging over the initial phase of the light wave ( ore over the

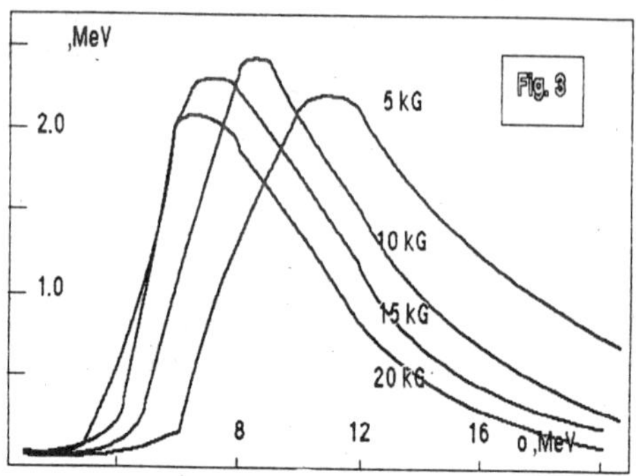


Fig.3 The average increment in the electron energy, $\delta\varepsilon$ versus the initial energy for the different transversal magnetic field strengths. $I=10^{15}W/cm^2$.

moment of entrance of an electron into the focal volume ). The first order effect [4] vanishes after this averaging . So to find the averaged change of energy of a beam of electrons it is necessary to consider the 2nd order of perturbation theory [4~8].

In the case of $\gamma < \gamma_o$ electron oscillations are multiple and uniform, with a constant period $\sim \lambda\, \gamma^2$. The contributions of these oscillations cancel each other and the net acceleration is exponentially small.

The case of $\gamma > \gamma_o$ corresponds to the existence of a region where the fields acting in electrons are quasi stationary. In this case the increase in average electron energy is the following

$$\frac{\delta\varepsilon^{(2)}}{\varepsilon}=\left(\frac{eE_o}{mc\omega}\right)^2 * \left(\frac{\gamma_o}{\gamma}\right)^{\frac{4}{3}} * \frac{2\sqrt{\pi}}{3^{7/6}} * F(\vartheta) \qquad (8)$$

where $\gamma_o$ is an optimal electron relativistic factor, and $F(\vartheta)$ is a function of the focusing length aperture $\vartheta$ (that equals the ratio of the focusing length diameter D to it's focal length f). If the the focusing is not too strong ,

$$\vartheta < \vartheta_o = (\omega/\Omega)^{1/3} \qquad (9)$$

then $F(\vartheta)$ is constant ~1. As the laser intensity in the focus is of a given laser increases with the increase in the focusing strength, it is reasonable to use the strongly focused laser wave. If the focusing strength exceeds the critical value $\vartheta_o$ then the efficiency of acceleration decreases proportionally to $(\vartheta_o/\vartheta)^2$ (for cylindrical focusing). So the focusing aperture

$$\vartheta_o = (\omega/\Omega)^{1/3} \qquad (10)$$

is the optimal one [8].

The highest laser intensity up to which the perturbation theory is valid is determined by the condition $eE\lambda = m$. Under this condition and for $\gamma = \gamma_c$ we may obtain $\delta\gamma^{(2)} \sim \gamma_o$.

Qualitatively, Eq.(14) may be treated as 'an acceleration in the cross-directed static electric and magnetic fields E and $H+H_o$ during the time $\sim \delta t \sim (\omega*\Omega^2)^{-1/3}$.

For $CO_2$ -laser ( $\lambda \sim 10^{-3}$ cm ) the condition of applicability of the perturbation theory $eE\lambda=m$ restricts the laser intensity I by $I_{max}=4*10^{14}$ W/cm$^2$. At higher intensities analytical calculations become rather difficult and the problem has been solved numerically.

## 5 THE RESULTS OF THE NUMERICAL SOLUTION OF THE PROBLEM

The numerical solution of the equations of motion has confirmed the described above qualitative description and our theoretical conclusions [ 4~8 ]. Fig. 3 shows the calculated energy increment of the electrons versus the injected electrons energy $\gamma_o$ for different values of the static magnetic field strength $H_o$. The intensity of the laser radiation ( $CO_2$-laser ) is the same($I=10^{15}$W/cm$^2$ ) for all the curves. The main feature of the curves is the strong increasing of the average energy increment when the injected electrons energy exceeds some threshold. This threshold corresponds to the condition $\gamma=\gamma_o$ (6).

In Fig.4 the average increment of the electron energy is described in its dependence on the laser intensity. The curve shows no strong saturation up to $I > 10^{16}$ W/cm$^2$, i.e. far beyond the limits of applicability of the perturbation theory. At $I \doteq 10^{16}$W/cm$^2$ $\delta\gamma \sim 1.3\ \gamma_0$.

The numerical calculations show that acceleration is accompanied by the " phase focusing " of the electrons. If incident electrons are uniformly distributed over the phases, their phase distribution near the point Z=0 has a wide peak near the light phase $\phi_0 = -\pi/4$. It should be noted that under this condition the first order acceleration [4] is maximal. This phenomenon is illustrated by the diagram at Fig.5.

The electron energy spectrum after interaction with laser light is described at Fig.6 . This diagram shows that after a single passage through the laser focus an originally mono energetic electron beam becomes broaden and shifted to the region of larger energies. The significant part of electrons are concentrated in a rather sharp peak at a high-energy wing of the spectrum. Under the conditions of Fig.6 ( $I=10^{15}$W/cm$^2$, $H_0$= 5 kG, $\varepsilon_0$= 11 MeV ) this peak has an energy shift 6 MeV and width ~ 2 MeV and it contains ~ 33% of all the electrons. Under these conditions the average acceleration of the electron beam as a whole is of about 2.2 MeV.

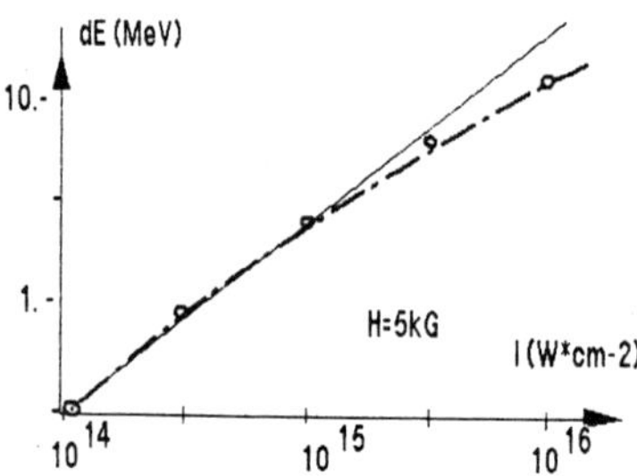


Fig.4 The average increment in the electron energy, $\delta\gamma$ versus the power density of the laser light. $\gamma_0$ =11 MeV, $H_0$= 5 kG.

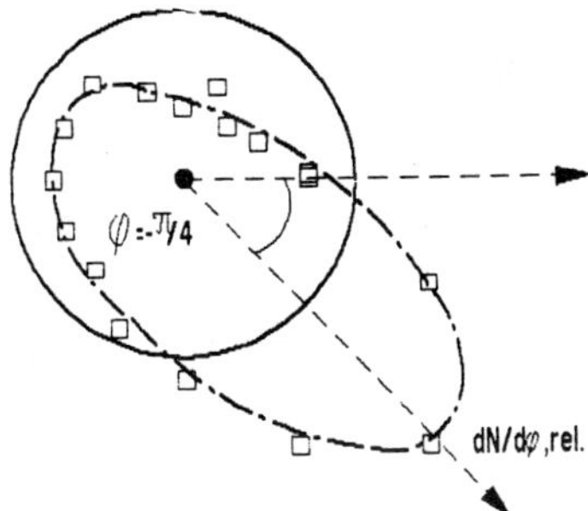


Fig.5 The electrons phase distribution at the Z=0 point. $H_o$=5 kG, $\gamma_o$=11 MeV, I=$10^{15}$W/cm$^2$. The solid line is the uniform phase distribution.

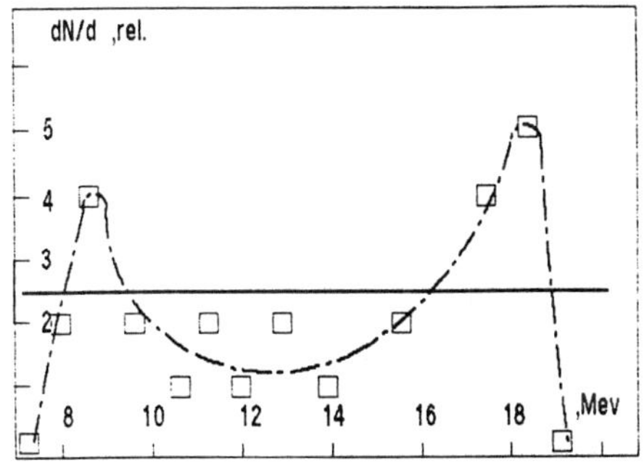


Fig.6 The accelerated electrons energy spectrum. $\gamma_o$=11 MeV, $H_o$=5 kG, I=$10^{15}$W/cm$^2$. The solid line is the uniform energy spectrum.

We guess the experimental realization of the method proposed would be easy and well reproducible. In fact to demonstrate the high efficiency of the acceleration, to see the existence of an optimal magnetic field $H_o$ and an optimal focusing conditions $\vartheta_o$, the dependence of the acceleration efficiency on the light intensity and the initial electron energy one needs nothing but a focused laser light , a magnetic field of available strength and an electron beam to be accelerated.

As the method proposed does not use any media like a plasma, no problems concerning the change of properties of accelerating medium under the action of the intense radiation arise. So this method allowes one to employ wery high intencity laser fields that are not compatible neither with the use of a plasma methods (as in the beat wave accelerator), nor with the inverse free electron laser scemese (due to their large dimentions and nonlinear effects in FEL).

REFERENCES

1. A.M.Sessler LASER ACCELERATORS . IEEE Trans. on Nucl. Sci., Vol. NS-30, No. 4, 1983, pp. 3145-3149

2. T.Katsouleas, C.Joshi, J.M.Dawson, E.F.Chen, C.Claiton, W.B.Mori, C.Darrow, D.Umstadter PLASMA ACCELERATORS. in Proc. Laser Acceleration of Particles ( Malibu, California, 1985 ), AIP No.130, pp. 66-99.

3. C.Joshi LASER ACCELERATORS. IEEE Trans. on Nucl. Sci., Vol. NS-32, No. 5, Pt. 1, pp. 1576-1581

4. V.V.Apollonov, A.I.Artemyev, Yu.L.Kalachev, A.M.Prokhorov, M.V.Fedorov ELECTRON ACCELERATION IN A STRONG LASER FIELD AND A STATIC TRANSVERSE MAGNETIC FIELD. JETP Lett., Vol. 47, No. 2, pp. 91-94 (1988)

5. V.V.Apollonov, A.I.Artemyev, Yu.L.Kalachev, A.M.Prokhorov, M.V.Fedorov . Sov. Phys. JETP, Vol. 70, N5, pp.846-852, (1990)

6 V.V.Apollonov, A.I.Artemyev, Yu.L.Kalachev, A.M.Prokhorov, M.V.Fedorov
Optical and acoustical review. Vol.1, N1, pp. 1-16. (1990)

7. V.V.Apollonov, A.I.Artem'ev, Yu.L.Kalachev, A.M.Prokhorov, A.G.Suzdaltsev, M.V.Fedorov. Zh.Eksp. i Teor.Fiz. (Soviet Series) Vol. 97,, N5, p.1498-1502. (1990)

8. V.V.Apollonov, A.I.Artem'ev, Yu.L.Kalachev, A.M.Prokhorov, M.V.Fedorov ACCElERATION OF ELECTRONS CROSSING A LASER FOCUS IN VACUUM IN THE PRESENCE OF A STATE TRANSVERSE MAGNETIC FIELD. (Unpublished)

**Research Trends in Physics: Coherent Radiation Generation and Particle Acceleration**
Editorial Board: J.M. Buzzi, A. Prokhorov (Editor-in-Chief), P. Sprangle, and K. Wille
*La Jolla International School of Physics*, The Institute for Advanced Physics Studies, La Jolla, California

# Acceleration of Electrons by a Focused Laser Radiation Combined With a Static Homogeneous Magnetic Field - A Search of an Optimum Geometry

**V.V. Apollonov, A.G. Suzdaltsev, and M.V. Fedorov**

General Physics Institute
Russian Academy of Sciences
117942 Moscow, Russia

## A B S T R A C T

A scheme of charged particles acceleration by means of a focused laser radiation in the presence of a constant magnetic field is optimized within the first order of the theory of perturbations with respect to the radiation field. The light is found to be focused with a cylindrical lens. The necessary angle between the magnetic field and the optical plane of the lens, the parameters of focusing are pointed out. Under optimum conditions the maximum energy gain is about $eE\lambda\gamma$.

 

During some last years a rather large attention is paid to the problem of acceleration of charged particles by means of intense laser radiation [1-9] . The two main groups of propositions have been done: acceleration by the electromagnetic fields generated in a laser plasma and acceleration by a focused laser radiation in the presence of some static fields from external sources. Such schemes as inverse free electron laser [4,5] (IFEL) and inverse gyrotron [7] (IG) belong to the second group. An evident disadvantage of these devices is a necessity to maintain high intensity of laser radiation within rather large volumes, along all the length of a device. The schemes offered in refs. [8,9] are free from this defect. They are based on interaction of electrons with radiation focused into a small caustic with an additional static and homogeneous magnetic field imposed. These schemes provide a sufficiently high efficiency of acceleration. They differ from the IFEL and IG both in size and in the physics of acceleration. A size of the laser focus is usually much shorter than the length of a wiggler or one of a gyrotron. In contrast with IFEL and IG the systems considered in refs.[8,9] are very far from cyclotron resonance: the Larmor frequency is assumed to be much smaller than the light frequency. However it should be noted that the conventional description of the radiation field within the caustic $E=E(r)\cos(\omega t-kr)$ used in refs. [8,9] is not perfectly correct because such a field does not satisfy the Maxwell equations. In this paper we will try to improve the theory describing the laser field by a superposition of plane waves. As it will be shown such an improved description affects essentially the final results and provides a deeper understanding of the physics of acceleration. Moreover in refs. [8,9] only two limiting cases were considered: a static homogeneous magnetic field was assumed to be either parallel or perpendicular to the laser optical axis. The present work is free from such a restriction and one of the main solved problems is a search of an optimum mutual orientation of a static magnetic field and an optical axis of the radiation. In this paper we will consider only the first order perturbation theory with respect to the electron-laser interaction. As usually [8,9] this result vanishes being averaged over the field phase or over the initial coordinates of electrons. Therefore to obtain the average nonzero gain of energy it is necessary to extend calculations upto the second order (compare with the similar problem considered in refs. [8,9]). Such a solution will be described elsewhere. But now we will confine ourselves with the first-order calculations. It should be emphasized that even these simplest calculations are nontrivial and they provide an information about the maximal energy spread of initially monoenergetic electron beam after crossing the laser focus.

Let the static magnetic field $\vec{H}$ be directed along the z axis and the rather radiation in the focus be given by a superposition of monochromatic plane waves with the same frequency and the wave vectors $\vec{k}$, $|\vec{k}|=\omega$ . Let us assume that for each of these plane waves the electric field strength lies in the $(x,z)$ plane. Introducing the spherical angles $\theta$ and $\phi$ of the wave vectors $\vec{k}$ in the frame with the z axis parallel to $\vec{H}$ we will write the electric field strength in the laser focus in the following form

$$\vec{E}(\vec{r},t)=\mathrm{Re}\int d\theta d\phi\ E_o(\theta,\phi)\ (\vec{e}_x\cos\theta\ -\vec{e}_z\sin\theta\cos\phi)\times$$

$$\times\exp\{i\omega[(x\ \cos\phi+y\ \sin\phi)\sin\theta+z\ \cos\theta-t]+i\phi_o\}, \tag{1}$$

where $\vec{e}_x$ and $\vec{e}_z$ are the unit vectors in the x- and z directions. Preexponential factor in eq. (1) is found from two following requirements: the field is polarized in the $(x,z)$ plane and all the plane waves in the superposition (1) are transversal. The function $E_o(\theta,\phi)$ defines the spatial structure of $\vec{E}(\vec{r},t)$ in the focal region.

Two opposite and the most featuring limits are the case of axially symmetrical distribution $E_o(\theta,\phi)=E_o(\theta)$ and the case of flat focusing by an infinitely long cylindrical lens

$E_o(\theta,\phi)=E_o(\theta)\delta(\phi)$. In the latter case assuming that $E_o(\theta)$ is given by the Gaussian function

$$E_o(\theta)=E_o\exp(-\theta^2/\Delta\theta^2)/(\Delta\theta\sqrt{\pi}) \tag{2}$$

with $\Delta\theta \ll 1$ we reduce eq. (1) to the well known form [10]. To find the energy gain of an electron crossing the laser focus we will use the well known formula for the work produced by the field ($|e|=-e$) [11]

$$\Delta\varepsilon = -\int_{-\infty}^{+\infty} e\vec{v}(t)\vec{E}(t,\vec{r}=\vec{r}(t))dt, \tag{3}$$

where $\vec{r}(t)$, $\vec{v}(t)$ are the electron radius-vector and velocity. Exact calculation of $\Delta\varepsilon$ (3) is hardly possible in an analytical form because it requires exact solution of equations of motion which is not known for the configuration of the fields under consideration. This is the reason why we use the perturbation theory with respect to the electron-light interaction.

In the zeroth order there is no an energy gain, $\Delta\varepsilon=0$. To calculate the first order energy gain $\Delta\varepsilon^{(1)}$ we substitute the unperturbed solutions $\vec{r}^{(o)}(t)$ and $\vec{v}^{(o)}(t)$ into the right hand side of eq. (3). $\vec{r}^{(o)}$ and $\vec{v}^{(o)}$ are the solutions of the equations of motion for the electron moving in static homogeneous magnetic field $\vec{H}$. They are well known and can be written in the form

$$\begin{cases} x^{(o)}=R\sin(\Omega t)+x_o, \\ y^{(o)}=-R\cos(\Omega t)+y_o, \\ z^{(o)}=v_{oz}t+z_o, \end{cases} \tag{4}$$

$$\begin{cases} v_x^{(o)}=R\Omega\cos(\Omega t)=v_\perp\cos(\Omega t), \\ v_y^{(o)}=R\Omega\sin(\Omega t)=v_\perp\sin(\Omega t), \\ v_z^{(o)}=v_{oz}=\text{const}. \end{cases} \tag{5}$$

Here $R=v_\perp/\Omega$ is the Larmor radius of the particle rotation in the $(x,y)$ plane, $\Omega$ is its cyclotronic frequency, $\Omega=ecH/\varepsilon_o$, $v_\perp=\left(v_x^{(o)2}+v_y^{(o)2}\right)^{1/2}$ =const is the constant (in the zeroth order) modulus of the $(x,y)$ plane projection of the particle velocity. Unperturbed energy of the particle $\varepsilon(t)$ is constant $\varepsilon^{(o)}(t)=\varepsilon_o$, where $\varepsilon_o$ is the initial electron energy; $x_o$ , $y_o$ and $z_o$ are the initial coordinates (at t=0). Let us assume that the electron is ultrarelativistic i.e. it's initial energy is much larger than $mc^2$ , where m is the electron rest-mass, $\varepsilon_o \gg mc^2$, or $\gamma=\varepsilon_o/mc^2 \gg 1$ where $\gamma$ is the relativistic Lorentz factor. Below we put c=1.

Substituting $\vec{r}^{(o)}(t)$ (4) and $\vec{v}^{(o)}(t)$ (5) into the right hand side of eq. (3) and using the definition (1) of $\vec{E}(\vec{r},t)$ (1) we obtain

$$\Delta\varepsilon^{(1)}=e\int d\theta d\phi\ E_o(\theta,\phi)\int dt\ (v_x^{(o)}\cos\theta-v_z^{(o)}\sin\theta\cos\phi)\times$$

$$\times\mathrm{Re}\ \exp\{i[\Phi_s+\omega R\ \sin(\Omega t-\phi)\sin\theta+\omega t(v_{oz}\cos\theta-1)]\}, \tag{6}$$

where $\Phi_s=\phi_o+\omega(x_o\sin\theta\cos\phi+y_o\sin\theta\sin\phi+z_o\cos\phi)$. Below we put $x_o=$ $=y_o=z_o=0$, $\Phi_s=\phi_o$. Expanding a periodic function of time t

$$\exp(i\omega R\ \sin\theta\ \sin(\Omega t-\phi))=\sum_{n=-\infty}^{+\infty} J_n(\omega R\sin\theta)\exp\{in(\Omega t-\phi)\} \tag{7}$$

into the Fourier series we reduce eq. (6) to the form

$$\Delta\varepsilon^{(1)}=e\int d\theta d\phi\ E_o(\theta,\phi)\int dt(v_\perp\cos(\Omega t)\cos\theta-v_{oz}\sin\theta\ \cos\phi)\times$$

$$\times\mathrm{Re}\sum_n J_n(\omega R\ \sin\theta)e^{i\phi_o}\ \exp\{it[\omega(v_{oz}\cos\theta-1)+n\Omega]-in\phi\}, \tag{8}$$

where $J_n$ are the Bessel functions.

If $E_o(\theta,\phi)$ does not depend on $\phi$ (a usual 3-dimensional focusing) integration over $\phi$ turns $\Delta\varepsilon^{(1)}$ into zero because of the oscillating factor $\exp(-in\phi)$.

Therefore in the first order in $E_o$ the energy gain is not equal to zero identically only if $E_o(\theta,\phi)$ is a very sharp function of $\phi$ :

$$E_o(\theta,\phi)=E_o(\theta)\delta(\phi). \tag{9}$$

As it was mentioned above expression (9) imitates an ideal flat focusing created by an infinitely long cylindrical lens. Of course in reality any cylindrical lens is finite and the distribution of $E_o$ over $\phi$ is not infinitely narrow. Applicability conditions of the model (9) are estimated below.

Within the frame of the model (9) $\Delta\varepsilon^{(1)}\neq 0$. Integration over t in eq. (8) is accomplished simply and gives rise to the

$\delta$-functions like

$$\delta(\omega(1-v_{oz}\cos\phi)-n\Omega). \tag{10}$$

Integrating $\Delta\varepsilon^{(1)}$ (8) over $\theta$ and using the well known recurrence formulas for the Bessel functions we can reduce eq. (8) for $\Delta\varepsilon^{(1)}$ to the form

$$\Delta\varepsilon^{(1)}= e\lambda\sum_{n_{min}}^{n_{max}} E_o(\theta_n)\cos\phi_o J_n(\omega R\sin\theta_n)\times \left[\frac{v_\perp \cos\theta_n}{v_{oz}\sin\theta_n}\frac{n}{\omega R\sin\theta_n}-1\right], \tag{11}$$

where $\theta_n$ are the solutions of equation

$$\omega(1-v_{oz}\cos\theta) = n\Omega . \tag{12}$$

When $\theta=\theta_n$ the argument of the $\delta$-functions (10) turns into zero. One can easily find from eq. (12) an explicit expression for $\theta_n$

$$\theta_n= \arccos\left[\frac{1}{v_{oz}}\left[1-n\frac{\Omega}{\omega}\right]\right] \approx \left[2\frac{n\Omega-\omega(1-v_{oz})}{\omega v_{oz}}\right]^{0.5} . \tag{13}$$

The latter approximate expression in eq. (13) is true when $\theta_n \ll 1$. $n_{min}$ and $n_{max}$ in eq. (11) are the lower and upper borders of the interval $[n_{min}, n_{max}]$ in which eq. (12) has a solution

$$n_{min}= \frac{\omega}{\Omega}(1-v_{oz}), \quad n_{max}= \frac{\omega}{\Omega}(1+v_{oz}). \tag{14}$$

Let's assume

$$n_{max} \gg n_{min} \gg 1 \tag{15}$$

That means that $v_{oz}$ is close to the speed of light i.e. $v_\perp \ll v_{oz} \approx 1$. In this case unperturbed trajectory is a spiral which radius is much smaller than step $D=v_{oz}/\Omega \approx \Omega^{-1}$. Equation (12) and its solution (13) can be interpreted as the condition of a resonance at the n-th harmonics of the cyclotronic frequency. Condition $n_{min} \gg 1$ means that the system is very far from the cyclotronic resonance: $\omega \gg \omega_{res}=\Omega/(1-v_{oz})$. There is no resonance also at all the lower harmonics of the cyclotronic frequency with $n<n_{min}$ . However for sufficiently large n within an interval $n_{min} \le n \le n_{max}$ at which the resonances at very high harmonics of $\Omega$ are realized. These are the resonances of very high orders which

give possibility to grow the electron energy $\Delta\varepsilon^{(1)}$.

Let's estimate the sum (11) assuming that the radiation has a Gaussian distribution on $\theta$ angle around some value $\theta_o \ll 1$ with a very small width $\Delta\theta$ ($\Delta\theta \ll \theta_o$)

$$E_o(\theta) = \frac{E_o}{\sqrt{\pi}\,\Delta\theta}\exp\left\{-\left(\frac{\theta-\theta_o}{\Delta\theta}\right)^2\right\}. \tag{16}$$

Optimum values of $\theta_o$ and $\Delta\theta$ will be found below. $\theta_o$ is the angle between the direction of the static magnetic field H (the z axis) and the optical axis of the focusing system. The distribution (16) is a generalization of the definition (2) and coincides with it at $\theta_o=0$. By the way one of the two particular situations considered earlier [8,9] corresponds to $\theta_o=0$. As it will be shown below this situation is very far from the optimum one.

It is known that at $n \gg 1$, $\xi \gg 1$ $J_n(\xi)$ reaches its maximum when the argument is about the order of the index n:

$$\xi_n \approx n, \tag{17}$$

where

$$\xi_n \equiv \omega R \sin\theta_n . \tag{18}$$

Far from these values $\xi_n$ and n, $J_n(\xi_n)$ is exponentially small. Rigorously considering eq. (17) as a precise equality $\xi_n = n$ we can find that its solution exists only if $v_{oz}=0$. This case is not considered here in particular because the condition $v_{oz}=0$ contradicts the formulated above assumption about a large interval of n $[n_{min}, n_{max}]$ where eq. (12) has its solution (13), $n_{max} \gg n_{min}$. Using the definitions of $\theta_n$ (13) and $\xi_n$ (18) it is easy to see that if $\gamma v_\perp \ll v_{oz}$ $\xi_n \ll n$ for any n and hence eq. (17) ($\xi_n \approx n$) has no solutions either. On the contrary, if

$$\frac{v_\perp}{v_{oz}} > \gamma^{-1} \tag{19}$$

such an approximate solution exists and is given by

$$n_o \approx \frac{\omega}{\Omega}\left(\frac{v_\perp}{v_o}\right)^2 \approx 2n_{min}, \tag{20}$$

where $v_o = \left(v_\perp^2 + v_{oz}^2\right)^{1/2} \approx 1$. In the latter approximate equation in (20) it is taken into account that $\gamma \gg 1$, $v_\perp \ll v_o \approx 1$. Let us introduce an angle

$$\chi = \operatorname{arctg}\frac{v_\perp}{v_{oz}} = \arcsin\frac{v_\perp}{v_o} \approx v_\perp \,, \tag{21}$$

which is a geometrical characteristic of the spiral motion in a magnetic field. Eq. (19) shows that although the angle is small it must exceed $\gamma^{-1}$ because otherwise $J_n(\xi_n)$ is exponentially small at any n. In terms of $\chi$ eq. (20) is written as

$$n_o \approx \frac{\omega}{\Omega}\chi^2. \tag{22}$$

Substituting $n_o$ (20) into the definition of $\theta_n$ (13) we obtain $\theta_{n_o} = \theta_o$. At relatively small deviations of n from $n_o$ (22) the angle $\theta_n$ (13) can be expanded in a power series in $|n-n_o|/n_o \ll 1$

$$\theta_n = \chi\left[1 + \frac{n-n_o}{n_o} + \left(\frac{n-n_o}{n_o}\right)^2 + \ldots\right]. \tag{23}$$

At large n and $\xi$, $\xi \sim n$ the Bessel $J_n(\xi)$ functions can be expressed in terms of the Airy function $\Phi$ [11]

$$J_n(\xi) \approx \frac{1}{\sqrt{\pi}}\left(\frac{2}{n}\right)^{1/3}\Phi\left[\left(\frac{2}{n}\right)^{1/3}(n-\xi)\right], \tag{24}$$

This representation is valid under the following conditions

$$\left|1-\frac{\xi}{n}\right| \ll 1, \quad n \gg 1. \tag{25}$$

Starting from the definitions of $\xi_n$ (18) and $\theta_n$ (13) it is easy to see that the Taylor expansion of $\xi_n - n$ has a form

$$\xi_n - n \approx -\frac{(n-n_o)^2}{n_o} + \ldots \,. \tag{26}$$

Using this expansion we can reduce eq. (24) to

$$J_n(\omega R \sin\theta_n) \approx \frac{1}{\sqrt{\pi}}\left(\frac{2}{n_o}\right)^{1/3}\Phi\left[\frac{(n-n_o)^2}{2^{2/3}\, n_o^{4/3}}\right]. \tag{27}$$

Using the known asymptotic of $\Phi(y)$ in the region $y \gg 1$ we find that $J_n(\xi_n)$ is very small if $|n-n_o| \gg n_o^{2/3}$. Hence $J_n(\xi_n)$ is not small in a small vicinity $\delta n$ of n close to $n_o$ (22) and the width of this region is about the order of

$$\delta n \sim n_o^{2/3}. \tag{28}$$

Taking into account that the argument of the Airy function (27) is not negative at any n (see eq. (26)) we find that in the region $|n-n_o| < \delta n \ll n_o$ $J_n(\xi_n)$ is not negative and is a monotonously decreasing function of $|n-n_o|$.

At last, remembering that $\Phi(y)\sim 1$ when $|y|<1$ we find that the maximum of $J_n(\xi_n)$ achieved at $n\approx n_o$ is about the order of

$$J_n(\xi_n)\Big|_{n\sim n_o} \sim n_o^{-1/3}. \tag{29}$$

The factor $\cos\phi_o$ in eq. (11) is determined by an initial phase of the field. In the bunch of electrons the phase $\phi_o$ is different for different electrons. In the first order the maximal effect of acceleration occurs for the particles with $\cos\phi_o=1$. This is the situation considered below.

Now we can write the energy gain $\Delta\varepsilon^{(1)}$ as

$$\Delta\varepsilon^{(1)}=e\lambda \sum_{n_{min}}^{n_{max}} E_o(\theta_n)J_n(\xi_n)\left[tg\theta_o\ ctg\theta_n \frac{n}{\xi_n} -1\right]. \tag{30}$$

Eq. (13) establishes a direct one to one correspondence between the dependencies of $J_n(\xi_n)$ on n and $\theta_n$. One can say that as a function of the angle $\theta_n$ $J_n(\xi_n)$ has its maximum at $\theta=\theta_{n_o}=\theta_o$. The width of its dependence on $\theta_n$ corresponding to the width $\delta n$ (28) of the dependence on n is equal to

$$\delta\theta \approx \frac{\theta_o}{n_o^{1/3}} \ll \theta_o . \tag{31}$$

The function summed over n in eq. (11) for $\Delta\varepsilon^{(1)}$ contains also the Gaussian function $E_o(\theta_n)$ (16) and a preexponential factor. Forgetting for a moment the preexponential factor and its role it's possible to assume that the sum (30) has its maximum if the maxima of the curves $J_n(\xi_n)$ and $E_o(\theta_n)$ (16) coincide, $\theta_o=\chi$ and if their widths are about the same order of magnitude:

$$\Delta\theta \approx \Delta\theta_{opt}=(\theta_o\Omega/\omega)^{1/3}. \tag{32}$$

Now we will try to calculate $\Delta\varepsilon^{(1)}$ (30). From the condition $n_o\gg 1$ we can conclude that $\delta n\gg 1$ also. The function summed in eq. (30) weakly changes from n to n+1 and hence the sum (30) can be replaced by the corresponding integral over n. In its dependence on n the Gaussian function $E_o(\theta_n)$ (16) at $\theta_o=\chi$ has a form

$$E_o(\theta_n)= \frac{1}{\Delta\theta\sqrt{\pi}}\exp\left[-\frac{\theta_o^2}{\Delta\theta^2}\left(\frac{n-n_o}{n_o}\right)^2\right]. \tag{33}$$

It's width is equal

$$\Delta n=\left(\frac{\Delta\theta}{\theta_o}\right)n_o . \tag{34}$$

Preexponential factor under the sign of sum in eq. (30) vanishes at $n=n_o$. Let's expand it into the Taylor series in powers of $n-n_o$

$$tg\theta_o ctg\theta_n \frac{n}{\xi_n} -1=-\frac{n-n_o}{n_o} + \frac{5}{2}\left(\frac{n-n_o}{n_o}\right)^2 + \dots . \tag{35}$$

The first term of this expansion (35) does not give its contribution to $\Delta\varepsilon^{(1)}$ (30) because whereas the functions $J_n(\xi_n)$ (27) and $E_o(\theta_n)$ (33) are even. The sum (30) is determined by the second term of the expansion (35). All the higher order terms give only small corrections because $\delta n<n_o$ and $\Delta n<n_o$. As a result eq. (30) for $\Delta\varepsilon^{(1)}$ is reduced to

$$\Delta\varepsilon^{(1)}=\frac{5}{\pi 2^{2/3}}\,\frac{eE\lambda}{\Delta\theta\; n_o^{1/3}}\int\limits_{n_o-\infty}^{n_o+\infty} dn\;\Phi\left[\frac{(n-n_o)^2}{2^{2/3}\; n_o^{4/3}}\right]\exp\left[-\frac{\theta_o^2(n-n_o)^2}{\Delta\theta^2\; n_o^2}\right]\times$$

$$\times\left(\frac{n-n_o}{n_o}\right)^2 . \tag{36}$$

In a general case analytical calculation of the integral (36) is hardly possible. It can be estimated only in two opposite limits: $\delta n\ll\Delta n$ and $\delta n\gg\Delta n$ (or, correspondingly, $\delta\theta\ll\Delta\theta$ and $\delta\theta\gg\Delta\theta$).

1) Let us assume $\delta n\ll\Delta n$, or $\delta\theta\ll\Delta\theta$ or

$$\Delta O\gg\left(\theta_o\frac{\Omega}{\omega}\right)^{1/3} \tag{37}$$

In this case the Airy function in the integrand of eq.(38) is much sharper than the exponent. The latter may be replaced by one and the integral from the Airy function multiplied by $(n-n_o)^2$ can be estimated as

$$\int dn\; n^2\;\Phi\left[\left(\frac{n}{2n_o^2}\right)^{2/3}\right]\sim\Phi_{max}(\delta n)^3\sim(\delta n)^3\sim n_o^2, \tag{38}$$

and this estimate yields

$$\Delta\varepsilon^{(1)}\sim eE\lambda\left(\frac{\Omega}{\theta_o^2\omega}\right)^{1/3}\frac{1}{\Delta\theta}. \tag{39}$$

2) Let us assume now that vice versa $\delta n\gg\Delta n$ or $\delta\theta\gg\Delta\theta$ or

$$\Delta\theta \ll \left[\theta_o \frac{\Omega}{\omega}\right]^{1/3}. \qquad (40)$$

Now the Airy function in eq. (36) is smooth compared to the exponent that makes it possible to substitute $\Phi_{max} \sim 1$ instead of $\Phi$. Residual integral is easily calculated to give

$$\Delta\varepsilon^{(1)} \sim eE\lambda \frac{(\Delta\theta)^2}{\theta_o^{5/3}} \left(\frac{\omega}{\Omega}\right)^{2/3}. \qquad (41)$$

The dependencies of $\Delta\varepsilon^{(1)}$ on $\Delta\theta$ in the described two cases are depicted on fig. 1. Decrease of $\Delta\varepsilon^{(1)}(\Delta\theta)$ at growing $\Delta\theta$ in the region (37) occurs because in this case the width $\delta n$ of the region of $n-n_o$ giving the main contribution into the integral (38) is fixed and independent of $\Delta\theta$ whereas the maximum magnitude of the field $E_o(\Delta\theta)$ in this region decreases proportionally to $1/\Delta\theta$ . One can say that with increase of the angular width $\Delta\theta$ with the fixed amplitude $E_o$ in eq. (14) the field is spread over the larger and larger number of angular components (modes) whereas the number of modes contributing effectively into the energy gain remains constant (under the condition (37)) resulting in the decrease of $\Delta\varepsilon^{(1)}$ with increasing $\Delta\theta$ .

In the region of small $\Delta\theta$ (40) decrease of $\Delta\varepsilon^{(1)}(\Delta\theta)$ with decreasing $\Delta\theta$ is explained by a small magnitude of the preexponential factor in eq. (36) and by diminishing of the region $\Delta n$ (34) which gives the main contribution into the integral (36). It should be noted that in this case some additional increase of $\Delta\varepsilon^{(1)}$ (36) could be provided by a small displacement of the maximum $\theta_o$ in the Gaussian distribution $E_o(\theta)$ (16) with respect to $\chi$ (under the restriction $|\chi-\theta_o|\leq\delta\theta$ (31)).

In this case $\Delta\varepsilon^{(1)}$ could become a little bit larger than that given by eq. (36) because of a shift of the maximum of the narrow function $E_o(\theta)$ (16) into a region where the preexponential factor in eq. (36) is not so small. However the arising integrals are rather complicated and that is why here we do not dwell upon their calculation and do not investigate in detail the described chance to provide an additional increase of the energy gain $\Delta\varepsilon^{(1)}$ in the case of small $\Delta\theta$ (40).

From the two asymptotics (39) and (41) of the curve $\Delta\varepsilon^{(1)}(\Delta\theta)$ one can find approximately the maximal energy gain (the dotted line on fig. 1) achieved in an intermediate case under the condition (32) and characterized by the magnitude

$$\Delta\varepsilon^{(1)}_{opt} \sim eE_o\lambda/\chi \; . \qquad (42)$$

Different curves on fig. 1 correspond to different values of the constant magnetic field H. When H decreases the maxima of the curves $\Delta\varepsilon^{(1)}(\Delta\theta)$ move to the region of small $\Delta\theta$ because of decreasing $\Delta\theta_{opt}$ (32). The width of the maxima ( also of the order of $\Delta\theta_{opt}$ ) are also diminishing with decreasing H. However the altitude of the maxima given by the eq. (42) is independent of H and remains constant. A limiting transition $\Delta\theta\to 0$ with H=const inevitably makes $\Delta\theta$ smallor than $\Delta\theta_{opt}$ (32). In this case $\Delta\varepsilon^{(1)}(\Delta\theta)$ is given by eq. (41) in accordance with which $\Delta\varepsilon^{(1)}(\Delta\theta)\to 0$ when $\Delta\theta\to 0$. This result agrees with the well known fact that in an ideal plane wave far from the cyclotron resonance an electron does not absorb energy of the electromagnetic field at all.

Restrictions on $\theta_o=\chi$ at which the result (42) is correct are

$$1/\gamma < \theta_o \ll 1 \; . \qquad (43)$$

On the lower boundary of this region at $\theta_o\approx 1/\gamma$ eq. (42) yields the limiting estimate of the electron energy gain

$$\Delta\varepsilon^{(1)}_{max} \sim eE_o\lambda\gamma \; . \qquad (44)$$

Eq. (44) shows that $\Delta\varepsilon^{(1)}_{max}$ is linear with respect to $\gamma$ and therefore the maximum absolute energy gain is the larger the higher the initial electron energy is. The relative energy gain is independent of $\gamma$

$$\left[\frac{\Delta\varepsilon^{(1)}}{\varepsilon_o}\right]_{max} \sim \frac{eE_o\lambda}{m}. \qquad (45)$$

As for the criterion of applicability of the used perturbation approach it can be found from the condition that the relative energy gain is small which is fulfilled as long as $eE_o\lambda < m$, i.e. up to $E\sim 10^{9}$ V/cm for $\lambda\approx 10^{-3}$ cm.

The derived result (45) seems to be independent of the constant magnetic field H and the latter could seem to be unnecessary for the considered scheme of acceleration. However in

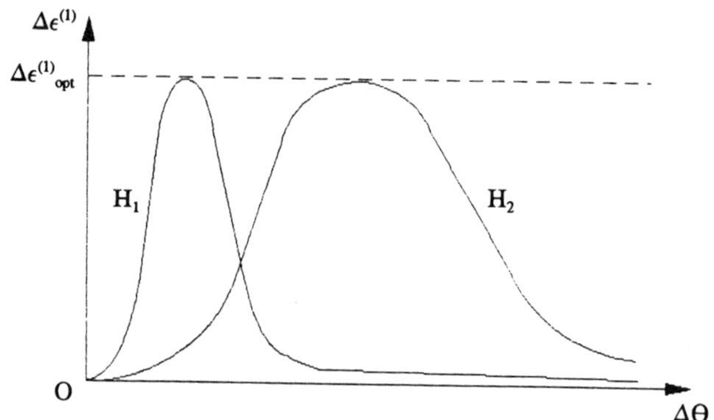


Fig. 1. The energy gain $\Delta\varepsilon^{(1)}$ versus the angular width of the laser radiation $\Delta\theta$ at different values of magnetic field: $H_1 < H_2$.

fact this conclusion is not correct: the magnetic field H enters into the definition of the optimum angular width $\Delta\theta_{opt}$ (32) at which $\Delta\varepsilon^{(1)}$ is given by eqs. (42), (44). Decreasing H (with all other parameters fixed including $\Delta\theta$) results in a decreasing $\Delta\theta_{opt}$ (32), which inevitably becomes smaller than $\Delta\theta$, $\Delta\theta \gg \Delta\theta_{opt}$. In this case $\Delta\varepsilon^{(1)}$ is given by eq. (39) in a accordance with which $\Delta\varepsilon^{(1)} \to 0$ when $H \to 0$. This result agrees with that known fact that without a static magnetic field an electron does not absorb photons from any superposition of plane electromagnetic waves, and in the first order there is no acceleration at all. Dependencies of $\Delta\varepsilon^{(1)}$ on H at different fixed values of the angular spread $\Delta\theta$ are depicted on fig. 2. When $\Delta\theta$ is diminishing the maxima of the curves $\Delta\varepsilon^{(1)}(H)$ move to the region of small H. The widths of the maxima of these curves at $\Delta\theta \to 0$ are decreasing whereas the height remains constant.

An interesting feature of the derived results is that formally the point $(H=0, \Delta\theta=0)$ on the plane $(H, \Delta\theta)$ proves to be essentially singular. $\Delta\varepsilon^{(1)}(H, \Delta\theta)$ tends to zero on any path approaching to the point $(H=0, \Delta\theta=0)$ except the curve determined by equation

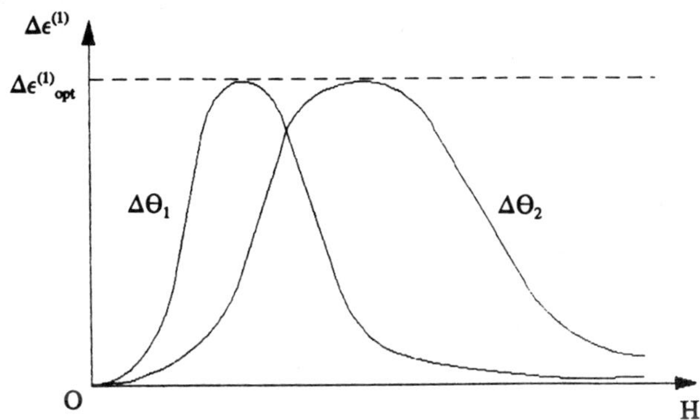


Fig. 2. The energy gain $\Delta\varepsilon^{(1)}$ versus the magnitude of magnetic field H at different values of the angular width of radiation: $\Delta\theta_1 < \Delta\theta_2$.

$\Delta\theta = \Delta\theta_{opt}(H)$ (32). On this curve $\Delta\varepsilon^{(1)} = \Delta\varepsilon^{(1)}_{opt}$ (42) and it depends neither on H nor on $\Delta\theta$, $\Delta\varepsilon^{(1)}_{opt} = const \neq 0$. This situation is reflected on fig. 3. As one can see from this picture the map of $\Delta\varepsilon^{(1)}(H, \Delta\theta)$ has a ridge of constant height $\Delta\varepsilon^{(1)}_{opt}$ (42) but of varying width. Approaching to the zero point ($H = \Delta\theta = 0$) the ridge becomes more and more narrow. But actually one should remember that the region of very small $\Delta\theta$ is nonphysical because it corresponds to an ideally plane wave, infinitely extended in space in transversal directions. Actually even without focusing the laser beam has a finite transversal size corresponding to finite values of $\Delta\theta \neq 0$. Therefore in fact the point $\Delta\theta = 0$, $H = 0$ in inachieveable and its described singularity hardly can have any practical consequences.

The optimum condition on $\Delta\theta$ (32) determines also both optimum transversal ($d_{opt}$) and longitudinal ($L_{opt}$) dimensions of the laser caustic

$$d_{opt} \sim \frac{\lambda}{\Delta\theta_{opt}} \sim \left(\frac{\lambda^2 D}{\chi}\right)^{1/3}, \tag{46}$$

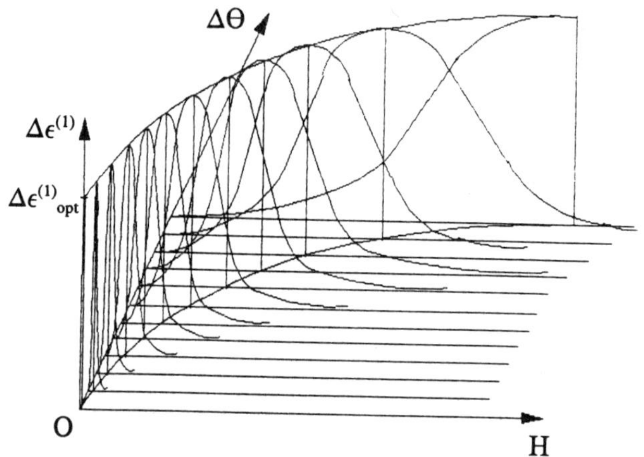


Fig. 3. The map of the energy gain $\Delta\varepsilon^{(1)}$ versus the magnetic field H and the angular width of radiation $\Delta\theta$. Figs. 1 and 2 are profiles of this surface at constant H and $\Delta\theta$ correspondingly.

$$L_{opt} \sim \frac{d^2_{opt}}{\lambda} \sim \left(\frac{\lambda D^2}{\chi^2}\right)^{1/3}, \tag{47}$$

where $D \approx 1/\Omega$ is the step of the spiral the electron moves along in the constant magnetic field.

As it was already stated the origin of the electron acceleration in our scheme is in realization of resonances at very high harmonics of the cyclotronic frequency whereas the system is very far from the resonance at the main cyclotronic frequency $\omega \gg \omega_{res} = \Omega/(1-v_{oz})$. The conditions of very high multipleness degree of the resonances and of a large number of harmonics giving contribution to the energy gain are: $n_{max} \gg n_{min} \gg 1$, where $n_{max}$ and $n_{min}$ are given by equations (14). Owing to these conditions $\theta_o \ll 1$ and $n_o \gg 1$, where $n_o$ (22) is that n at which $J_n(\xi_n)$ has its maximum. Inequality $n_o \gg 1$ excludes too small $\theta_o$

$$\theta_o \gg (\Omega/\omega)^{1/2} \sim (\lambda/D)^{1/2}. \tag{48}$$

This restriction together with eq. (47) show that under the conditions of optimum acceleration the length $L_{opt}$ of the laser focus is much shorter than the step of the spiral along which the electron moves in the field $\vec{H}$.

In other words $L_{opt}$ is much shorter than the distance traversed by the electron along the z direction during one cyclotron period $1/\Omega$. Hence the electron-light interaction is very local. In this respect the proposed scheme differs very much from the IG, where the length of interaction is assumed to be much longer than D. In terms of the Larmor radius $R=\chi D=\theta_o D$ inequality (48) takes the form $\theta_o \gg \lambda/R$ which yields

$$d_{opt} \ll R \ . \qquad (49)$$

From this inequality we conclude that the transverse dimension of the focus is much smaller than the Larmor radius R.

Such a local character of the electron-laser interaction makes it possible in principle to use quite a different way to calculate the energy gain $\Delta\varepsilon^{(1)}$. In the calculations done above we have expanded periodical functions into the Fourier series and very high harmonics have been shown to give the main contributions to $\Delta\varepsilon^{(1)}$. Corresponding characteristic time and space intervals $\sim 1/n\Omega$ are very small namely because of large n. For this reason the integrals over the time t determining $\Delta\varepsilon^{(1)}$ in eq. (6) can be calculated by means of expanding the integrand in powers of t instead of expansion in the Fourier series. This is approach used in refs. [8,9] where another geometry of the fields has been considered. An analysis of the present problem by means of expansion in powers of t as well as the correspondence between such a time evolution approach and the considered here Fourier spectrum expansion will be described elsewhere.

Remind up to now we have used in fact been considering the assumption about ideal plane focusing where the field $\vec{E}(\vec{r},t)$ is independent of one of the transverse coordinates (y). In this case $d_{opt}$ from eq. (46) is the waist of the focus in the other (x) transversal direction whereas its size in the y-direction ($d_y$) is infinitely large. However actually $d_y \neq \infty$. The question is how large must be $d_y$ and how narrow must be $E_o(\theta,\phi)$ in its dependence on $\phi$ to justify the used approximations and the derived results. From eq. (8) for $\Delta\varepsilon^{(1)}$ it's clear that the function $E_o(\theta,\phi)$ can be approximated by $\delta(\phi)$ if its width $\Delta\phi$ satisfies a condition

$$n_o\Delta\phi \ll 1, \tag{50}$$

where $n_o$ is determined by eq. (22). Under this condition the integral over $\phi$ in eq. (8) does not turn into zero. From the definition (1) of $\vec{E}(\vec{r},t)$ we find

$$d_y \sim \lambda/(\theta_o\Delta\phi). \tag{51}$$

Using the definition of $n_o$, $n_o \approx R\varkappa/\lambda$ from the condition (50) we find a searched restriction on $d_y$

$$d_y \gg R. \tag{52}$$

As it could be expected in the problem under consideration the focusing may be considered as an ideally plane one if its y-dimension is larger than the electron Larmor radius.

At last it is interesting to compare the obtained estimates for $\Delta\varepsilon^{(1)}_{opt}$ (42) and $\Delta\varepsilon^{(1)}_{max}$ (44) to those obtained in ref. [9] in the first order but for a different geometry. In our notations a geometry considered in ref. [9] corresponds to the field $\vec{H}$ directed along y-axis, $\varkappa=0$, $v_y=0$, $R \gg L$ . Such a comparison shows, the ratio $\Delta\varepsilon^{(1)}_{max}$ (44) to one of the work [9] is determined by a factor

$$\gamma\left(\frac{\Omega}{\omega}\right)^{1/3} = \left(\gamma^2\frac{eH}{m\omega}\right)^{1/3} \approx \left(\frac{\gamma}{\gamma_{cr}}\right)^{2/3}, \tag{52}$$

where $\gamma_{cr}=(m\omega/eH)^{1/2}$ is a crucial value of the relativistic Lorentz factor introduced in refs. [8,9] at which the scheme considered earlier [9] provides the most effective acceleration. At $\gamma>\gamma_{cr}$ the ratio (53) is larger than 1 and the efficiency of the present scheme is larger than one with perpendicular direction of the magnetic field [9].

Now we will give some numerical estimates. At $\lambda=10^{-3}$ cm ($\omega=2\times10^{14}$ Hz), $\gamma=30$, H=20 kG ($\Omega=10^{10}$ Hz), $\theta_o=0,1$, $E=10^8$ V/cm we get $\Delta\theta_{opt}\sim10^{-2}$, D=0.4 cm, $n_o\cong200$, R=0.04 cm, $d_{opt}\approx1.3\times10^{-2}$ cm, $L_{opt}\approx0.22$ cm, $d_y=0.1$ cm, $\Delta\varepsilon_{opt}\approx0.03\varepsilon_o\approx mc^2=0.5$ meV. The effective rate of acceleration is $\Delta\varepsilon_{opt}/L_{opt}\sim2.5$ MeV/cm. The required laser intensity is $I=2\times10^{13}$ W/cm and its power is $P=3\times10^{10}$ W i.e. for example laser producing 100 ps pulses with the energy about 3 J in each pulse can be used.

Note this situation is realistic in the experimental sence but is far below the perturbation limit. Increasing the intensity we will obtain growth of the particle energy gain as a square root of the intensity.

Let's formulate once more the optimum conditions of the acceleration scheme described. First the laser radiation with linear polarization is to be focused with a cylindrical lens the "optical plane" of the lens being perpendicular to the polarization plane. Second the small angle $\theta_0$ between the directions of propagation of the laser radiation (i.e. optical plane of the lens) and of the static magnetic field H is to be approximately equal to the angle $\chi$ defining the small ratio of the transverse and longitudinal components of the electron velocity in the magnetic field, $\chi$ must be about few units of $\gamma^{-1}$. Third the magnitude of the magnetic field is to be as large as it is possible to provide the most efficient use of the radiation energy. Forth for given energy $\gamma$, wavelength $\lambda$ and magnetic field H the focusing properties of the lens ($\Delta\theta_{opt}$, $d_{opt}$, $d_y$, $L_{opt}$) are to obey the appointed equations. Or vice versa for a given lens and given $\gamma$ and $\lambda$ the magnetic field $H\sim\Omega$ is to satisfy the optimum condition (32).

From the given above analysis we can make a conclusion that the offered scheme can provide a rather efficient acceleration of electrons. If the optimum configuration is held the energy gain is proportional to the initial energy that makes it possible to hope to use the given method for particles with a rather large initial energy. The found optimum configuration seems to be realizable experimentally and the demands to the laser energetic performance are quite acceptable. To investigate the possibility of repeated acceleration by means of consequent pulses of a train of laser pulses it is necessary to find the energy gain averaged over the initial phase $\phi_0$. Such a problem is to be solved within the second order of the perturbational theory that will be described elsewhere.

## REFERENCES

1. Tajima T., Dawson J.M., 1979, Phys. Rev. lett., 43, 267.
2. Clayton C.E., Joshi C., Darrow C., Umstudter D., 1985, Phys. Rev. lett., 54, 2343.
3. Katsouleas T., Dawson J.M.,1983, Phys. Rev. lett. 51, 392.
4. Joshi C.,1985, IEEE Trans. Nucl. Sci. NS-32, 1576.
5. Varfolomeev A.A., Lachin Yu.Yu.,1986, Zh. Eksp. Theor. Fiz., 56, 2122.
6. Apollonov V.V., Kalachev Yu.L., Prokhorov A.M., Fedorov M.V., 1986, Pis´ma Zh. Eksp. Theor. Fiz., 44, 61.
7. Caspers F., 1989, CERN PS/OP/Note 89-17, 10.04.89.
8. Apollonov V.V., Artemyev A.I., Kalachev Yu.L., Suzdaltsev A.G., Fedorov M.V.,1990, Opt. Acoust. Rev. 1, 1.
9. Apollonov V.V., Artemyev A.I., Kalachev Yu.L., Suzdaltsev A.G., Fedorov M.V., 1990, Zh. Eksp. Theor. Fiz., 97, 1498.
10. Karlov N.V., 1983, Lectures on Quantum Electronics, (Moscow, Nauka).
11. Landau L.D., Lifshits E.M., 1988, Theory of Field (Moscow, Nauka).

**Research Trends in Physics: Coherent Radiation Generation and Particle Acceleration**

*La Jolla International School of Physics*, The Institute for Advanced Physics Studies, La Jolla, California

# Excitation of Nonlinear Wake Field in a Plasma for Particle Acceleration

**B.N. Breizman, T. Tajima, D.L. Fisher, and P.Z. Chebotaev**

University of Texas at Austin
Institute for Fusion Studies
Austin, TX 78712

and

Institute of Nuclear Physics
630090 Novosibirsk, Russia

## ABSTRACT

Excitation of large amplitude wake fields in a plasma for acceleration of particles is theoretically and computationally considered. The wake electric field can be generated either by short laser pulses or charged particle (electrons) beam pulses. We treat both cases from a unified point of view and compare them. In two (or three) dimensional investigations, the wake causing agencies are treated as rigid, while in the one dimensional cases the feedback of the wake field on the driving pulse is accounted for fully kinetically and relativistically. We elucidate transverse and longitudinal wavebreaking effects, nonlinear wake field effects, pulse shaping, multiple pulses, the coherency length of wake fields and comparison of laser and electron beam pulses.

## INTRODUCTION

A promising way of achieving super high accelerating gradients for future linear colliders is the use of a plasma-based accelerator in which a plasma wave is driven either by a short laser pulse or a relativistic electron beam. These schemes are known now as the laser wake field accelerator (LWFA) and the plasma wake field accelerator (PWFA). Some of the early ideas on this have been introduced by Veksler in 1956 (see Ref. 1). In Ref. 2 a laser pulse was introduced to induce a wake, as the laser pulse is capable of creating a fast phase velocity. Some later papers went back to the original charged particle ideas. The electron beam version of PWFA was proposed in the theoretical paper,[3,4] and its basic principles have also been tested experimentally.[5] Recently, an enhanced accelerating gradient has been obtained in a new experiment,[6] and there are at least two more experiments[7,8] proposed with substantially increased parameters.

 

To address an experimental situation, theory of either the LWFA or the PWFA is desirable to be both three dimensional and nonlinear. Three dimensional effects are especially important when the wake field is excited by an electron beam, since the transverse dimensions of the driving beam are typically less than its length. In addition, the self fields of an ultrarelativistic electron beam have predominantly transverse components. The nonlinearity of a plasma is also an important issue in the whole problem, since there is a nonlinear limitation on the amplitude of the plasma wave excited by the beam. Though the above arguments had long been commonly accepted, the theory developed so far has been often limited by using either a linear approximation or a one-dimensional model. The two objectives of this paper are to develope a simple fluid model for analytical and numerical study of the nonlinear dynamics of a three dimensional wake field and to present related fully self-consistent computer simulations for a one dimensional kinetic model.

When solving the general wake field problem, one has to take into account both the nonlinearity of the plasma and the deformation of the driver. However, in the ultrarelativistic limit the driver may be assumed to be "rigid", i.e. the back effect of the excited field on the driver can be neglected (at least as a first approximation).

The nonlinear limitation on the accelerating gradient stems from Langmuir wave breaking. As it was shown for the one-dimensional case in Ref. 9, the wave breaking is characterized by a finite threshold amplitude of the electric field. In contrast to this, the wave will always eventually break even for small amplitude waves in the three-dimensional case (see Ref. 9). With decreasing wave amplitude the wave breaking is delayed in time, but it does not disappear. In the context of the problem, it means that that the length of the "laminar wake", where the field structure is suitable for particle acceleration, depends on the accelerating gradient. As the gradient increases, the length become shorter, hence, the question arises as how large a gradient can be achieved with a acceptable length of the laminar wake.

This particular question may be addressed within the fluid description of plasma electrons on the background of immobile plasma ions. As the velocity of the driver is very close to the velocity of light, all the fields and currents in a plasma are assumed to depend on the longitudinal coordinate $z$ and time $t$ in a combination $z - ct$, like in a travelling wave. This assumption is consistent with the fact that the plasma is unperturbed far ahead the leading edge of the driver because the group velocities of both electromagnetic and Langmuir waves are less than the driver velocity. An additional assumption is an axial symmetry of the problem.

In the three dimensional model the fluid approximation is used and the driver acts on the plasma; however, there is no feedback on the driver. This "rigid" driver approximation which allows an important three dimensional analysis is valid over short distances. The one dimensional kinetic model gives the self-consistent interaction of the driver with the plasma at the cost of reduction in dimensions. This simulation can test the rigidity approximation

The fluid model that we use reduces a physically three dimensional problem to a mathematically one dimensional one for which a numerical solution can be obtained easily. A simple fluid code which solves this problem can be further combined with a much more sophisticated kinetic code to describe the dynamics of the driving beam on a time scale determined by the distortion of the driver. This second time scale is much longer than the period of plasma wave. The combined code will certainly be less time consuming than a fully kinetic code.

## BASIC EQUATIONS

### Electron Driver

Derivation for an electron beam driver will be initially investigated, with modifications then given for a laser driver. A set of equations to be solved includes a relativistic equation of motion for plasma electrons

$$\frac{\partial \mathbf{p}}{\partial t} + (\mathbf{v}\nabla)\mathbf{p} = -e\left(\mathbf{E} + \frac{1}{c}\left[\mathbf{v} \times \mathbf{H}\right]\right) , \tag{1}$$

wherein $\mathbf{v} = c\mathbf{p}/\sqrt{p^2 + m^2c^2}$, a continuity equation

$$\frac{\partial n}{\partial t} + \operatorname{div}(n\mathbf{v}) = 0 \tag{2}$$

and Maxwell equations:

$$\operatorname{rot}\mathbf{H} = \frac{4\pi}{c}\mathbf{j} + \frac{1}{c}\frac{\partial \mathbf{E}}{\partial t} , \tag{3}$$

$$\operatorname{rot}\mathbf{E} = -\frac{1}{c}\frac{\partial \mathbf{H}}{\partial t} . \tag{4}$$

A total current $\mathbf{j}$ in Eq. (3) is the sum of a plasma electron current

$$\mathbf{j}_p = -en\mathbf{v} \tag{5}$$

and a current of a driving beam $\mathbf{j}_b$. For a rigid axisymmeric beam, $\mathbf{j}_b$ is directed along the $z$-axis of a cylindrical coordinate system $(r, \varphi, z)$ and is a given function of $r$ and $t - z/c$. For the particular computational runs we take a gaussian radial profile of the beam. It is convenient to rewrite Eqs. (1)-(4) in a dimensionless form by introducing new (dimensionless) variables that are marked by primes and defined by the following transformations:

$$t \Rightarrow \sqrt{\frac{m}{(4\pi n_0 \theta^2)}}\, t' \ ; \ \mathbf{r} \Rightarrow \sqrt{\frac{mc^2}{(4\pi n_0 e^2)}}\, \mathbf{r}' \ ; \ \mathbf{p} \Rightarrow mc\mathbf{p}' \ ; \ \mathbf{v} \Rightarrow c\mathbf{v}' \ ;$$

$$n \Rightarrow n_0 n' \ ; \ \mathbf{j} \Rightarrow -en_0 c\mathbf{j}; \ ; \ \mathbf{E} \Rightarrow -\sqrt{4\pi n_0 mc^2}\,\mathbf{E} \ ; \ \mathbf{H} \Rightarrow -\sqrt{4\pi n_0 mc^2}\,\mathbf{H}' \ . \tag{6}$$

Here, $n_0$ is the unperturbed plasma density and the fields are normalized to the wavebreaking value $E_0 \equiv m\omega_p c/e$. In what follows we omit the prime at a dimensional variable. By assuming an axisymmetric travelling wave dependence $(r;\ t - z/c)$ for all the functions we eliminate $z$-derivatives from Eqs. (1)–(4). We also introduce new unknown functions $V_z$, $V_r$ and $N$ that are related to $v_z$, $v_r$ and $n$ by equations

$$V_z \equiv \frac{v_z}{1 - v_z}\,, \quad V_r \equiv \frac{v_r}{1 - v_z}\,, \quad N \equiv n(1 - v_z)\,.$$

Finally, Eqs. (1)–(4) take the form

$$\frac{\partial}{\partial t} V_r + V_r \frac{\partial}{\partial r} V_r = \sqrt{1 + 2V_z - V_r^2}\,\left[(1 + V_z - V_r^2)(E_r - H) + H + V_r E_z\right]\,, \tag{7}$$

$$\frac{\partial}{\partial t} V_z + V_r \frac{\partial}{\partial r} V_z = \sqrt{1 + 2V_z - V_r^2}\,\left[(1 + 2V_z)E_z + V_r H - V_r V_z (E_r - H)\right]\,, \tag{8}$$

$$\frac{\partial}{\partial t} N + \frac{1}{r}\frac{\partial}{\partial r}(rNV_r) = 0\,, \tag{9}$$

$$\frac{\partial}{\partial t}(E_r - H) = -NV_r\,, \tag{10}$$

$$\frac{\partial}{\partial r} E_z = NV_r\,, \tag{11}$$

$$\frac{\partial}{\partial r}\frac{1}{r}\frac{\partial}{\partial r}(rH) = \frac{\partial}{\partial t}(NV_r) + \frac{\partial}{\partial r}(NV_z) + \frac{\partial}{\partial r} j_b\,. \tag{12}$$

For the driving beam current, $j_b$, in Eq. (12) we choose

$$j_b = j_0 \begin{cases} \exp\left(-\dfrac{r^2}{2\sigma_r^2} - \dfrac{t^2}{2\sigma_-^2}\right)\,; & t \le 0 \\[2ex] \exp\left(-\dfrac{r^2}{2\sigma_r^2} - \dfrac{t^2}{2\sigma_+^2}\right)\,; & t \ge 0 \end{cases} \tag{13}$$

with the parameters $j_0, \sigma_r, \sigma_-$, and $\sigma_+$ that determine a beam-to-plasma density ratio ($j_0$), beam radius ($\sigma_r$) and the widths of the beam leading edge ($\sigma_-$) and trailing edge ($\sigma_+$) measured in units of the plasma collisionless skin depth.

By analyzing the solution of Eqs. (7)–(12) one can, in principle, find the optimum values for the parameters $j_0, \sigma_r, \sigma_-$, and $\sigma_+$ to maximize the plasma wake field. Since the complete analysis of the solution is rather difficult we will restrict ourselves with a more practical problem of outlining a relevant range of parameters for creating experimentally the wake field of the order of hundreds GeV/m.

### Laser Driver

We will model the laser pulse in the plasma by the ponderomotive force it produces, otherwise the electromagnetic fields of the laser pulse are ignored. This gives the "rigid" laser pulse with all the fields in the basic equations due only to the plasma. Effects of the feedback of wakefields on the laser pulse have been treated by a fluid approach, including the pump depletion in Ref. 10. A laser pulse with a radial profile will produce a radial and axial ponderomotive force

$$\mathbf{F}_p = -\frac{m}{2}\nabla\left(v_{\rm osc}^2\right) , \tag{14}$$

where $v_{\rm osc} = eE_{\rm laser}/m\omega_{\rm laser}$ is an electron oscillatory velocity in the laser field. A relativistic generalization can be done.

This force is to be included in the equation of motion for plasma electrons

$$\frac{\partial \mathbf{p}}{\partial t} + (\mathbf{v}\nabla)\mathbf{p} = -e\left(\mathbf{E} + \frac{1}{c}[\mathbf{v}\times\mathbf{H}]\right) + \mathbf{F}_p \tag{15}$$

where the fields are only those due to the plasma.

The normalized force becomes

$$\mathbf{F} \Rightarrow \sqrt{4\pi n_0 mc^2 e^2 \mathbf{F}} , \tag{16}$$

Then the dimensionless equations (7) and (8) become

$$\begin{aligned}\frac{\partial}{\partial t}V_r + V_r\frac{\partial}{\partial r}V_r = \sqrt{1+2V_z-V_r^2}\,\big[&(1+V_z-V_r^2)(E_r-H+F_r)\\ &+H+V_r(E_z+F_z)\big] ,\end{aligned} \tag{17}$$

$$\begin{aligned}\frac{\partial}{\partial t}V_z + V_r\frac{\partial}{\partial r}V_z = \sqrt{1+2V_z-V_r^2}\,\big[&(1+2V_z)(E_z+F_z)+V_rH\\ &-V_rV_z(E_r+H+F_r)\big] .\end{aligned} \tag{18}$$

The rest of the dimensionless equations (9)–(12) are as before, except that the beam current $j_b$ is equal to zero.

## LINEAR THEORY

### Electron driver

A sufficient applicability condition for the linear theory is given by inequality $j_0 \ll 1$ for the electron beam or $F \ll 1$ for the laser pulse. With this condition being satisfied one can put $N = 1$ in Eqs. (10)–(12) to split off Eq. (9) and to rewrite the remaining equations in the form

$$\frac{\partial}{\partial t} V_r = E_r \ , \tag{19}$$

$$\frac{\partial}{\partial t} V_z = E_z \ , \tag{20}$$

$$\frac{\partial}{\partial r} E_z = V_r \ , \tag{21}$$

$$\frac{\partial}{\partial t} (E_r - H) = -V_r \ , \tag{22}$$

$$\frac{1}{r} \partial r \, (rH) = \frac{\partial}{\partial t} E_z + V_z + j_b \ . \tag{23}$$

We now substitute $E_r$, $E_z$ and $V_r$ in Eqs. (22) and (23) from Eqs. (19)–(21) to obtain

$$H = \frac{\partial}{\partial r} \left( \frac{\partial^2}{\partial t^2} V_z + V_z \right) \tag{24}$$

$$\frac{1}{r} \frac{\partial}{\partial r} (rH) = \frac{\partial^2}{\partial t^2} V_z + V_z + j_b \ . \tag{25}$$

It follows from Eqs. (24) and (25) that a vector potential $A$ given by

$$\frac{\partial^2}{\partial t^2} V_z + V_z = A \tag{26}$$

satisfies the equation

$$\frac{1}{r} \frac{\partial}{\partial r} r \frac{\partial}{\partial r} A - A = j_b \ , \tag{27}$$

After solving this equation the longitudinal electric field $E_z$ can be found from Eqs. (20) and (26):

$$E_z(r;t) = \int_{-\infty}^{t} A(r;t') \cos(t - t') dt' \ . \tag{28}$$

Equation (28) gives a sinusoidal wake field behind a driving beam

$$E_z(r;t)]E \cos(t + \psi) \tag{29}$$

with an amplitude $E$ that depends on radius as

$$E = \sqrt{\left( \int_{-\infty}^{+\infty} A(r;t') \cos(t') dt' \right)^2 + \left( \int_{-\infty}^{+\infty} A(r;t') \sin(t') dt' \right)^2} \ . \tag{30}$$

For the current given by Eq. (13) the amplitude takes the form

$$E = j_0 R(r;\sigma_r)\sqrt{(Z_1(\sigma_-;\sigma_+))^2 + (Z_2(\sigma_-;\sigma_+))^2}\ , \tag{31}$$

wherein the following notations are introduced

$$R(r;\sigma_r) \equiv \left|\sigma_r^2 \int_0^\infty \frac{1}{k^2+1}\exp\left(\frac{k^2\sigma_r^2}{2}\right) J_0(kr)k\,kdk\right|\ . \tag{32}$$

$$Z_1(\sigma_-;\sigma_+) \equiv \sqrt{\frac{\pi}{2}}\left(\sigma_-\exp\left(-\frac{\sigma_-^2}{2}\right) + \sigma_+\exp\left(-\frac{\sigma_+^2}{2}\right)\right)\ , \tag{33}$$

$$Z_2(\sigma_-;\sigma_+) = \sigma_{-1}^2 F_1\left(1;\frac{3}{2};-\frac{\sigma^2}{2}\right) - \sigma_{+1}^2 F_1\left(1;\frac{3}{2};-\frac{\sigma_+^2}{2}\right)\ . \tag{34}$$

Here $J_0$ is a Bessel function and ${}_1F_1$ is a confluent hypergeometric function.

It follows from Eqs. (31) and (32) that the electric field $E$ as a function of r reaches a maximum value on the beam axis. The dependence of $E(0)$ on $\sigma_r$ is characterized by a monotonic function $R(0;\sigma_r)$ that is shown in Fig. 1. Since it is typical for an electron beam that both $\sigma_-$ and $\sigma_+$ are larger than $\sigma_r$, we restrict $\sigma_r$ by a condition

$$\sigma_r \leq \min(\sigma_-;\sigma_+)\ . \tag{35}$$

To maximize the wake field under this condition we will assume in what follows that

$$\sigma_r = \min(\sigma_-;\sigma_+)\ . \tag{36}$$

As Eq. (31) is symmetric with respect to $\sigma_-$ and $\sigma_+$, it is sufficient for our purpose to consider the case $\sigma_+ \leq \sigma_-$. Shown in Fig. 2 is the dependence of the wake field on $\sigma_+$ with $\sigma_+$ given by Eq. (36) for the symmetric beam ($\sigma_- = \sigma_+$) and for the beam with a very long leading edge ($\sigma_- \Rightarrow \infty$). The positions of the maxima are almost the same in both cases ($\sigma_+ = \sigma_r \approx 1.3$), while the maximum values of $E$ differ by a factor of 2. Namely, $E_{max} \approx 0.8 j_0$ for the symmetric beam and $E_{max} \approx 0.4 j_0$ for the asymmetric one. There is a simple qualitative argument for the existence of the maxima in Fig. 2. At small values of $\sigma_+$ the wake field is less than the maximum value because of a smaller number of particles in a driving beam. At a large values of $\sigma_+$ the field decreases because the beam switches on adiabatically so that the beam space charge can be neutralized by the plasma. By extrapolating the results of linear theory up to $j_0 = 0.5$ we would obtain a dimensionless wakefield to be 0.4 for the symmetric beam and 0.2 for the asymmetric beam.

With the normalization (6) we have for the above quantities in their dimensional form $j_0 = 0.5\,enc$, $E_{max} = 0.4\,m\omega_p c/e$, and $E_{max} = 0.2 m\omega_p c/e$.

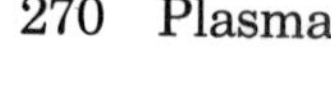


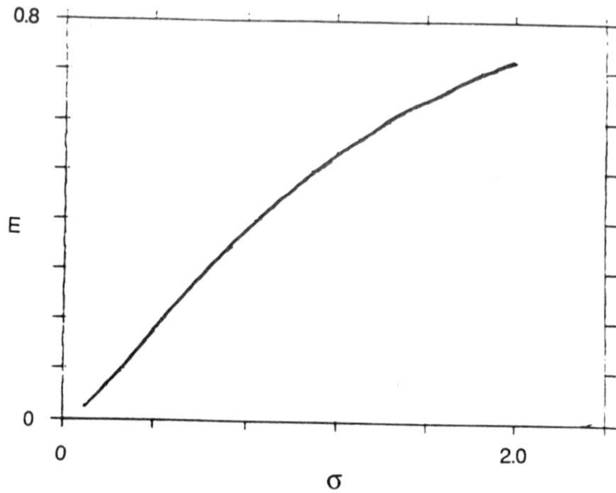


1. From equations (26) and (27) the dependence of the normalized wake field $E$ at $r = 0$ as a function of given by $R(0, \sigma_r)$. $E$ is normalized to $m\omega_p c/e$ and $\sigma$ to $c/\omega_p$.

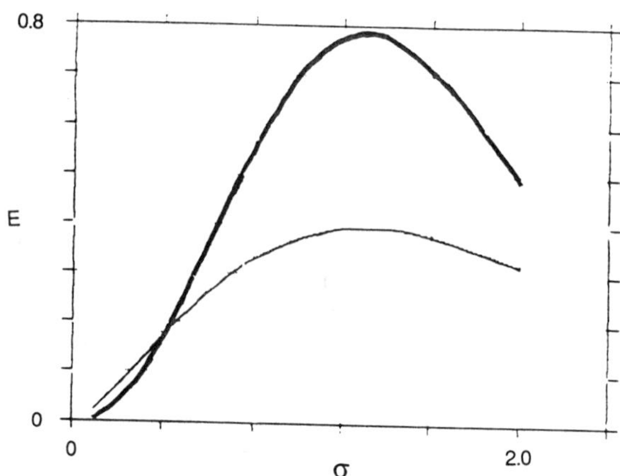


2. From equation (31) the dependence of the normalized wake field on normalized $\sigma_r$ for a symmetric beam with $\sigma_- = \sigma_+$ (thick line) and for a beam with $\sigma_- \to \infty$ (thin line).

At a plasma density $10^{15}\,\mathrm{cm}^{-3}$ these numbers correspond to 1200 MeV/m and 600 MeV/m respectively. An optimal beam radius is equal to 0.22 mm for this density. The above values determine the characteristic scales for the parameters in our nonlinear numerical simulations.

Laser driver

In the linear approximation for the laser pulse Eqs. (17) and (18) become

$$\frac{\partial}{\partial t} V_r = E_r + F_r \ , \tag{37}$$

$$\frac{\partial}{\partial t} V_z = E_z + F_z \ . \tag{38}$$

We now define the electric potential $\psi$ and the ponderomotive potential $\phi$ such that

$$E_r = \frac{\partial \psi}{\partial r} \ , \qquad E_z = \frac{\partial \psi}{\partial t} \ , \tag{39}$$

$$F_r = \frac{\partial \phi}{\partial r} \ , \qquad F_z = -\frac{\partial \phi}{\partial t} \ . \tag{40}$$

Since the force is derivable from a potential, there is no magnetic field in the linear approximation. From these and the original equations we get

$$\frac{\partial^2}{\partial t^2} \psi + \psi = \phi \ . \tag{41}$$

The solution for Eq. (41) is

$$\psi = \int_{-\infty}^{t} \phi(t'; r) \sin(t - t') dt' \ . \tag{42}$$

The amplitude of the potential $\psi$ depends on the ponderomotive potential. Assuming $f(t)$ to be an even function of $t$, we obtain

$$\mathrm{ampl} = \max \int_{-\infty}^{\infty} \phi(t'; r) \cos(t') dt' \ . \tag{43}$$

For a ponderomotive potential with a gaussian radial and longitudinal profiles

$$\phi = C \exp\left(\frac{-t^2}{2\sigma_z^2}\right) \exp\left(\frac{-r^2}{2\sigma_r^2}\right) \tag{44}$$

the amplitude becomes

$$\mathrm{ampl} = \sqrt{2\pi}\, \sigma_z C \exp\left(-\frac{\sigma_z^2}{2}\right) \ . \tag{45}$$

The calculation of the force from this form of the laser pulse is valid only for nonrelativistic velocities of the plasma electrons. If the amplitude of the force is such that either the velocity of the electrons due to the electric field of the laser pulse or the velocity of the electrons in the plasma wave becomes comparable to the speed of light, then the force given above will not come from the gaussian form of the laser pulse. The corrections to the nonrelatistivic ponderomotive force have not been considered here.

## COMPUTATIONAL RESULTS

Equations (7) to (12) from the 2-D problem with a rigid driver were solved numerically with a fluid code. We also used a fully relativistic 1-D kinetic code (see e.g. Ref. 11) to study the effects of particle velocity spread which is important for studying wave breaking. In this section we present the numerical results for both electron and laser drivers with various longitudinal profiles and intensities. The radial profile of the driver for all 2-D runs is chosen to be gaussian with $\sigma_r = \min(\sigma_-;\sigma_+)$ for the electron driver and $\sigma_r = \sigma_z$ for the laser driver. For the longitudinal gaussian half width $\sigma$ we used the optimal value from linear theory of 1.3 $c/\omega_p$ for the electron pulse and 1 $c/\omega_p$ for the laser pulse. In the case of a rigid driver relativistic factor$\gamma$ was not specified ($v = c$); for the one-dimensional kinetic case we put $\gamma = 20000$.

Initially we shall look at electron drivers starting with the single pulse driver. In Figs. 3-5 we present the results for various densities of the driving beam. In each of the figures the radial profiles of the wake field and focusing force correspond to the right end point of the longitudinal profiles. Fig. 3 corresponds to a beam peak density which is 20% of the plasma electron density. Although the plasma wave look linear (sinusoidal), it is actually in the weakly nonlinear regime as is evidenced by the form of the focusing force. Since we are dealing with accelerating relativistic particles, the focusing force is defined as the radial electric field minus the magnetic field. In the linear region, the focusing force's sign is always opposite the slope of the wakefield, however, in Fig. 3b we see that the focusing force becomes positive after about 2.4 radii. The radial profile of the longitudinal field in Fig. 3 has a gaussian-like form as expected in the linear regime.

Next we look at a beam density that is 40% that of the plasma density (Fig. 4). There are now indications of nonlinearity in both the longitudinal and radial structure. The longitudinal field now has a steepened trailing edge and doubling the current from 20% to 40% increased the maximum amplitude by more than a factor of two. The amplitude also drops off due to nonlinear phase mixing of the harmonics. The structure of the focusing force is definitely nonlinear showing regions of large focusing and defocusing. Also the radial profile of the longitudinal wakefield has a blunted top and steep dropoff after about one radius due to nonlinearity.

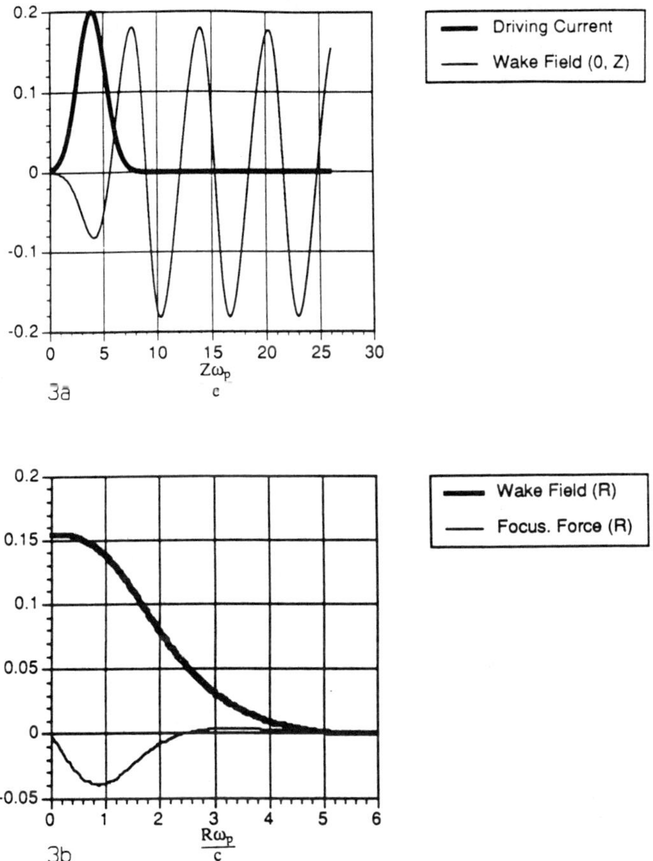


3. Single gaussian electron beam, 2D fluid code. 3a — normalized beam current and normalized wake field for a beam peak current of $0.2\,nec$. 3b — normalized radial profiles of the wake field and the focusing force (radial electric field minus magnetic field) at the right end point of the wake field longitudinal profile. The wake field and the focusing force are normalized to $m\omega_p c/e$.

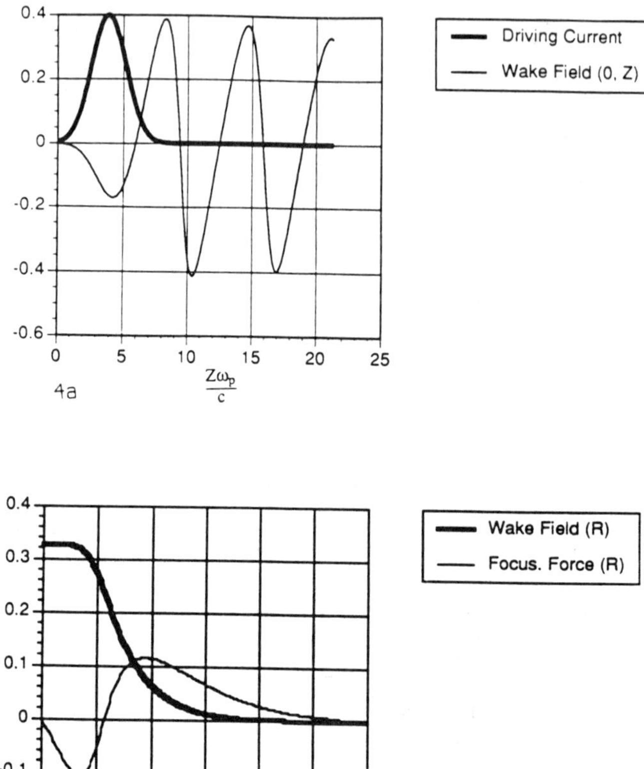


4. Same as Fig. 3 for beam current of 0.4.

The last 2-D single gaussian bunch (Fig. 5) has a beam density that is 60% that of the plasma density. The longitudinal wakefield has a steep trailing edge with a decreasing amplitude due again to nonlinearity. Although the longitudinal field looks nonlinear it does not appear to be close to wave breaking. However, if we look at the radial profile there is a very sharp change in the focusing force and in the radial profile of the longitudinal field at about three-fourths of a radii. The wave is on the verge of wave breaking in the radial direction.

Now we look at the velocity distributions with the one dimensional kinetic code. Since the 1-D code has an infinite radius, we expect the maximum wakefield amplitude to be larger than that for a finite beam radius of the same beam to plasma density ratio. Fig. 6 shows the phase space and plasma wakefield for a beam density of 20%. The initial distortion in the first 10 $c/\omega_p$ is due to startup and is to be neglected since it is unphysical. We see that the density perturbation is slightly peaked showing a weak nonlinearity. However, there is no wavebreaking and heating of the plasma and the plasma wave has a sinusoidal structure with no decrease in amplitude. The peak amplitude of the longitudinal oscillation is as expected larger than that for the 2-D case with a finite radius.

Fig. 7 shows a very nonlinear plasma oscillation which almost immediately starts to break. The beam density is 60% that of the plasma. In the phase space plot we see a very pointed top and a heating of plasma electrons. The electrons with large momentum are trapped by the wake field and are accelerated by the longitudinal field. The initial amplitude of 1.4 mwpc/e is above the nonrelativistic wave breaking limit of $1.0 m\omega_p c/e$.

Figure 8 shows the 2-D results for a slowly ramped pulse composed of two half gaussian, one with a wide width and the other with a sharp drop. The maximum current is 60% with a half width of about 17 plasma wavelengths. This shape gives an enhancement by a factor of 2 over the linear theoretical prediction of $0.24 m\omega_p c/e$ for the peak amplitude of the wake field. We see from Fig. 8c that inside the pulse the focusing force is always negative and the longitudinal field is very low. In the wake field there is the usual nonlinear flattening of the radial profile of the wakefield and the nonlinear distortion of the focusing force.

Figure 9 shows the 2-D results of 5 bunches all spaced one plasma wavelength apart. The beam density of each bunch is 10% that of the plasma. This shows that following bunch amplifies the pulse in a linear fashion, the peak amplitude growing linearly with the number of bunches which have amplified the wave. After the final bunch the amplitude reduces somewhat as the wave is now in the nonlinear regime. The radial profile also shows as before the nonlinear behavior of distorting the focusing force and radial profile of the longitudinal electric field. Careful inspection of the wakefield shows a progression from the linear regime of a sinusoidal wake field to that of the nonlinear wave with a steepened profile.

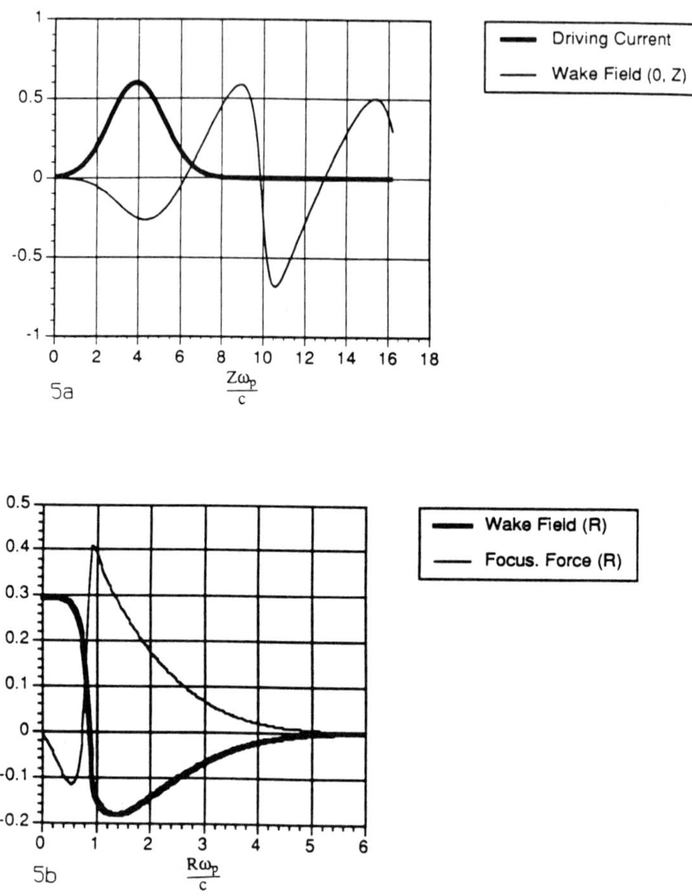


5. Same as Fig. 3 for beam current of 0.6.

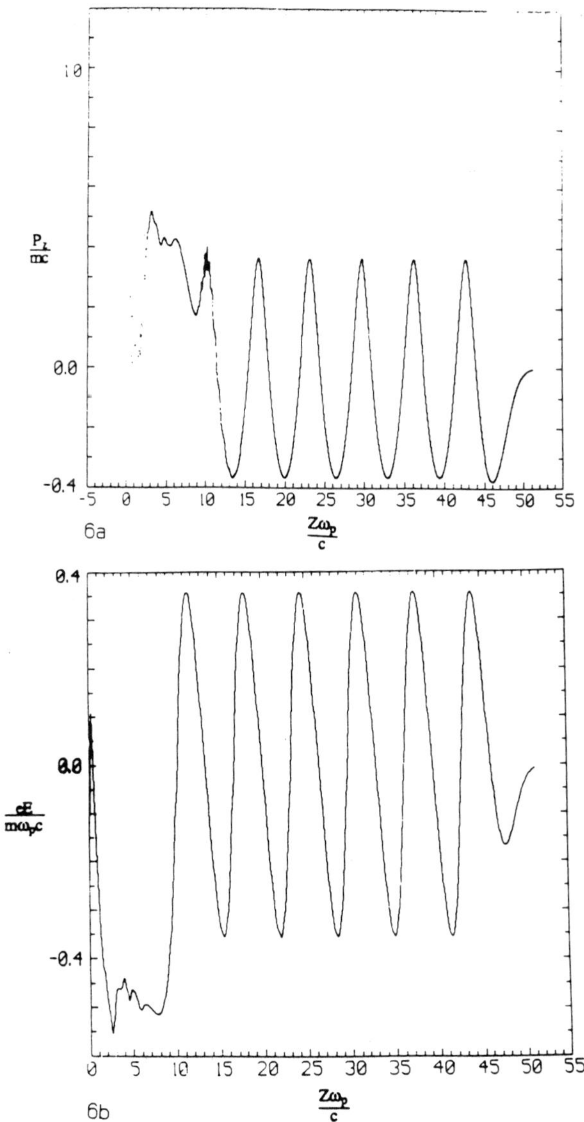


6. Single gaussian electron beam, 1D kinetic code with normalized beam current of 0.2. 6a — phase space diagram and 6b — normalized wake field.

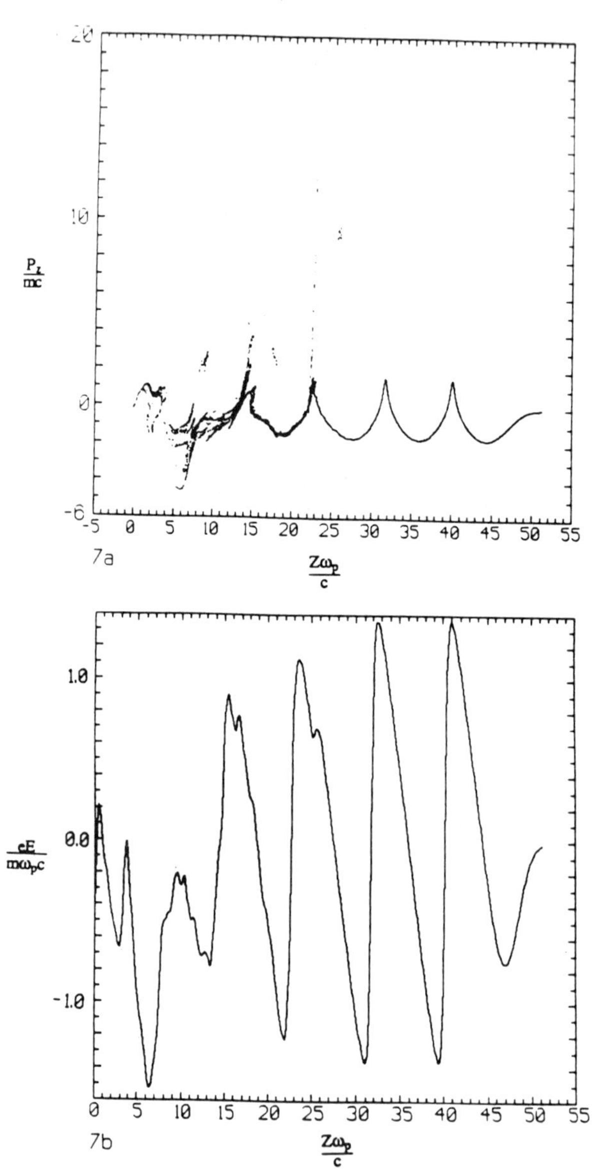


7. Same as Fig. 6 for beam current of 0.6.

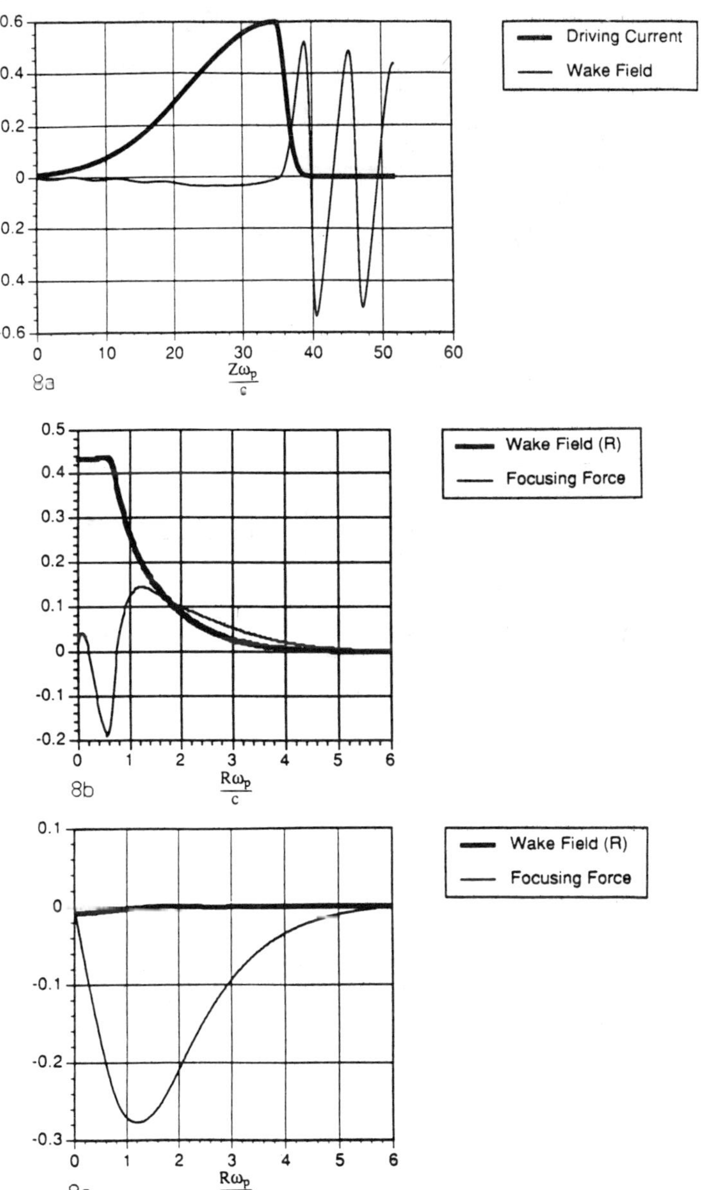


8. Slowly sloped electron beam with normalized peak current of 0.6, 2D fluid code. 8a — normalized driving beam and wake field. 8b — normalized radial profile of the wake field and focusing force at the right end point of the wake field longitudinal profile. 8c — normalized radial profile of the wake field and focusing force at the peak of the driving beam.

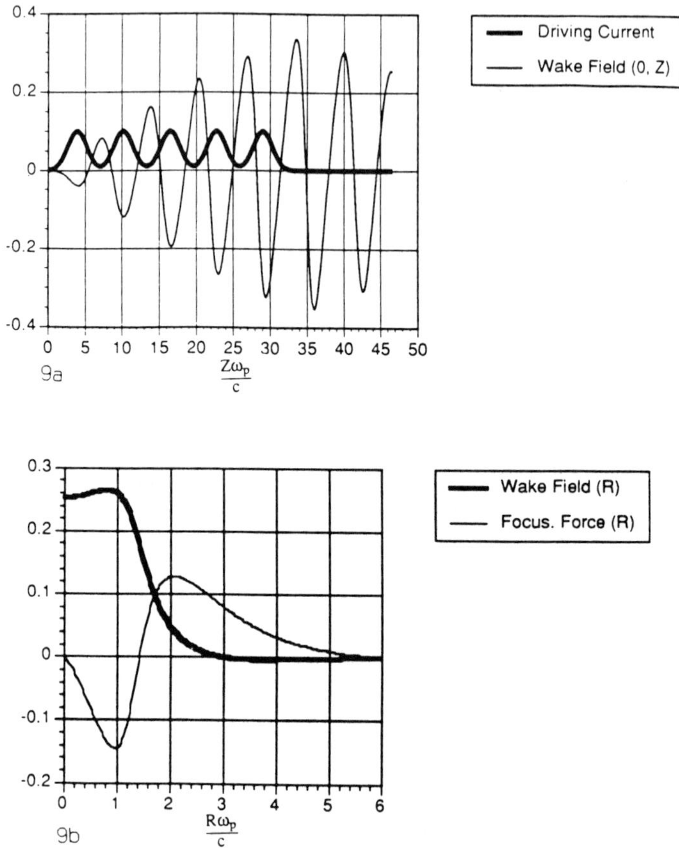


9. Five electron beams spaced one plasma wavelength apart each with current of 0.1. 9a — normalized driving beam current and wake field. 9b — normalized radial profiles of the wake field and focusing force at the right end point on wake field longitudinal profile.

Now we look at the cases for the laser pulse. In Fig. 10 we have the ponderomotive potential of an amplitude $0.1mc^2$ with a half width $\sigma$ of 1. This corresponds to the laser intensity $I$ of $0.1I_0$, where $I_0 = (\frac{\omega}{\omega_p})^2 nmc^2$. This gives a wake field amplitude of $0.15m\omega_p c/e$ consistent with linear theory, however, from the radial profile of the focusing force we observe that we are entering the nonlinear regime.

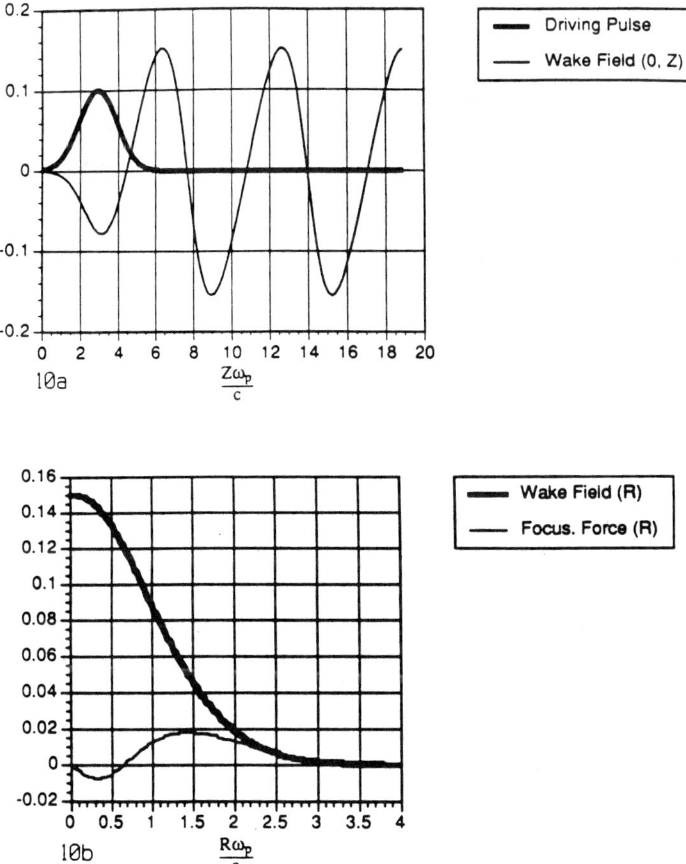


10. Gaussian laser pulse with normalized ponderomotive potential amplitude of $0.1mc^2$. 2D fluid code. 10a — normalized driving pulse and wake field. 10b — normalized radial profile of wake field and focusing force at the right end point on the wake field.

Fig. 11 now shows definite nonlinearity in both the nonsinusoidal longitudinal wake field and in the radial profiles. The normalized amplitude of the ponderomotive potential of the laser force is now $0.3mc^2$ in this case. As discussed in the theory section, the assumption that this force models a gaussian pulse is in question since we are now in the relativistic region.

For the last 2-D case we have three laser pulses of normalized amplitude $0.1mc^2$ spaced one wavelength apart. As in the multiple electron bunch case we see an amplification of the wake field starting in the weakly nonlinear regime and ending in the strongly nonlinear regime. From the radial profiles we see that the wave is near wavebreaking.

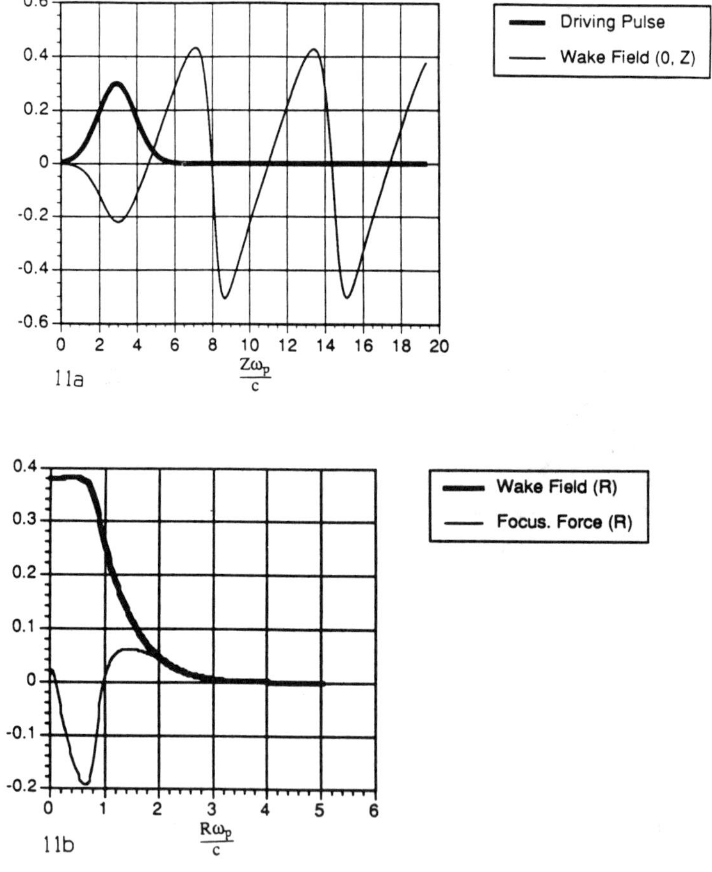


11. Same as Fig. 10 for a normalized pulse amplitude of $0.3mc^2$.

For the 1-D cases we now look at a normalized ponderomotive potential of $0.1mc^2$ in Fig. 12. As expected the amplitude is larger than that of the finite radius case and is in the weak nonlinear region. We observe no wavebreaking or heating in this region.

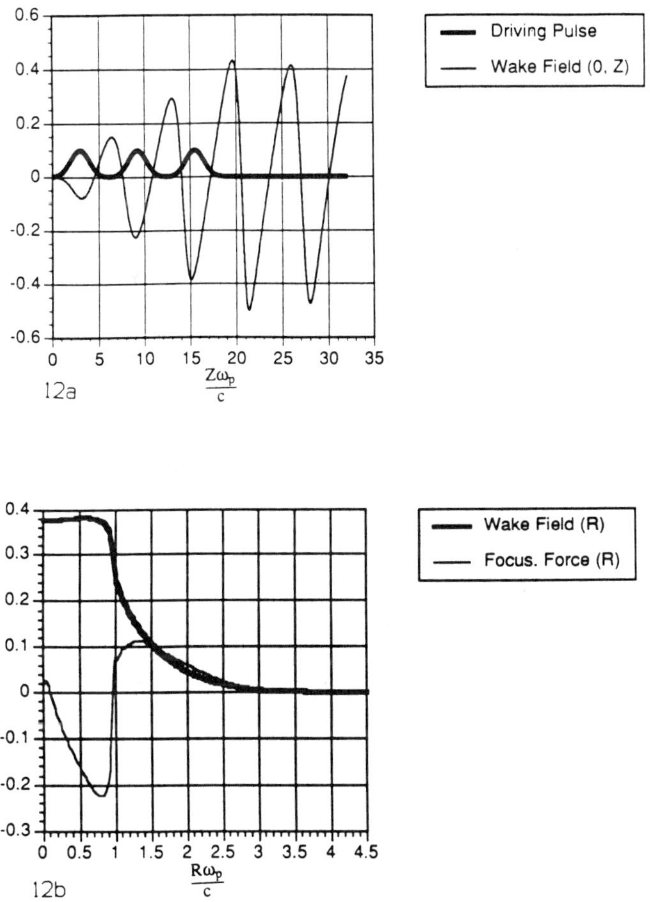


12. Same as Fig. 10 for three laser pulses spaced one plasma wave length apart, each with normalized amplitude $0.1mc^2$.

Figure 13, with a normalized potential of $0.6mc^2$, shows definite nonlinearity in both the phase space and in the wakefield. The phase space diagram shows the wave starting to break after about 2 oscillations, with the harmonics starting to modulate the wave. After about 4 oscillations we see the transition to chaos and the heating of the plasma. Correspondingly we see the amplitude of the

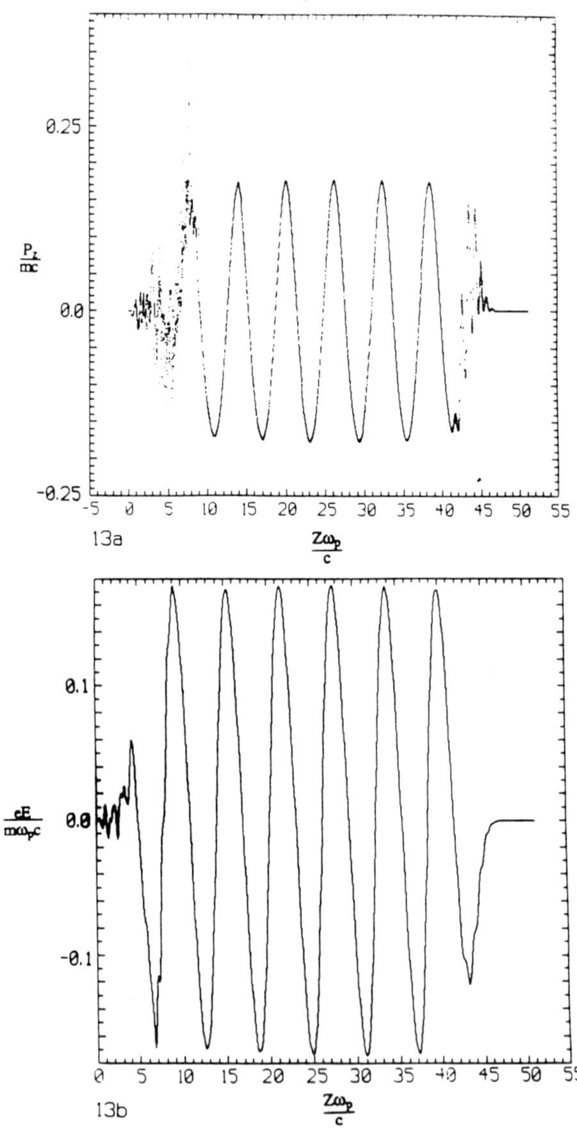


13. Single gaussian laser pulse with normalized amplitude of $0.1mc^2$, 1-D kinetic code.13a — phase space plot. 13b — normalized wake field.

wakefield decrease due to nonlinearity and noise introduced due to the heating. Also observed are electrons trapped by the wakefield and accelerated.

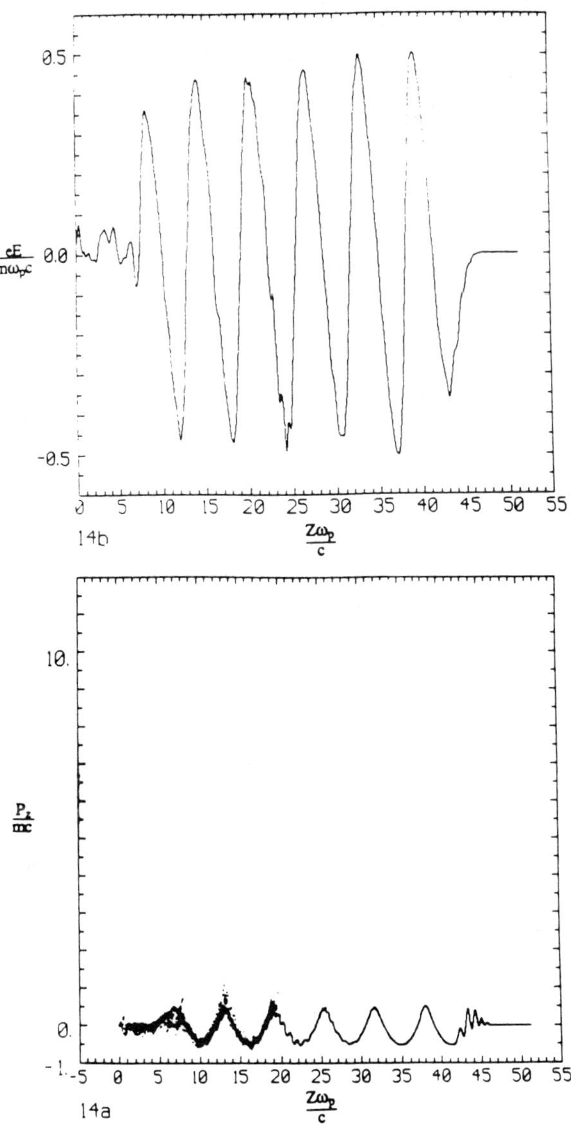


14. Same as Fig. 13 for a pulse amplitude of $0.6mc^2$.

## CONCLUSION

In this paper we considered the excitation of a nonlinear plasma wave for the wakefield acceleration of high energy particles. By using the two dimensional fluid model we calculate the maximum amplitude of a wake field driven by either "rigid" electron beams or a given laser pulse and study the dynamics of the wave steepening and approach to wavebreaking. We also studied a one dimensional kinetic model in which depletion of a driver is taken into account and which is capable of describing a plasma motion after wave breaking. We observed that in a two-dimensional problem the wave steepening and breaking can occur first in the radial profile and not in the longitudinal profile. The length of the "laminar" wave after the driver decreases with increasing amplitude of the wake field. However, one can produce the wake field close to the wave breaking limit and have it suitable for particle acceleration and focusing on a distance of a few wavelengths behind the driver. In particular the accelerating field of $0.6m\omega_p c/e$ is feasible. The amplitude and the structure of the excited field depends substantially on the shape of the driver. We observed a nonlinear amplification of the wake field by a factor of 2 for a driving electron beam with a gentle slope leading edge and a sharply cut trailing edge. It is shown that the field can be strongly enhanced by employing multiple drivers with a spacing equal to one wavelength. The electron beam driver and a laser driver of a similar shape produce comparable wake fields when

$$\frac{\langle v_{\mathrm{osc.}}^2 \rangle}{2c^2} \approx \frac{j_b}{enc} \,. \tag{46}$$

A simple fluid code presented in this paper can be further incorporated into a more sophisticated three dimensional kinetic code to describe the dynamics of the driver and accelerated particles over the distance comparable to the length of acceleration. This longer distance typically exceeds the plasma wavelength by several orders of magnitude and the driver does not remain rigid over such a length, though it can be assumed rigid over a few wavelengths. The combined code would be much faster than a fully kinetic code alone since a lot of time can be saved by calculating the fields on the intermediate steps from the fluid code with a locally rigid driver.

## ACKNOWLEDGMENTS

The work was supported in part by the U.S. Department of Energy grant DE-FG05-80ET-53088, and in part by the University Research Association subcontract to UTA.

## REFERENCES


1. V.I. Veksler, Proceedings of CERN Symposium of High Energy Accelerators and Pion Physics, Vol.1, CERN, Geneva (1956)

2. T. Tajima and J. M. Dawson, Phys. Rev. Lett. **43**, 267 (1979)

3. P. Chen, J.M. Dawson, R.W. Huff, and T. Katsouleas, Phys. Rev. Lett. **54**, 693 (1985).

4. R.D. Ruth, A. Chao, P.L. Morton, and P.B. Wilson, Particle Accelerators **20**, 171 (1985).

5. J.B. Rosenzweig, D.B. Cline, B. Cole, H. Figueroa, W. Gai, R. Wonecny, J. Norem, P. Schoessow, and G. Simpson, Phys. Rev. Lett. **61**, 98 (1988).

6. A. Ogata, in *Proceedings of International Topical Conference on Research Trends in Coherent Radiation Generation and Particle Accelerators*, February 11-13,1991, La Jolla, California.

7. B.N. Breizman, P.Z. Chebotaev, I.A. Koop, A.M. Kudryavtsev, V.M. Panasyuk, Yu.M. Shatunov, A.N. Skrinsky, and F.A. Tsel'nik, in *Proceedings of the 8th International Conference on High-Power Particle Beams* (BEAMS–90), July 2–5, 1990, Novosibirsk.

8. C. Joshi, private communication (1991).

9. J.M. Dawson, Phys. Rev. **113**, 383 (1959).

10. W. Horton and T. Tajima, Phys. Rev. A **34**, 4110 (1986).

11. T. Tajima, "Computational Plasma Physics" (Addison-Wesley, Redwood City, 1989).

**Research Trends in Physics: Coherent Radiation Generation and Particle Acceleration**

*La Jolla International School of Physics*, The Institute for Advanced Physics Studies, La Jolla, California

# Frequency Upconversion of Intense Femtosecond Pulse in Density-Modulated Plasmas

**M.C. Downer and Wm. M. Wood**

Department of Physics
University of Texas at Austin
Austin, TX 78712

## ABSTRACT

The new plasma diagnostic technique of spectral blue shifting of intense, femtosecond pulses is used for the first time to quantitatively analyze the ionization of noble gases under the influence of intense, femtosecond illumination. The two processes of strong-field tunnelling ionization and electron impact ionization are found to play competing roles on these time scales.

## INTRODUCTION

Highly ionized plasmas approaching atmospheric density are a potential future source of coherent x-rays, and a potential medium for charged particle acceleration. Ionization by intense, femtosecond pulses holds promise for precise control of initial plasma conditions (temperature, ionization state, density) which are critical to these applications. At the same time, new, quantitative, experimental diagnostics -compatible with the high gas density and ultrafast time scale - are needed to measure the ionization and subsequent plasma dynamics which give rise to these conditions. A number of authors have shown, both experimentally and theoretically, that a laser pulse which rapidly ionizes a gas experiences a frequency blueshift caused by the creation of a free-electron plasma[1,2]:

$$\Delta\omega = -\frac{\omega_0}{c}\int_0^z \frac{\partial n}{\partial t}(l)\, dl \qquad (1)$$

Here, $\omega_0$ is the angular frequency of the light, 'l' is the longitudinal distance over which the interaction occurs; $n = (1 - \omega_p^2/\omega_0^2)^{1/2}$ is the index of refraction of the medium through which the pulse travels, and is found using the Drude model ('$\omega_p$' is the plasma frequency.) In experiments which used pulse durations long with respect to ion-ion collision times[2], plasma expansion and recombination following the ionization contributed to and complicated the phase modulation induced on the pulse. In a previous publication[3], we showed for the first time that *femtosecond* pulses, tightly focused to intensities above the ionization threshold in air experience a much simpler, "pure" blue shift, i.e. with no trace of components red shifted from the original pulse spectrum, indicating qualitatively that the phase modulation on the laser pulse is caused entirely by an ionization process. Furthermore, self focusing and spectral broadening caused by nonlinear optical interactions with neutral gas were suppressed in gases at pressures less than 5 atmospheres with sufficiently tight focusing, suggesting that plasma density grows rapidly near the beginning of the pulse. This qualitative result suggested that quantitative, time-resolved measurement and analysis of the spectral shifts of femtosecond laser pulses might serve as a simple diagnostic for measuring detailed growth of the ionization front, distinct from the subsequent dynamics, offering in addition time resolution over a broad range of gas pressures and gas species. In a wider context, such frequency upshifts might also provide quantitative diagnosis of plasma density oscillations in plasma wakefield and beat-wave particle accelerator schemes, as recently pointed out by several authors[4].

The purpose of this article is to demonstrate, for the first time, the quantitative use of this plasma diagnostic technique in analyzing the breakdown of noble gases caused by intense, femtosecond illumination. This will be done in three stages: 1) We report systematic experimental observations of the self- blueshifting of laser pulses after ionizing 1-5 atm. pressure samples of each of the noble gases He, Ne, Ar, Kr, and Xe, which reveal a universal, reproducible pattern in the shape of the blueshifted spectra. Specifically, with increasing laser intensity, gas pressure, and atomic number, the self-blueshifted spectra develop from a near replica of the incident pulse spectrum into a complex structure consisting of two spectral peaks: a narrow peak shifted between 5 and 10 nm towards the blue from the original spectrum, whose position is independent of pulse energy and pressure above some threshold energy; and a broad, low peak shifted further towards the blue whose position and width depend strongly on the gas pressure and the laser pulse energy. 2) We report time-resolved spectral shifts of a weak probing pulse which show different temporal evolution for each of these two spectral features. 3) Finally, we propose a quantitative, *ab initio* model which relates these two spectral features to two competing ionization mechanisms: collisionless tunnelling ionization, predicted to dominate early in the ionizing pulse profile, and electron impact ionization, predicted to dominate in the intense maximum of the pulse profile.

## EXPERIMENT

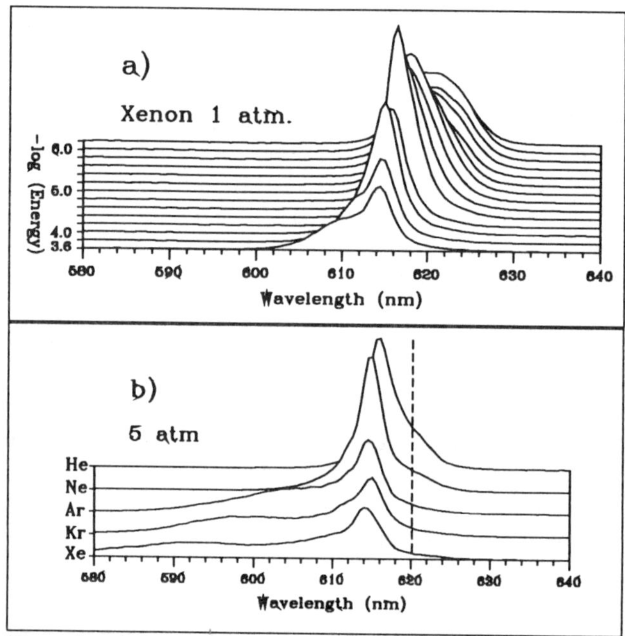


**Figure 1:** **a)** Spectra after interaction with 1 atm. pressure Xe as a function of pulse energy. Pulse energy is increasing towards the bottom in steps of (x $10^{0.2}$). **b)** Spectra after interaction of 0.25 mJ pulses in 5 atm. pressure of each of the noble gases studied.

In our experiments, 100-femtosecond laser pulses with center wavelength of 620 nm and energies less than or equal to 0.4 mJ are focused at f/5 into a glass cell containing a pressure between 1 atm. and 5 atm. of He, Ne, Ar, Kr, or Xe gases. The minimal focal spot size is measured as 7.0 microns in diameter for the intensity to reach 1/e of the central intensity maximum. Light transmitted through the focal region is collected and analyzed by a spectrometer. The spectra of pulses are recorded for varying gases, pressures, and pulse energies. Typically, less than 1% of the pulse energy is lost in ionizing the gas. Figure 1a shows the results of varying the laser pulse energy on the measured spectrum after breakdown in Xe gas at 1 atm. of pressure. As the pulse energy is increased, the spectrum peak moves towards the blue until the pulse energy reaches 0.1 mJ (-log(Energy) = 4.0), after which the position of the peak appears to change very little. As the energy increases further, a shoulder appears on the blue side of the spectrum which becomes broader and shifted further towards the blue. Figure 1b shows the resulting spectra after interaction with each of the different gases at 5 atmospheres of pressure using a pulse energy of 0.25 mJ. The center of the unshifted pulse spectrum is indicated by a vertical line, showing that a blue shift occurs in all cases. However, the shape of the blueshifted spectrum depends strongly on gas species. In Ar, Kr, and Xe, the narrow, less-shifted peak and the broad, blue shoulder are clearly discernible. In fact, in Kr and Xe, the shoulder is actually a separate peak. In He and Ne, on the other hand, the blue shoulder is absent. Instead, a different feature - a "red shoulder" corresponding approximately to the unshifted spectrum - showing that a significant amount of the initial pulse energy remains unshifted. Time-resolved pump-and-probe experiments show that each of these spectral features has a characteristic temporal evolution within the pump pulse.

To obtain time-resolved spectra of the blueshifting process, a probe beam with polarization perpendicular to the ionizing "pump" pulse and approximately 4% of the pump energy is recombined with the pump after some time delay, and allowed to traverse the focal region. Because the Drude polarization does not depend upon the vector character of the electric field, the probing pulse experiences the same spectral shift as the pump pulse when the two are coincident in time. By arranging for the probe to arrive before, during, and after the pump pulse, the blue shifting, and therefore the ionization, can be mapped out as a function of time. Figure 2a shows a series of probe spectra after ionization in 1 atm. Xe gas corresponding to different time delays between the probe and the pumping beams. There are 27 fs between spectra; time increases towards the top of the figure, with negative times corresponding to the probe pulse arriving before the peak of the pump pulse. Coincidence of the pump and probe pulses is indicated by an emboldened spectrum, and is denoted "0" on the time scale. A quantitative evaluation of the temporal behaviour of the two features (the narrow, less-shifted peak and the broad, bluer shoulder) can be had by plotting the area under the spectra in small regions around 615 nm, 610 nm, and 600 nm. The values at 615 nm will reflect the time behaviour of the process giving rise to the narrow, less shifted peak. Those values at 610 nm and 600 nm will reflect the temporal behaviour of the process giving rise to the broad, blue shoulder. These plots are shown in figure 2b for 1 atm. Xe gas. The values at 610 nm and 600 nm display the same temporal behaviour, with maxima at approximately 100 fs before the maximum of the ionizing pulse. This indicates that the process giving rise to the broad, blue shoulder/peak occurs in the early part of the pulse. The values plotted for 615 nm show a maximum approximately 25 fs after the maximum of the pump pulse, indicating that the process causing the narrow, less-shifted peak occurs near or slightly after the maximum of the ionizing pulse, and definitely occurs later than that process giving rise to the blue shoulder. We now propose a simple, *ab initio*, ionization model which accounts for 1) the narrow, less shifted peak and the broad blue shoulder/peak observed in Ar, Kr, and Xe, 2) their respective temporal behaviour, and 3) the "unshifted" portions of the spectra and the *absence* of the blue shoulder observed in He and Ne.

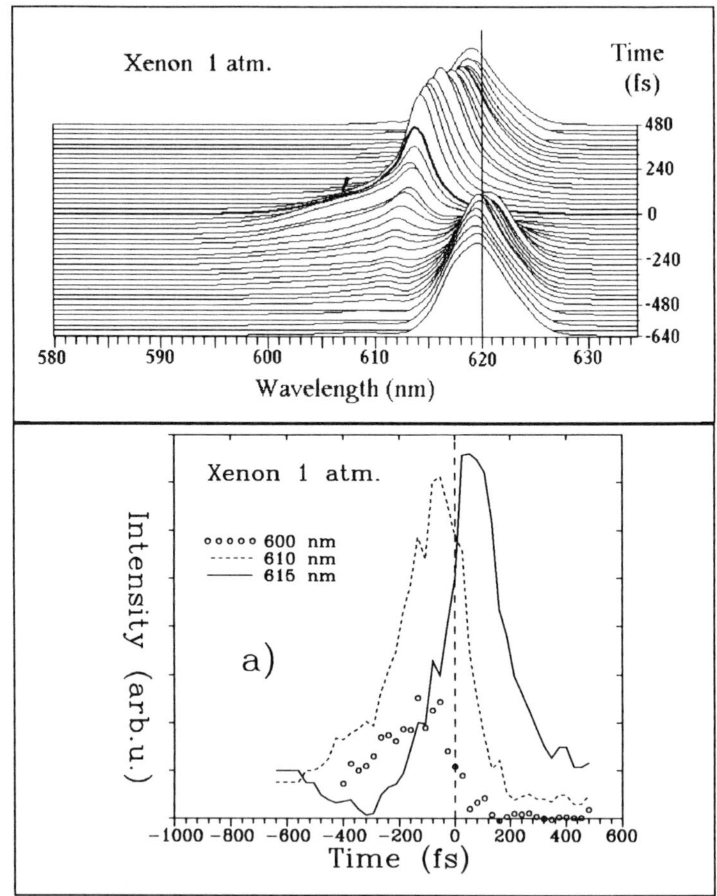


**Figure 2: a)** Time resolved spectra in xenon at 1 atm. using pump pulse energy of 0.22 mJ. There are 27 fs between spectra; time increases towards the top of the figure, with negative times corresponding to the probe pulse arriving before the pump pulse. Coincidence of the pump and probe pulses is indicated by an emboldened spectrum. **b)** Energies at 600 nm, 610 nm, and 615 nm as functions of time for time-resolved spectra in 1 atm. Xe.

## MODEL

In the presence of a strong light field, the ionization of the atoms can be modelled using the strong-field tunnelling theory due to Keldysh or Ammosov[5], which yield similar results for the lower stages of ionization. The coupled equations governing the densities $N_i$ of atoms of a particular ionization state 'i' are written in the following way:

$$\frac{dN_i}{dt} = (P_{i-1} N_{i-1} - P_i N_i) + (N_e \sigma_{i-1} v_e N_{i-1} - N_e \sigma_i v_e N_i) \qquad (2)$$

The first term on the right hand side of the equation involving the $P_k$ represents the tunnelling probability per unit time from the Keldysh or Ammosov theories for the ions changing from k-times ionized to (k+1)-times ionized. The second term on the right hand side of the equation represents the collisional probability per unit time for ionization via electron impact, where $N_e$ is the free electron density, $v_e$ is the r.m.s. electron velocity, and $\sigma$ is the cross-section for the process. Intermediate resonant states can be ignored in the equations[6]. The density of free electrons and the resulting index of refraction as a function of time calculated from these equations allows the calculation of the phase modulation and blueshift induced on the laser pulse. The model using only the tunnelling ionization terms $P_k$ proves adequate for describing the observed blue shifting of the laser light in He and Ne gas breakdown, and for Ar gas at low intensities and pressures (where there is no indication of the presence of the blue "shoulder.") Because tunnelling ionization occurs almost entirely during the first half of the laser pulse, only the leading edge of the pulse becomes blueshifted; the trailing part of the pulse remains unshifted, with the resulting spectral "red" shoulder observed in the He and Ne gas breakdown.

In pressures above 1 atm. of the heavier gases Ar, Kr, and Xe, the "red" shoulder in the spectrum after ionizing breakdown disappears, and the *entire* pulse spectrum is shifted at the higher pulse energies (the absence can be seen in figure 1b.) The process of ionization via electron impact, as described by the second term in equation (2), can be used to explain this phenomenon. At or above a light intensity of $10^{15}$ W/cm$^2$, the excursion amplitude $eE/m\omega^2 \sim 3$ nm for the electrons "wiggling" in the laser field at a frequency corresponding to $\lambda$=620 nm. is greater than the mean spacing, ~ 2 nm, between atoms or ions in a gas at STP. The cross sections for electron impact ionization ($\sigma \sim 4 \times 10^{-16}$ cm$^2$ in Ar) in the noble gases is maximal for electron energies in the range 100 eV - 500 eV [7], very close to the quiver energy of the electrons during the *peak* of the laser pulse. Because the density $N_e$ and r.m.s. velocity 'v' of free electrons generated by the tunnelling ionization is near its maximum at the pulse peak, the collisional process will occur at its maximal rate, $N_e\sigma v$, during the peak of the laser pulse. Estimates using typical values for the terms in equation 2 reveal that in both He and Ne at 5 atm. pressure, the time $\tau = (N_e\sigma v)^{-1} \sim 200$ fs between ionizing collisions is significantly *longer* than the light pulse duration; therefore collisional ionization effects play only a minor role in these gases, and no index change and therefore no shifting occurs during the second half of the pulse. In Ar, Kr, and Xe at 5 atm. pressure, however, these collision times, $\tau \sim< 30$ fs, are significantly *shorter* than the pulse duration; as a result, collisions play an important role in the ionization process during and after the peak of the laser pulse in these gases, resulting in the *entire* spectrum shifting.

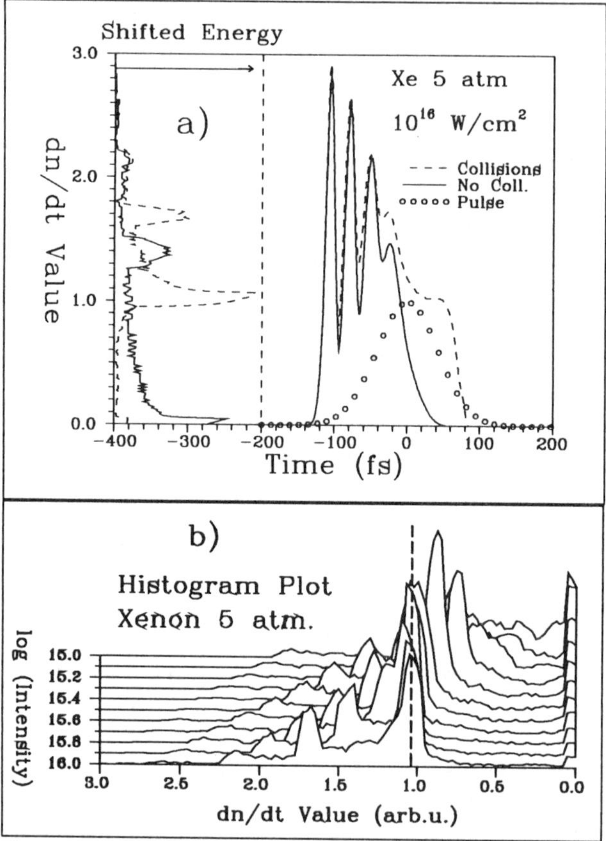


**Figure 3:** **a)** Histogram plot for breakdown of 5 atm. pressure Xe gas. Solid curve shows calculations without collisions. Dashed curve includes collisions. **b)** Spectra after interaction of 0.25 mJ pulses in 5 atm. pressure of each of the noble gases studied.

The existence of the two features in the shifted spectra outlined above can be explained in terms of a contrast between the strong-field and the collisional ionization rates. Because the blue shift is proportional to the *derivative* of the index of refraction, this contrast can be seen most readily by plotting the derivative as a function of time, calculated from the model. It is expected that there should be two primary values which the derivative of the index takes on during the presence of the laser pulse. Figure 3a shows a plot of the derivative versus time calculated for 5 atm. pressure of xenon gas, both with (dashed) and without (solid) the collisional terms included. The vertical axis which has values for dn/dt is also used as the abscissa of a histogram plot of energy versus value of the derivative. This plot is an indication of how much of the light pulse energy is shifted by the amount corresponding to the value of the derivative. The solid histogram plot shows only a single non-zero peak which is caused by the oscillating features in the derivative during the early part of the pulse. (These "oscillations" are the result of transitions between completion of the ionization of one level of electrons, e.g. neutral atoms becoming singly ionized, and a brief "wait" as the pulse intensity becomes strong enough to begin ionizing the next level of electrons.) The dashed curve reveals that when collisions are included in the model, the peak caused by the oscillating features moves to a higher dn/dt value, and a *second* peak appears at a lower value of dn/dt, due to collisional ionization during and after the maximum of the laser pulse.

A plot of such histograms as a function of laser pulse energy is shown in figure 3b for 5 atm. pressure Xe, and reveals that at each energy, the histogram displays two primary peaks: one caused by the strong-field ionization at the beginning part of the pulse, and the second caused by the collisional ionization occurring at or after the peak of the pulse. The position of the peak caused by the collisional terms in the ionization equations as a function of energy shows the same saturation behaviour that the narrow, less-shifted peak displays in the observed spectra. Above a certain "threshold" energy, the position of this collisional peak in the model is very stable, as indicated by a vertical dashed line in the diagram. In contrast to this, the position of the peaks in the histogram plots that correspond to the strong-field ionization display a very strong dependence on the laser pulse energy. This behaviour is qualitatively the same as that of the broad, bluer shoulder or peak observed in the spectral data. Similar results are found in Xe at lower pressures, and in Ar and Kr gases. These correspondences between the collisional ionization in the model and the narrow less-shifted peak in the observed spectra on the one hand; and between the strong-field ionization and the broad blue shoulder/peak on the other, can be drawn further by considering the temporal behaviour of the two features. Of the two peaks in the histogram model, the strong field ionization, at a time *before* the maximum of the laser pulse, gives rise to the larger value of dn/dt, which matches the behaviour of the broad, blue shoulder/peak; the collisional ionization, at a time *after* the maximum of the laser pulse, gives rise to the peak at the smaller value of dn/dt, which matches with the behaviour of the narrow, less shifted peak observed in the data. This temporal behaviour is also observed in Xe gas at lower pressures, and in Ar and Kr gases.

## CONCLUSION

In summary, then, we see that the technique of spectral blueshifting of femtosecond laser pulses provides a powerful diagnostic of plasma evolution. Such experiments significantly complement standard diagnostics of strong-field ionization (ion yield or photoelectron spectroscopy) by providing femtosecond time resolution and by permitting experiments in much denser plasmas. As an example, a simple, ab initio model of gas ionization in the presence of an intense laser pulse, using the competing mechanisms of strong-field tunnelling ionization and electron impact ionization, has been used to provide an explanation of the two common features observed in the blueshifted spectra: the narrow, less-shifted peak, and the broad, bluer shoulder/peak. The broad, blue shoulder/peak is caused by the primarily strong-field ionization which occurs in the leading part of the laser pulse. The narrow, less-shifted peak is caused by the collisional ionization process which occurs during and immediately after the maximum of the laser pulse. In addition, the model accounts for the "red" shoulder observed in the shifted spectra after breakdown in He and Ne gases; lack of collisional ionization in these gases causes the second part of the laser pulse to remain unshifted. Extensions of this technique to laser-induced plasma density oscillations appear feasible.

## ACKNOWLEDGMENTS

This research was supported by the National Science Foundation (Grant DMR8858388), the Robert A. Welch Foundation (Grant F-1038), and the Air Force Office of Scientific Research (Contract F49620-89-C-0044).

## REFERENCES

1. S.C. Wilks, J.M. Dawson, and W.B. Mori, Phys. Rev. Lett. **61**, 337. (1988)
   E. Esarey, A.Ting, P.Sprangle, Phys. Rev. A **6**, 3526. (1990)
2. E. Yablonovitch, Phys. Rev. A **10**, 1888. (1974)
   E. Yablonovitch, Phys. Rev. Lett. **32**, 1101. (1974)
3. Wm M. Wood, G.B. Focht, M.C. Downer, Opt. Lett. **13**, 984. (1988)
4. P. Sprangle, E. Esarey, A. Ting and G. Joyce, Appl. Phys. Lett. **53**, 2146 (1988)
   S.C. Wilks, J.M. Dawson, W.B. Mori, T.Katsouleas, Phys.Rev.Lett. **62**, 2600. (1989)
   T. Osuga, K. Mima, and H. Takabe, Jap. J. Appl. Phys. **28**, 1474. (1989
5. L.V. Keldysh, Sov. Phys. JETP **20**, 1307. (1965)
   M.V. Ammosov, N.B.Delone, and V.P.Krainov, Sov. Phys. JETP **64**, 1191. (1986)
6. P.B. Corkum, N.H.Burnett, and F.Brunel, Phys. Rev. Lett. **62**, 1259. (1989)
7. H.Tawara and T.Kato, Atomic Data and Nuclear Data Tables **36**, 167-353. (1987)

**Research Trends in Physics: Coherent Radiation Generation and Particle Acceleration**
Editorial Board: J.M. Buzzi, A. Prokhorov (Editor-in-Chief), P. Sprangle, and K. Wille
*La Jolla International School of Physics*, The Institute for Advanced Physics Studies, La Jolla, California

# The European Beat Wave Programme

**A. Dyson and A.E. Dangor**

Blackett Laboratory
Imperial College
London SW7, U.K.

## 1. Introduction

In the laser beat wave accelerator a large amplitude relativistic plasma wave is generated by the beating together of two laser frequencies whose difference frequency is close to the plasma frequency. The resonantly excited plasma wave offers the potential to accelerate plasma particles to relativistic energies over distances a short as several tens of meters. This is because a plasma is fully ionised and so can support accelerating fields orders of magnitude greater than possible in current accelerator structures. Another application of these plasma waves is the final focussing of particle beams in high energy accelerators.

In this paper we review the laser beat wave programmes at the Rutherford Appleton Laboratory (RAL) in the U.K. and at Ecole Polytechnique (EP) in France (both programmes are partly supported by a CEC stimulation contract). In these programmes the plasma wave is generated using neodymium glass lasers giving laser light at $\omega_0$=1.064μm (YAG) and $\omega_1$=1.053μm (YLF). Other beat wave programmes (Clayton *et al.*, UCLA; Ebrahim *et al.*, NRC; Kitagawa *et al.*, Osaka) use carbon dioxide lasers at 9.6μm and 10.6μm. However, the advantage of the shorter wavelength is that a plasma wave of higher phase velocity is produced and this is more relavant for particle acceleration. The use of the shorter wavelengths does mean that a more uniform plasma is required for resonant excitation of a high amplitude plasma wave. For this reason multiphoton ionisation has been investigated at both RAL and EP.

The beat wave scheme does suffer the limitations of needing a precise plasma density and the destructive influence of instabilities due to the relatively long growth time of the plasma wave (200ps). These disadvantages are not present in the laser wakefield scheme (Tajima & Dawson 1979; Sprangle *et al.* 1988). Here a very short laser pulse (pulse length = half the plasma wavelength, i.e $\tau$=300fs for a typical plasma density of $10^{17}cm^{-3}$) travelling through a plasma leaves a plasma wave in its wake. Work has started at RAL on the development of a suitable very short pulse high power laser for use in plasma wakefield experiments.

In this review emphasis will be given to the experimental aspect of the beat wave programmes. We will not discuss the considerable amount of theoretical work that has been done except where relavent. We begin with a very brief summary of the basics of the beat wave.

## 2. Scalings and limitations of the laser beat wave.

The basic mechanism for the beat wave is illustrated in figure 1. The two laser lines $\omega_0,\omega_1$ beat together and the ponderomotive force of the beat envelope pushes the plasma electrons apart so generating the plasma wave. Provided the plasma is very underdense the phase velocity of the plasma wave $v_p$ is equal to the group velocity of the pump waves, i.e.

$$v_p = c\left(1-\frac{\omega_p^2}{\omega_0\omega_1}\right)^{1/2} \approx c$$

For our experiments this gives a value for relativistic factor associated with the plasma waves of $\gamma_p$=100. This should be compared to $\gamma_p$=10 for the experiments performed with $CO_2$ lasers.

The amplitude of the plasma wave $\delta n/n$ can be calculated from the wave envelope equation (Rosenbluth and Liu 1972; Karttunen and Salomaa 1986; Mora 1988; Gibbon and Bell 1988)

$$\frac{1}{\omega_p}\frac{d}{dt}\left(\frac{\delta n}{n}\right) = \frac{1}{4}\left(\frac{v_0}{c}\right)\left(\frac{v_1}{c}\right) + \frac{3}{16}i\left|\frac{\delta n}{n}\right|^2\left(\frac{\delta n}{n}\right)$$

$$- \frac{\omega_p-(\omega_1-\omega_0)}{\omega_p} - \frac{v_c}{2\omega_p}\left(\frac{\delta n}{n}\right)$$

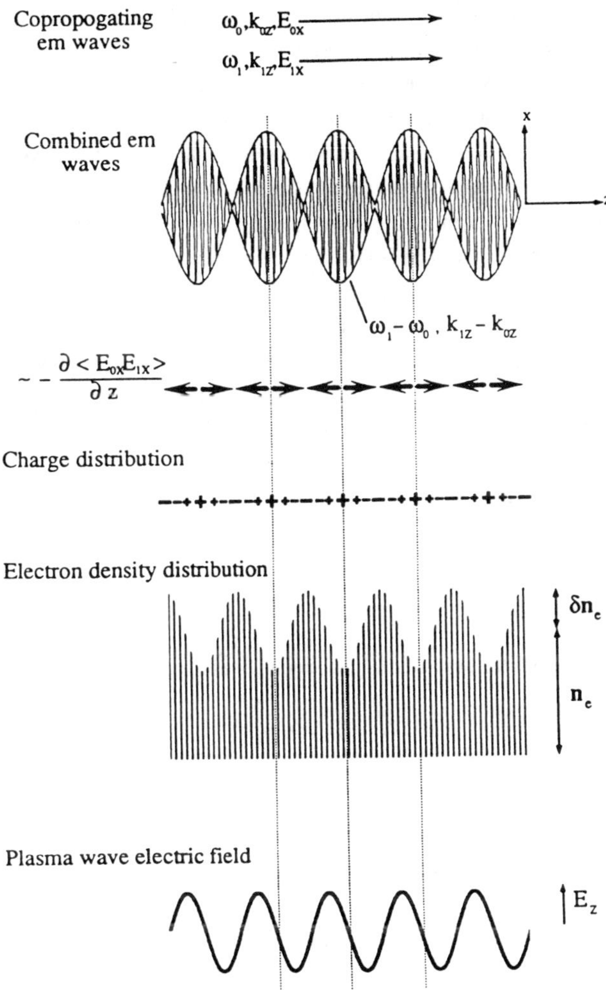


Fig 1. The beat wave concept.

where $v_0, v_1$ are the electron quiver velocities in the laser field $v_{0,1}=eE_{0,1}/m_e w_{0,1}$ and $\upsilon_c$ is normally taken to be the electron collision frequency. The first term on the r.h.s of this equation is the laser driving term. The second term on the r.h.s represents the relativistic detuning of the plasma frequency as a result of the relativistic mass increase of the plasma electrons as they move in the longitudinal field of the plasma wave. Thus as soon as the plasma wave begins to grow the system moves off resonance and the maximum plasma wave amplitude is limited. The third term on the r.h.s allows for the fact that the plasma frequency may not be at the exact difference frequency of the two laser beams. This is known as the linear detuning term. It is important that in any experiment this term is less than relativistic detuning so as not to limit the maximum possible plasma wave amplitude. The fourth term on the r.h.s represents damping of the plasma wave at a characteristic frequency $\upsilon_c$. Since the plasma wave takes time to grow any damping process which occurs on a faster time scale may severely limit its amplitude. In our experiments the modulational decay instability is considered to be a possible candidate for such limitation.

The maximum $\delta n/n$ due to relativistic detuning (2nd term on r.h.s.) is given by

$$\frac{\delta n}{n} = \left(\frac{16}{3}\left(\frac{v_0}{c}\right)\left(\frac{v_1}{c}\right)\right)^{1/3}$$

For constant laser intensities the time to reach this amplitude is

$$\omega_p \tau = \left(\frac{16}{3}\right)^{1/3}\left[\left(\frac{v_o}{c}\right)\left(\frac{v_1}{c}\right)\right]^{-2/3}$$

For our experiments with $I_{0,1}=10^{14}$Wcm$^{-2}$ we have a maximum value of $\delta n/n=7\%$ after 250ps, to be compared to the laser pulse duration of 200ps.

The maximum $\delta n/n$ due to linear detuning (3rd term on r.h.s.) is given by

$$\frac{\delta n}{n} = \left(\frac{v_0}{c}\right)\left(\frac{v_1}{c}\right)\Big/\left(\frac{\Delta n}{n}\right)$$

where

$$\frac{\omega_p - (\omega_1 - \omega_0)}{\omega_p} = \frac{1}{2}\frac{\Delta n}{n}$$

Comparing this with the relativistic detuning limit shows that $\Delta n/n$ must be less than 0.1% otherwise linear detuning will dominate. This indicates that very homogeneous plasmas are needed for beat wave experiments. Figure 2 shows the expected plasma wave amplitudes for relativistic detuning only and for linear detunings of 0.5% and 1%.

In an actual experiment the finite laser spot size gives a plasma wave of finite transverse dimension which further limits the maximum amplitude. This has two causes. First the transverse ponderomotive force of the laser beam will push out plasma electrons from the focal region taking the system off resonance (Mora 1988). Second the ponderomotive force due to the plasma wave itself also drives off plasma electrons. For our experiments where the spot size is 300μm diameter these two effects give a limit to $\delta n/n$ comparable to relativistic detuning.

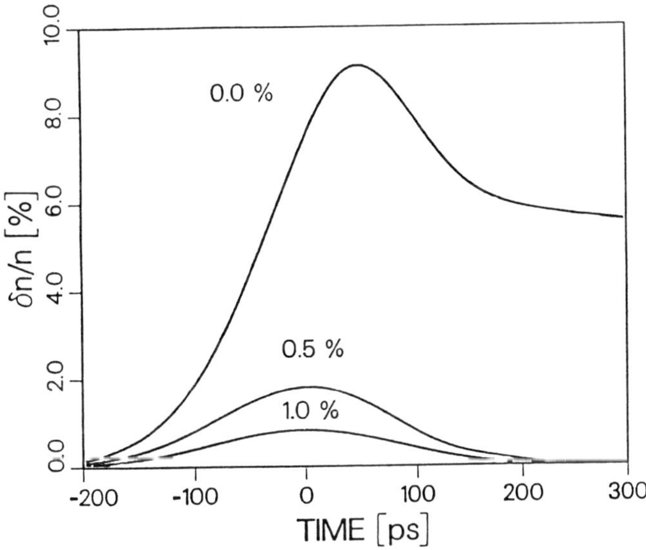


Fig. 2. The calculated plasma wave amplitude for relativistic detuning only ($\Delta n/n=0$) and linear detuning of $\Delta n/n=0.5\%,1\%$.

The longitudinal electric field of the plasma wave is given by

$$E_p = \frac{\omega_p m_e c}{e}\left(\frac{\delta n}{n}\right) \cong n^{1/2}(\delta n/n)$$

where the plasma density n is in $cm^{-3}$. Thus for our experiments for which $n=1.1\ 10^{17} cm^{-3}$ and the relativistic detuning limit gives $\delta n/n=10\%$ we have

$$E_p = 3.10^7 V/cm$$

This should be compared to $E_p=10^5 V/cm$ in conventional particle accelerators. An actual accelerator would consist of several stages since the plasma wave travels slightly slower than c so the accelerated electrons will outrun the wave.

A numerical solution to the two dimensional envelope equation (Gibbon and Bell 1988) shows that focussing of the laser beam occurs due to the casade process (the interaction between the plasma wave and the laser light to give sidebands on the laser light). This is important as it offers a means of confining the pump beams over distances greater than the Rayleigh length.

## 3. Plasma generation experiments

We now describe the experiments performed for uniform plasma production. As mentioned above only a very small plasma density variation can be tolerated if the plasma wave is to grow to the maximum amplitude allowed by relativistic detuning. For this reason multiphoton ionisation of static fill gas was investigated as pinch plasmas were found to be too irreproducible. The first experiments were performed using a frequency doubled 1.053μm laser beam at 0.53μm. However, this has the disadvantage that in a beat wave experiment an additional high power frequency doubled beam is required. Thus multiphoton ionisation with a 1.053μm beam has also been investigated.

3.1 *Multiphoton ionisation at 0.53μm*

In the experiments at RAL a frequency doubled beam of typically 10J in 100ps was focussed into hydrogen fill gas to a spot of 400μm (c.f. plasma wavelength of $\lambda_p$=100μm), giving a peak intensity of about $10^{14}$Wcm$^{-2}$. The plasma was diagnosed by observing Thomson scattered light from this same laser beam. A grating spectrometer and streak camera were used to monitor the scattered light. Spectra were obtained at four different scattering angles for each shot to give a wide range of the scattering parameter $\alpha$ (Evans and Katzenstein 1969). The length of the plasma generated was determined by collecting the scattered light at different positions along the axis of the beam. The scattering techniques are described in detail in Dangor *et al*. (1988).

The measured plasma density is shown as a function of fill presure in figure 3. This clearly shows that 100% ionisation is obtained over the pressure range investigated for beat wave experiments. The scattered spectra from different axial positions along the beam are shown in figure 4. The identical seperation of the wings, which is $\alpha\ n^{1/2}$, shows that the plasma is uniform over a length of at least 8mm, the Rayleigh length of the optics used. The electron velocity distribution is measured as maxwellian with a temperature of 8eV.

These results show that multiphoton ionisation generates a highly reproducible quiescent plasma, uniform over a least the Rayleigh length of the focussing optics used (Dangor *et al*. 1987). The plasma can be produced with a prescribed density by fixing the pressure and temperature of the hydrogen fill gas in the interaction chamber.

This technique for plasma production has been refined at EP with the development of a special interaction chamber (Amiranoff et al. 1990). The chamber can be baked to 150°C to reduce outgasing. The fill pressure can be set to better than 0.1% and the temperature of the fill gas regulated to 0.2°C. By these means the gas density can be set and maintained to 0.1% for several hours, as required for beat wave experiments. Further, a novel alignment technique has been developed which does not require the use of an alignment microballoon placed inside the vessel. This eliminates any need to let the vessel up to air.

The multiphoton ionisation experiments are similar to those performed at RAL described above. A longer pulse length of 600ps was used and focussed to a much tighter focal spot of 50μm. However, the intensity was kept at about $10^{14}$Wcm$^{-2}$. The Thomson scattered specta obtained show a strong variation of the plasma density over the time of the laser pulse. This corresponds to a drop in the average density in the focal region of 5%/100ps. This is due to the

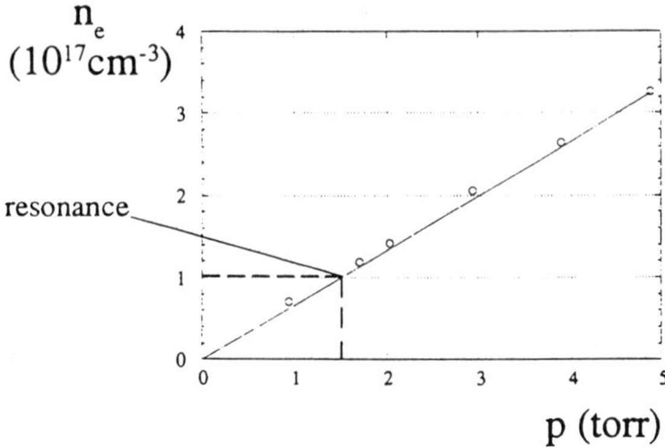


Fig. 3. The measured plasma density against hydrogen fill pressure for multiphoton ionisation at 0.53μm.

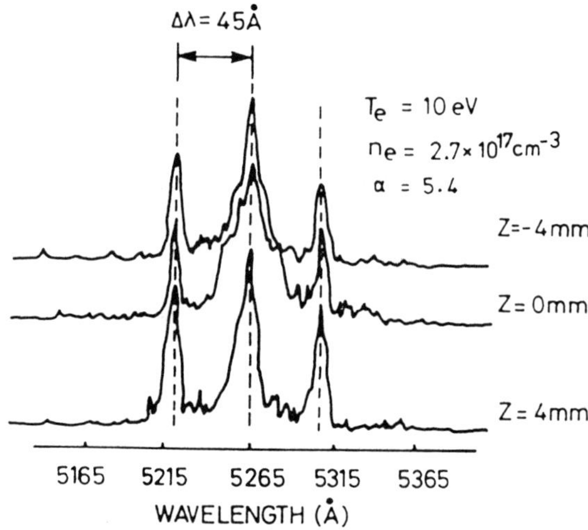


Fig. 4. Spectral profiles of the scattered light at different axial positions for multiphoton ionisation at 0.53μm.

much stronger ponderomotive force of the focussed beam driving the plasma outwards. It should be noted that here the spot size is smaller than the plasma wavelength.

### 3.2 *Multiphoton ionisation at 1.053μm*

Several experiments have been performed at RAL using a beam of typically 50J in 200ps focussed to a 400μm spot giving a peak intensity of 5 $10^{14}$Wcm$^{-2}$ (Dangor *et al.* 1990a). The experimental arrangement was essentially the same as for the experiments at 0.53μm. A streak photograph of a typical scattered spectrum is shown in figure 5. It can be seen that the spectrum in the forward direction (15°) is assymetric with a dominant upshifted wing (c.f. symmetric Thomson scattered spectra obtained at 0.53μm, figure 4). The position of this blue wing corresponds to 100% ionisation of the fill gas. The magnitude of this wing is estimated to be three orders of magnitude larger than expected from a thermal plasma. The scattered spectra in the backward direction (165°) are symmetric but non maxwellian. These results imply the existence of large turbulence at the plasma frequency.

**Streak Camera Record of Scattered Spectra in Forward(15°) and Backward (165°) Directions**

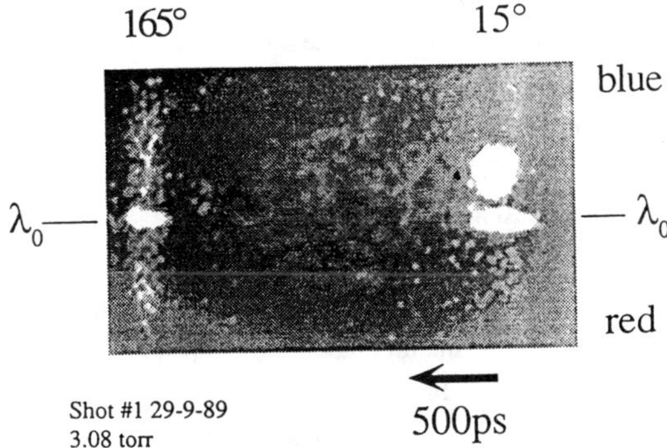


Fig. 5. Streak photograph showing the scattered light for multiphoton ionisation at 1.053μm. Note assymetry in forward direction (15°).

The anomalous nature of these scatering spectra has yet to be explained. However, the results indicate that the laser pump beams in a beat wave experiment themselves generate the plasma without recourse to a seperate 0.53μm preionising beam.

## 4. The beat wave experiment

In this experiment the Vulcan laser at RAL was operated to give beams at 1.064μm and 1.053μm. Each beam was 50J in 200ps focussed with f/20 optics to a focal spot of 400μm. The plasma wave was diagnosed by monitoring the sidebands generated on a third copropagating beam at 0.53μm. This beam was apertured to f/100 and blocks of corresponding size were placed in each of the pump beams. This ensured that the probe and pump beams had almost no common path in any of the quartz optical elements. This was necessary to reduce the sidebands on the probe beam produced by interaction with phonons in quartz induced by the pump beams (Dyson *et al.* 1989). Only the upshifted sideband on the pump was monitored in order to maintain sufficient rejection of the unshifted light with the triple monochromator used.

It is interesting to note that phonon generation in quartz is identical to that of the beat wave in a plasma. A similar process gives rise to sidebands in air if the pump beams have a common air path (Dangor *et al.* 1988). This interaction is strong as the beat frequency is close to a vibrational transition in the nitrogen molecule. Both these effects must be avoided to observe the beat wave.

It is found that the upshifted sideband on the probe light varies strongly with fill pressure. An enhancement of more than one order of magnitude is observed at resonant density indicating the presence of a plasma wave generated by the beat wave. The results are shown in figure 6. The plasma wave amplitude is estimated to be 2% corresponding to a longitudinal electric field of about 5MeV/cm (Dangor *et al.* 1990b). This is similar in magnitude to that observed by Clayton *et al.* (1985) in a beat wave experiment using a $CO_2$ laser.

The amplitude of the plasma wave is smaller than that expected for relativistic detuning, see figure 3. Further as shown in figure 7 the lifetime of the sideband is only 50ps compared to the theoretically expected value of 150ps. This implies that some instability is limiting the growth of the plasma wave. One possible instability is the modulational (Mora 1988).

A 1d non-relativistic particle simulation code to simulate the modulational instability has been written by Bell (1989). In the code the plasma wave is driven by the ponderomotive force

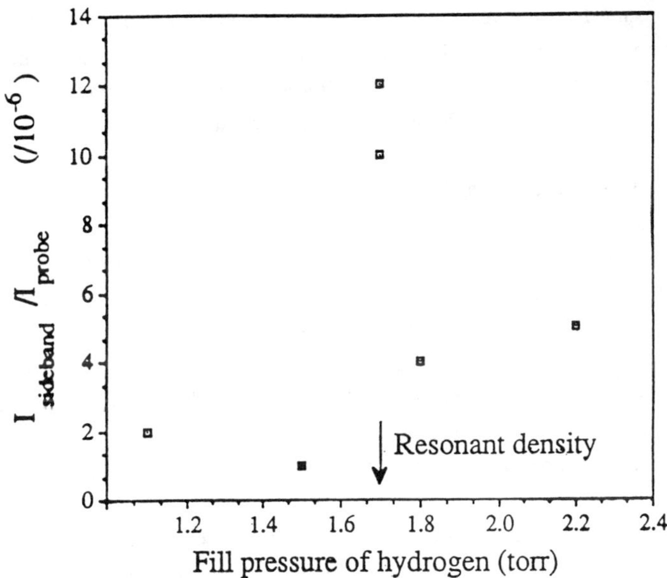


Fig. 6. Ratio of the sideband to the copropagating probe at 0.53μm for various fill pressures of hydrogen in the beat wave experiment. The order of magnitude enhancement indicates a plasma wave.

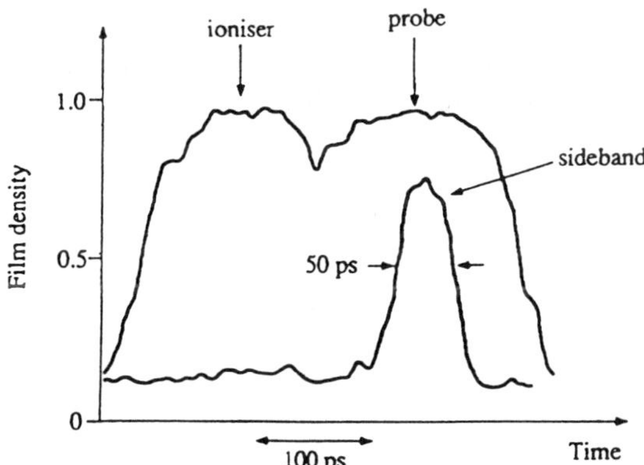


Fig. 7. Temporal evolution of the sideband at resonance in the beat wave experiment.

which was assumed to be given and ion motion is allowed. The code limits the plasma wave amplitude to our experimentally observed value with the same plasma wave duration. Future experiments using Thomson scattering to monitor ion waves and the evolution of the electon density distribution will help to further investigate the instability.

## 5. Discussion and future work

The beat wave has two shortcomings. These are that the plasma density needs to be precisely tuned and the detremental influence of instabilities. In order to minimise the effects of instabilities the plasma wave should be driven to maximum amplitude as rapidly as possible. Thus the growth time should be of the order of the pump beam duration. Using the equations given earlier and matching the pulse duration we see that $dn/n \quad 1/\tau_{pulse}{}^{1/2}$ and the necessary beam energy $E \quad I\tau_{pulse} \quad 1/\tau_{pulse}{}^{1/2}$. Thus using shorter pulses gives larger plasma wave amplitudes. A pulse width of 20ps requires a beam energy of 35J for matching and would give a plasma wave of 25%.

In the wakefield the plasma wave is excited on the timescale of the laser pulse and no strict resonance condition exists (Sprangle *et al.* 1988). For optimum driving of the plasma wave the pulse duration should be of the order of half the plasma period i.e. $\omega_p\tau=\pi$. The resulting plasma wave amplitude is $\delta n/n=(1/2)(v/c)^2$. If the focal spot is taken to be at least the plasma wavelength then $\delta n/n \quad E/\tau_{pulse}{}^3$ . This is necessary to avoid any two dimensional effects limiting the plasma wave amplitude. For a 1μm laser pulse of 300fs duration (corresponding to a plasma density of 5 $10^{16}cm^{-3}$) a laser energy of 35J would give $\delta n/n=40\%$. An ongoing programme at RAL is developing the laser necessary for such experiments.

## REFERENCES

Amiranoff, F. *et al.* 1990 J. Appl. Phys. (accepted for publication)
Bell, A. R. & Gibbon, P. 1988 Plas. Phys. and Contr. Fus. **30**, 1319
Clayton, C. E. *et al.* 1985 Phys. Rev. Lett. **54**, 2343
Dangor, A. E. *et al.* 1987 IEEE Trans. on Plas. Sci. **PS-15**, 161
Dangor, A. E. *et al.* 1988 J. Appl. Phys. **64**, 6182
Dangor, A. E. *et al.* 1989 J. Phys. B: At. Mol. Opt. Phys. **22**, 797
Dangor, A. E. *et al.* 1990 Physica Scripta (accepted for publication)
Dyson, A *et al.* 1989 J. Phys. B: At. Mol. Opt. Phys. **22**, L231
Ebrahim, N. A. *et al.* 1986 IEEE Trans. Nucl. Sci. **32**, 3539
Evans, D. E. & Katzenstein, J 1969 Rep. Prog. Phys. **32**, 207
Gibbon, P & Bell, A. R. 1988 Phys. Rev. Lett. **61**, 1599
Karttunen, S. J. & Salomoa, R. R. E. 1986 Phys. Rev. Lett. **29**, 701
Kitagawa, Y. *et al.* 1990 Phys. Rev. Lett. (submitted for publication)
Mora, P. 1988 Revue Phys. Appl. **23**, 1489
Rosenbluth, M. N. & Liu, C. S. 1972 Phys. Rev. Lett. **29**, 701
Sprangle, P. *et al.* 1988 Appl. Phys. Lett. **53**, 2146
Tajima, T. & Dawson, J. M. 1979 Phys. Rev. Lett. **43**, 267

**Research Trends in Physics: Coherent Radiation Generation and Particle Acceleration**
Editorial Board: J.M. Buzzi, A. Prokhorov (Editor-in-Chief), P. Sprangle, and K. Wille
*La Jolla International School of Physics*, The Institute for Advanced Physics Studies, La Jolla, California

# Laser Particle Acceleration Research at Chalk River Laboratories

**N.A. Ebrahim**

Chalk River Laboratories, AECL Research
Chalk River, Ontario, Canada K0J 1J0

ABSTRACT

At Chalk River we have established a facility to study various aspects of novel particle acceleration concepts. A number of important features have been incorporated into this experimental facility. The pulsed high pressure $CO_2$ laser system is capable of producing short pulses (300 - 500 ps) at two wavelengths (9.6 and 10.6 $\mu$m or 10.3 and 10.6 $\mu$m) at focal power densities in excess of $10^{14}$ W/cm$^2$, both of which are relevant to most particle acceleration schemes. The plasma medium at a density of $10^{16}$ - $10^{17}$ cm$^{-3}$ is produced by tunneling ionization of neutral hydrogen or argon gas by focused laser radiation, which also drives the relativistic electron plasma wave. A 3 GHz electron linear accelerator injects 10 MeV electrons in 30 ps microbunches with approximately $10^8$ - $10^9$ electrons in each microbunch. Extensive beam dynamics calculations with TRANSOPTR and TRANSPORT and computer modeling of the dipole magnets with POISSON and TOSCA were used to design a doubly achromatic, double-focusing beam line injection system and a broad-range, modified Browne-Buechner electron spectrometer with a multichannel detector system to detect accelerated particles from the laser plasma interaction region.

## 1. INTRODUCTION

Recently there has been a great deal of interest in studying the behaviour of relativistic large amplitude plasma waves because of their relevance to problems ranging from hot electron preheat in laser fusion targets through current drive in tokamaks to their potential applications in rapidly accelerating particles to relativistic energies, and in the final focusing of particle beams in high-energy particle accelerators. In the area of particle acceleration, there are at least three concepts for plasma based collective accelerators all of which involve plasma waves with phase velocities almost equal to the velocity of light, and plasma wave amplitudes which are a significant fraction ($\geq 10\%$) of the cold-plasma wavebreaking limit. In the laser plasma beatwave accelerator (LPBWA) concept proposed by Tajima and Dawson[1], the plasma waves are generated by the non-linear coupling of two intense laser beams of slightly different frequencies propagating through a low-density plasma. If the difference frequency of the lasers is chosen to match the plasma frequency, the ponderomotive force of the beatwave can resonantly build up the relativistic plasma wave. In the plasma wakefield accelerator (PWFA) concept the plasma waves are excited by a short, intense electron bunch propagating through a high density plasma[2]. The space charge force of the electron bunch displaces the plasma electrons and generates a wake of plasma oscillations with a phase velocity that is equal to the driving electron bunch velocity, which is very close to the velocity of light. An alternative to using an electron bunch, is to inject an extremely short but intense laser pulse into a low density plasma as is the case in the laser wakefield accelerator (LWFA) concept[3]. The ponderomotive force of the laser pulse envelope initially expels the plasma electrons both radially and axially, resulting in plasma oscillations as the returning electrons overshoot their initial positions. The acceleration of particles in the three concepts is identical. A trailing relativistic electron bunch injected into the potential well of the plasma wave at the appropriate phase remains synchronized to the wave and accelerated for a significant period of time.

In this paper we discuss a facility we have established at Chalk River to study novel concepts in particle acceleration and to assess the role of lasers and plasmas in future high energy accelerators[4]. There are four separate technologies that need to be developed and integrated in such an experimental facility. The first is a multi-atmosphere, multi-wavelength subnanosecond $CO_2$ laser facility for driving the plasma waves; the second is a plasma source for producing high density, relatively, homogeneous plasmas; the third is a short pulse, high energy linear accelerator to produce short electron bunches; and the fourth is appropriate injection beamlines, spectrometers and detection systems. We discuss a number of important features that have been incorporated into this experimental facility to give it the flexibility necessary to examine a range of problems in laser and plasma based acceleration concepts.

## 2 BASIC PRINCIPLES

### (i) Plasma Wave Generation

In the laser plasma beatwave acceleration concept, the plasma waves are generated by the non-linear coupling of two copropagating laser beams in a low density plasma. The force which acts on the plasma electrons (ponderomotive force) and causes them to bunch is directed along the propagation direction of the electromagnetic waves, is periodic and originates in a non-zero **v** x **B** force of the electromagnetic waves as shown in Figure (1). The ponderomotive force is given by[5]

$$\boldsymbol{F}_{NL} = -\frac{\omega_{pe}^2}{\omega_0\omega_1}\nabla\frac{\langle|\boldsymbol{E}_0+\boldsymbol{E}_1|^2\rangle}{8\pi} \tag{1}$$

where $\omega_{pe} = (4\pi n_e e^2/m_e)^{1/2}$ is the electron plasma frequency, $n_e$ is the plasma electron density, $\omega_0$ and $\omega_1$ are the laser frequencies and $\mathbf{E}_0$ and $\mathbf{E}_1$ are the electric fields of the two laser beams.

We can estimate the maximum possible amplitude of the plasma wave by considering the maximum possible bunched electron density or density fluctuation $\delta n_e$. For a background plasma electron density of $n_e$, the maximum possible density fluctuation $\delta n_e = n_e$ in Figure (1).

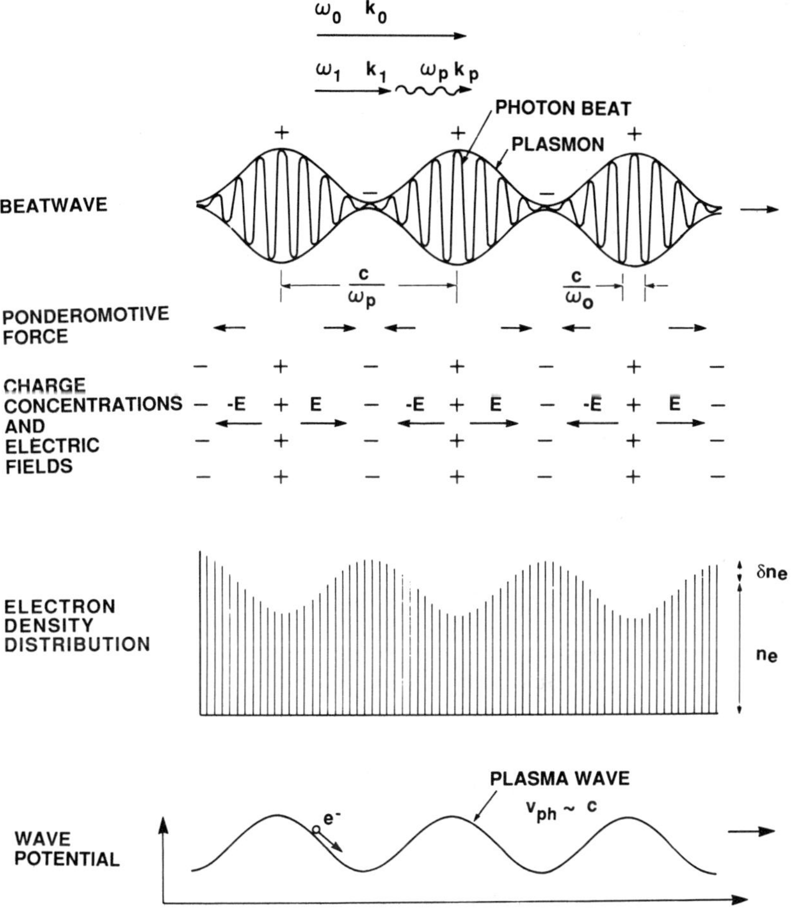


Fig.1. Schematic diagram of the beatwave, ponderomotive force, concentrations of the positive and negative charges, the associated electric fields, electron density distribution, and the ponderomotive wave in the plasma.

From Poisson's equation

$$\nabla \cdot \mathbf{E} = 4\pi\rho = -4\pi e\delta n_e = -4\pi e n_e \qquad \delta n_e \sim n_e \tag{2}$$

and

$$\nabla^2\phi \approx 4\pi e n_e \tag{3}$$

and

$$|e\phi| \sim \frac{4\pi n_e e^2}{m_e}\frac{m_e}{k^2} = \frac{\omega_{pe}^2}{c^2k^2}m_e c^2 \tag{4}$$

The maximum electric field

$$|\mathbf{E}_{max}| = |k\phi_{max}| = \frac{\omega_{pe} m_e c^2}{ce} = 0.94\sqrt{n_e} \qquad V/cm \tag{5}$$

where $n_e$ is the plasma electron density in $cm^{-3}$.

For a general case the amplitude of the density fluctuation $\epsilon = \delta n_e/n_e$ can be obtained from the following equation within the slowly varying envelope approximation[6-9]

$$\left[\frac{1}{\omega_{pe}}\frac{d}{dt} - i\frac{3}{16}|\varepsilon|^2 + i\frac{\Delta\omega}{\omega_{pe}} - \frac{\nu_c}{2\omega_{pe}}\right]\varepsilon = \frac{1}{4}\alpha_0\alpha_1 \tag{6}$$

where $\Delta\omega = \omega_{pe} - (\omega_0 - \omega_1)$ is the frequency detuning from the exact resonance, $\nu_c$ is the electron-ion collision frequency and $\alpha_1 = v_{01}/c = eE_1/m_e\omega_1 c$ is the normalized oscillatory velocity of an electron in the laser field $E_1$. The driver term on the right hand side of Equation (6) is defined in terms of the laser field strengths ($E_0$ and $E_1$) of the two laser beams. The second term on the left hand side, which gives a nonlinear frequency shift as the plasma wave amplitude increases, is due to the relativistic mass increase of the oscillating electrons as $m_e \to \gamma m_e$ and $\omega_{pe}^2 \to \omega_{pe0}^2/\gamma$, where $\omega_{pe0}$ is the plasma frequency with the rest mass, and $\gamma$ is the relativistic Lorentz factor defined by the mean electron velocity in the wave. The third term on the left hand side describes a linear detuning of the plasma frequency $\omega_{pe}$ from the exact resonance with the beat frequency $(\omega_0 - \omega_1)$. The fourth term on the left hand side is the linear damping term due to electron-ion collisions.

Equation (6) shows that the plasma wave amplitude initially grows linearly with time according to the relation

$$\frac{1}{\omega_{pe}} \frac{d\varepsilon}{dt} = \frac{1}{4} \alpha_0(t)\ \alpha_1(t) \tag{7}$$

In the absence of other effects, relativistic detuning limits the amplitude of the plasma wave. As the plasma wave grows, the plasma electrons increase in mass and the plasma frequency $\omega_{pe}$ no longer matches the driver frequency $(\omega_0 - \omega_1)$ and this dephasing causes the amplitude growth to slow, stop and then reverse[10]. The saturated amplitude is given by

$$\varepsilon_s = \left[ \frac{16}{3} \alpha_0(t)\ \alpha_1(t) \right]^{\frac{1}{3}} \tag{8}$$

and the time to saturation is determined from

$$\frac{\omega_{pe}}{4} \int_0^{\tau_s} [\ \alpha_0(t')\ \alpha_1(t')\ ]\ dt' = \left[ \frac{16}{3} \alpha_0(t)\ \alpha_1(t) \right]^{\frac{1}{3}} \tag{9}$$

Assuming a linear risetime $\tau$ for the laser intensity,

$$I(t) = I_0 \frac{t}{\tau}$$

$$\alpha(t) = \alpha_0 \left(\frac{t}{\tau}\right)^{\frac{1}{2}} \tag{10}$$

the time to saturation is given by

$$\omega_{pe}\tau_s = 4.87 \left[ \frac{\tau\ \omega_{pe}}{\alpha_0\ \alpha_1} \right]^{\frac{2}{5}} \tag{11}$$

The saturated electric field amplitude is then given by

$$E_s = 0.94 \sqrt{n_e}\ \varepsilon_s \quad V/cm \tag{12}$$

For maximum plasma wave amplitude, the time to saturation Equation (11) should be matched to the laser pulse duration. It is an advantage to choose the plasma density so that the exact frequency match between the driver and the plasma frequency occurs not at the beginning of the laser pulse but later on during the growth when the laser intensity is greater[10]. This detuning can increase the saturated amplitude by as much as 60%. It has also been suggested that the plasma density could be continuously adjusted to maintain exact resonance at all times during the laser pulse and a number of approaches have been suggested to achieve this in practice[11].

For resonant excitation of plasma waves (frequency $\omega_p$ and wavenumber $k_p$) with two copropagating laser beams of frequency $\omega_0$ and $\omega_1$ and wavenumbers $k_0$ and $k_1$ respectively, the following wavematching relations must be satisfied

$$\omega_p = \omega_0 - \omega_1 = \Delta\omega \qquad \boldsymbol{k_p} = \boldsymbol{k_0} - \boldsymbol{k_1} = \Delta \boldsymbol{k} \tag{13}$$

The phase velocity of the plasma wave $v_p$ is given by

$$v_p = \frac{\omega_p}{k_p} = \frac{\omega_0 - \omega_1}{k_0 - k_1} = \frac{\Delta\omega}{\Delta k} = v_g^{EM} \tag{14}$$

where the group velocity of the electromagnetic waves $v_g^{EM}$ is obtained from the dispersion relation for an electromagnetic wave in a plasma

$$\omega^2 = \omega_{pe}^2 + k^2c^2 \tag{15}$$

The group velocity is

$$v_g^{EM} = \frac{d\omega}{dk} = \frac{kc^2}{\omega} = c\left(1 - \frac{\omega_{pe}^2}{\omega^2}\right)^{1/2} \tag{16}$$

The phase velocity is then given by

$$v_p = c\left(1 - \frac{\omega_{pe}^2}{\omega^2}\right)^{1/2} \tag{17}$$

and the Lorentz factor $\gamma_p$ is defined as

$$\gamma_p = \left(1 - \frac{v_p^2}{c^2}\right)^{-1/2} = \left(\frac{\omega}{\omega_{pe}}\right) = \left(\frac{n_c}{n_e}\right)^{1/2} \tag{18}$$

where $n_c$ is the critical density for the laser beam (where the laser frequency equals the plasma frequency), which for

Table I. Parameters of the Chalk River Laser Acceleration Experiment

| | | |
|---|---|---|
| Laser Wavelengths | 10.25 $\mu$m<br>10.59 $\mu$m | 9.55 $\mu$m<br>10.55 $\mu$m |
| Resonant Electron Density ($n_e$) | $10^{16}$ $cm^{-3}$ | $10^{17}$ $cm^{-3}$ |
| Plasma Frequency | 5.69 x $10^{12}$ | 1.87 x $10^{13}$ |
| Phase Velocity ($\beta_p = v_p/c$) | 0.9995 | 0.995 |
| Lorentz Factor $\gamma_p$ | 32 | 10 |
| Electron Quiver Velocity ($v_o/c$) | 0.07 | 0.07 |
| Pulse risetime ($\tau_r$) | 250 ps | 250 ps |
| Pulse width ($\tau$) | 500 ps | 500 ps |
| Saturation time ($\tau_s$) | 130 ps | 65 ps |
| Density Fluctuations ($\delta n_e/n_e$) | 0.24 | 0.19 |
| Field Gradient | 2.4 GeV/m | 6.0 GeV/m |
| Dephasing Length ($L_p$) | 22 cm | 0.7 cm |
| Injection Energy | 5 - 10 MeV | 5 - 10 MeV |
| Maximum Energy Gain | 30 MeV | 40 MeV |
| Maximum Output Energy | 40 MeV | 50 MeV |

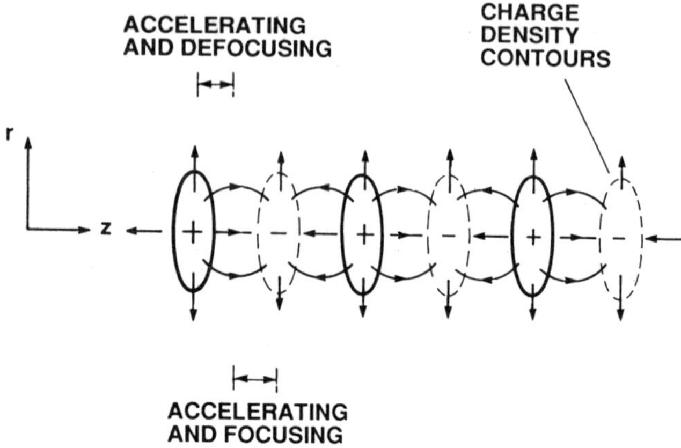


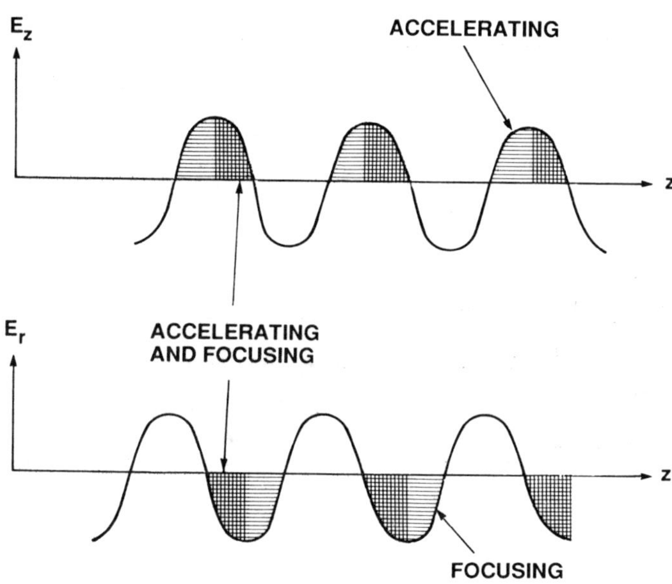


Fig.2. Charge density contours, axial and radial electric fields in a plasma wave of finite radial dimensions.

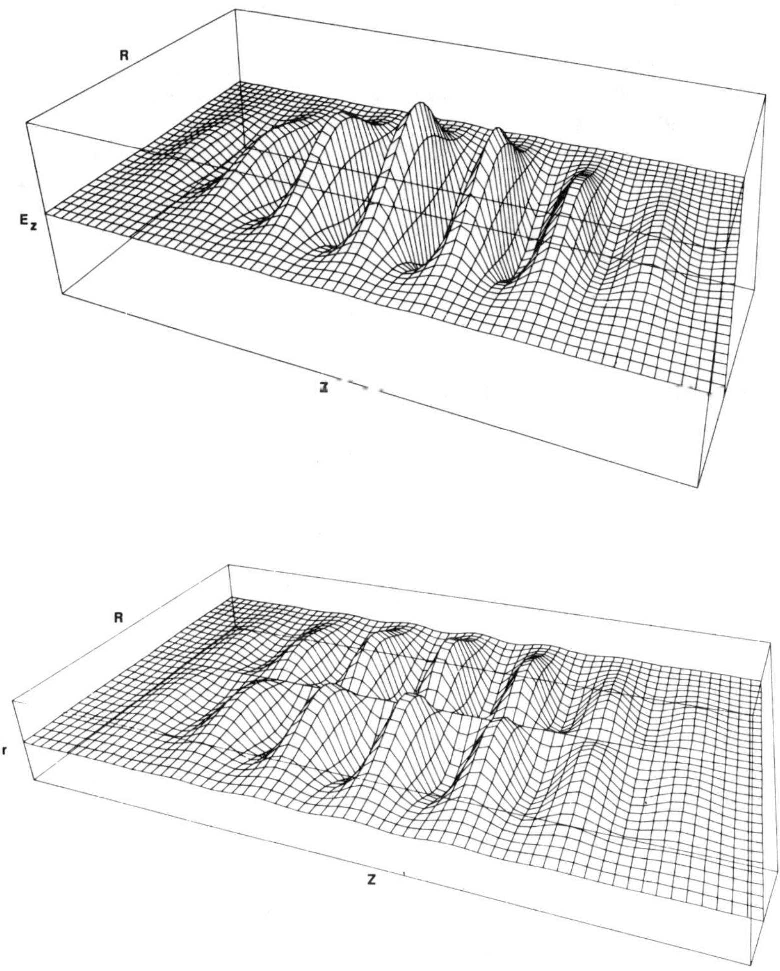


Fig.3. Profiles of the axial ($E_z$) and radial ($E_r$) fields of the plasma waves. The laser axis lies along the axial z-direction.

a $CO_2$ laser is $n_c = 10^{19}$ cm$^{-3}$.

An important feature of laser accelerators is the synchronism that exists between the group velocity $v_g$ of the laser pulse and the phase velocity of the plasma waves as shown in Equations (16) and (17). It follows that the trapped particles being accelerated are also in synchronism with the laser pulse, and particles are accelerated by the electrostatic wave until they outrun the wave. The maximum energy gain is

$$\Delta W \approx 4 \, \varepsilon_s \, \gamma_p^2 \, m_e \, c^2 \qquad (19)$$

The dephasing distance is

$$L_p = \frac{1}{2} \lambda_p \, \gamma_p^2 \qquad (20)$$

Equation (20) is nothing more than a statement that when a particle accelerates, it can move from the top to the bottom of the potential well of the plasma wave. In other words it can travel, at most, half a plasma wavelength in the (moving) wave frame.

Typical parameters for the Chalk River experiments calculated from Equations (7) through (20) are given in Table I for plasma densities of $10^{16}$ and $10^{17}$ cm$^{-3}$. Since the phase velocity of the plasma wave is somewhat lower at the higher electron densities as shown in Equation (17), the injection energy of the particles for trapping is also lower. Furthermore significant laser beam refraction effects are likely to occur at the higher densities as a result of the intensity threshold behaviour of the multiphoton ionization process and the propagation of the laser beam through such a medium, as discussed in the next section.

As a result of the finite radial extent of the plasma waves (on the order of the focal spot size of the laser beams), there will be radial fields[12,13] that will influence the transverse dynamics of the injected electron beam (Figure 2). Profiles of the axial and radial fields (Figure 2 and 3) show that the radial fields are focusing or defocusing depending on the particular phase. Ideally, particles should be injected in the accelerating and focusing phase in which case the accelerated beam is likely to be tightly focused. In the defocusing phase, on the other hand, the particles are likely to be scattered.

The presence of radial fields has a strong bearing on the conceptual design of the electron beam injection system. Ideally the electron beam spot size should be on the order of the laser beam spot size or less. Because of the long beam waist of the laser beam (on the order of 1.5 cm) it is desirable that the electron beam injection system also have a beam waist in that region. As discussed in the following sections, in the present facility the injection beam line is designed to produce an electron beam focal spot diameter of approximately 300 $\mu$m which matches the laser focal spot size.

### (ii) Plasma Generation

The long plasma column in these experiments is generated by ionization of hydrogen or argon gas by a focused laser beam. For the ionization of a hydrogen atom (ionization potential of 13.6 eV) with a $CO_2$ laser beam (photon energy of 0.117 eV) 117 photons are required. To produce a fully ionized plasma column 1.5 cm long, 300 $\mu$m in diameter with a uniform density of $10^{16}$ $cm^{-3}$, $10^{15}$ photons are required. For comparison, a 50 joule $CO_2$ laser pulse contains 3 x $10^{21}$ photons.

The nonresonant ionization of a uniform gas or partially ionized plasma by intense laser fields is separated into two regimes by the Keldysh tunneling parameter[14]

$$\gamma_K = \left(\frac{E_{ion}}{2\Phi_{pond}}\right)^{1/2} \tag{21}$$

where $E_{ion}$ is the ionization potential of the atom or ion and $\Phi_{pond}$ is the ponderomotive potential of the laser (average kinetic energy of an electron in the laser field)

$$\Phi_{pond}(eV) = \frac{e^2E^2}{4m_e\omega^2} = 9.33 \times 10^{-14}\, I\, \lambda^2 \tag{22}$$

where E is the electric field strength of the laser, I is

the focused laser irradiance in watts/cm$^2$ and $\lambda$ is the laser wavelength in microns.

For short wavelength lasers (ruby or neodymium) at moderate laser intensities we are in the $\gamma_K > 1$ regime, and ionization is described as a multiphoton ionization process where an atom or an ion simultaneously absorbs N photons

$$A^{n+} + N\hbar\omega \rightarrow A^{n+1} + e^- \tag{23}$$

where $N\hbar\omega > E_{ion}$.

With high laser intensity, long wavelength lasers (typical of $CO_2$ laser experiments) we are in the $\gamma_K < 1$ regime and ionization is described as a tunneling process.

Tunneling ionization in hydrogen gas has been used to produce plasmas for laser acceleration experiments with a $CO_2$ laser[15-16] and plasmas produced by multiphoton ionization with frequency-doubled 1.053 $\mu$m lasers have been extensively studied using Thomson scattering of laser light[17-18]. The latter studies suggest that uniform plasma columns can be produced by this technique.

There is however a problem with laser beam refraction that occurs at higher densities in plasmas produced by multiphoton ionization, as a result of plasma induced refractive index change. This effect limits the intensity of the focused laser beam to a value close to the threshold for multiphoton ionization. The conditions for significant laser beam refraction can be derived from considerations of Gaussian laser beam propagation in a medium with threshold intensity for ionization[19]. The intensity distribution I of a gaussian laser beam is given by

$$I = \frac{I_0}{(1 + z^2/z_r^2)} \exp-(2r^2/\omega^2) \tag{24}$$

and the wavefront radius of curvature

$$R(z) = z\,(1+\frac{z_r^2}{z^2}) \tag{25}$$

where z is the distance along the axis measured from the beam waist, $z_r^2 = (\pi\omega_0^2/\lambda)^2$, where $z_r$ is the Rayleigh range, $I_0$ is the intensity of the focused laser beam and

$$\omega(z) = \omega_0 \left( 1 + \frac{z^2}{z_r^2} \right)^{\frac{1}{2}} \tag{26}$$

z = 0 corresponds to the beam waist with spot radius $\omega_0$.

As the focused laser beam converges, the laser intensity eventually exceeds the threshold intensity $I_{th}$ for multiphoton ionization. Assuming that $I_0 >> I_{th}$ there will exist in the near field a threshold contour (with radius of curvature $R_0$) beyond which the gas will be ionized (Figure 4). From Equation (24), the curvature of the threshold contour in the limit $r_{th} \to 0$ is

$$\frac{d^2 z_{th}}{dr_{th}^2} = \frac{1}{R_0} = -\frac{2}{\omega_0^2}\frac{1}{(z_{th}/z_r^2)} \simeq -2\,\frac{z_r}{\omega_0^2}\frac{I_0/I_{th}}{(I_0/I_{th}-1)^{3/2}} \tag{27}$$

If $R_1$ is the wavefront curvature of the focused laser beam just to the left of the ionized region (Fig. 4), then from Equation (24)

$$\frac{1}{R_1} = \frac{z_{th}}{z_r^2}\frac{1}{(1+z_{th}^2/z_r^2)} = \frac{(I_0/I_{th}-1)^{1/2}}{z_r(I_0/I_{th})} \tag{28}$$

If $R_2$ is the wavefront curvature just inside the ionized region, then from the thin lens formula

$$\frac{1}{R_2} = \frac{n_2-n_1}{n_2}\frac{1}{R_0} + \frac{n_1}{n_2}\frac{1}{R_1} \tag{29}$$

where $n_1$ and $n_2$ are the refractive indices of the regions to the left and right of the threshold contour respectively. The index of refraction of light in a plasma is

$$n = \left[ 1 - \left(\frac{\omega_{pe}}{\omega_0}\right)^2 \right]^{\frac{1}{2}} \approx 1 - \frac{1}{2}\frac{n_e}{n_c} \tag{30}$$

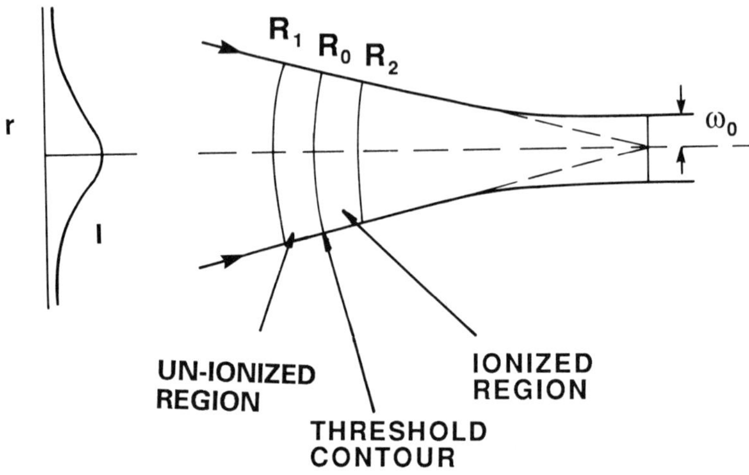


Fig.4. Focusing of a Gaussian laser beam. Axial direction of the laser beam is along the z-axis. $R_0$ is the radius of curvature of the threshold contour. $R_1$ is the curvature of the contour that lies in the un-ionized region. $R_2$ is the curvature of the contour that lies in the ionized region.

where $\omega_{pe}$ is the plasma frequency, $\omega_0$ is the light frequency, $n_e$ is the electron density and $n_c$ is the critical density (where the light frequency equals the plasma frequency). The change in the index of refraction across the threshold contour $\delta\eta \approx 1/2\ \delta n_e/n_c$, where $\delta n_e$ is the electron density increase as a result of multiphoton ionization. As the incident light propagates across the threshold contour, its radius of curvature $R_2$ increases (as a result of beam refraction) and becomes infinite when

$$\frac{\delta n_e}{n_c} = \frac{\lambda^2}{\pi^2 \omega_0^2} \tag{31}$$

Equation (31) is derived by setting $R_2 = \infty$ and $n_2 - n_1 \approx 1/2\ \delta n_e/n_e$, in Equation (29). For electron density change greater than that given in Equation (31), laser light propagating through the plasma will diverge thereby limiting the laser intensity to a value near the threshold for multiphoton ionization.

In the case where the laser light is focused onto a vacuum-gas interface, multiphoton ionization will still result in laser beam refraction, since the electron density builds up most rapidly on the axis at the peak of the Gaussian radial intensity profile[19]. The change in refractive index as a result of multiphoton ionization is given by $\delta\eta \approx 1/2\ \delta n_e/n_c$. For the plasma to modify the beam by one diffraction-limited beam-divergence the wave front must undergo a phase change of $\pi/2$. A phase change due to the plasma is given by

$$\delta\phi = \frac{2\pi\int\delta\eta\, dz}{\lambda} = \frac{\pi\int\delta n_e dz}{\lambda n_c} \tag{32}$$

From Equation (32) the distance over which the phase changes by $\pi/2$ as a result of multiphoton ionization is $L_{mpi} \approx 1/2 \lambda n_c/\delta n_e$. The diffraction length $L_{diff} = \pi\omega_0{}^2/\lambda$. For multiphoton ionization to significantly shorten the focal depth, $L_{mpi} < L_{diff}$ in which case

$$\frac{\delta n_e}{n_c} > \frac{1}{2}\frac{\lambda^2}{\pi\omega_0^2} \tag{33}$$

This expression gives similar upper limits on the plasma density for significant laser beam refraction to that derived in Equation (31).

Equation (31) shows that in $CO_2$ laser experiments with $\lambda = 10.6\ \mu m$ and a focal spot $\omega_0 \approx 100\ \mu m$, laser beam refraction from multiphoton ionization will be significant for plasma electron densities in excess of $10^{16}\ cm^{-3}$.

To avoid significant beam refraction effects, it is desirable that the plasma density not exceed $10^{16}\ cm^{-3}$. However, at these lower densities the dephasing length $L_p$ (Eq. 20) is considerably longer, and a longer plasma length is required to accelerate a particle to the maximum possible energy.

## 3 EXPERIMENTS

### (i) Laser Facility

The layout of the low energy short pulse driver for the $CO_2$ laser facility is shown schematically in Fig. (5). For simultaneous dual-wavelength operation it is essential that the two emissions be independently tunable, free from gain-competition effects, and well synchronized in time. Gross jitter problems are eliminated by using a common electrical discharge in the oscillator for the two cavities. The oscillator in this facility is a UV-preionized transversely-excited atmospheric-pressure (TEA) Lumonics 821 HP $CO_2$ laser with an active discharge length of 48 cm and a clear aperture of 3.3 cm x 2.8 cm. The two cavities are spatially separated to avoid gain competition. The variable delay in the two emissions, which arises from the different gains of the corresponding rotational lines is compensated for by two separate quasi-cw low-pressure gain sections, which allow a control of the pulse buildup time by varying the low-pressure section parameters[20]. Furthermore, the hybrid configuration allows the operation of both cavities in single longitudinal and transverse modes, resulting in smooth temporal and spatial pulse shapes.

The two low-pressure discharge tubes are identical in construction, with an active length of 60 cm and a bore diameter of 1.5 cm. The plasma tubes are operated with a trigger-transformer-delay unit (TTD) in conjunction with an

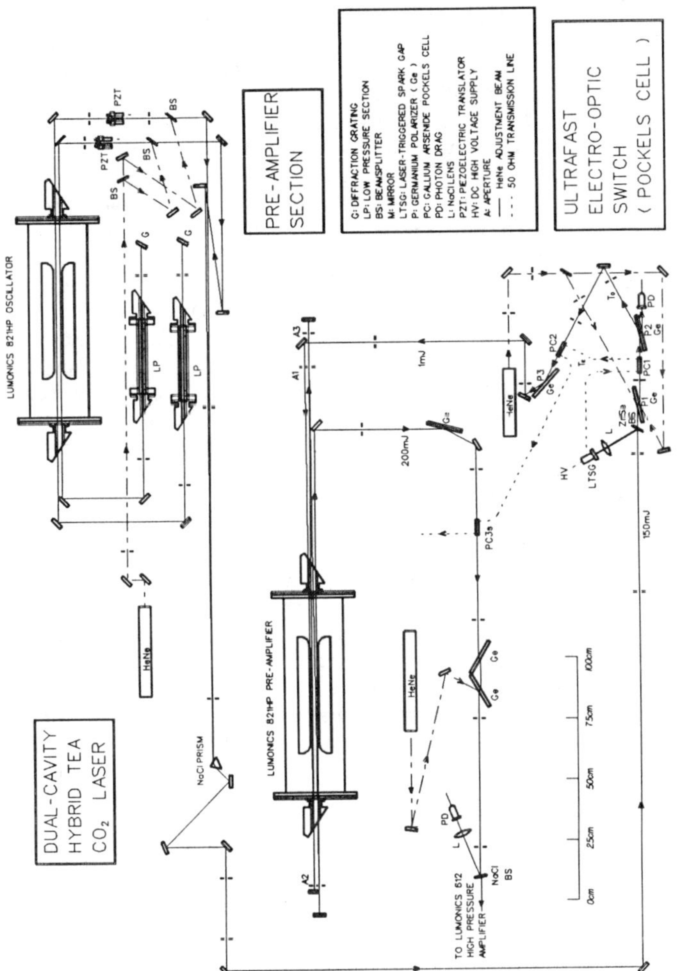


Fig.5. Schematic of the dual-cavity hybrid TEA $CO_2$ laser, the electro-optic (Pockel's cells) switch, and the pre-amplifier section.

auto-ignition coil high voltage pulse transformer. The low-pressure sections are timed to fire approximately 650 $\mu s$ before the main TEA discharge and are operated at pressures between 3 and 12 Torr of premixed gas with the $CO_2$:$N_2$:He ratio of 10:10:80 flowing continuously through the tubes. Both the main oscillator and the low-pressure tubes are sealed with anti-reflection-coated NaCl windows at Brewster's angle. The two cavities are formed with Littrow-mounted blazed gratings at one end and dielectric-coated 60% reflectivity 20-m radius of curvature germanium mirrors at the output end. The output mirrors are mounted on piezoelectric transducers (PZT's) to give fine control of cavity lengths. Adjustable intracavity apertures near the gratings are used to achieve the lowest transverse-mode operation. The two cavities are 2.7 m long with the corresponding longitudinal-mode spacing of approximately $\Delta\nu = c/2L \approx 55$ MHz. Figure (6) shows a photograph of the hybrid TEA oscillator system and the associated optical components and alignment beams.

The dual-cavity hybrid $CO_2$ oscillator produces 200 mJ, 100 ns single longitudinal and transverse mode pulses tuned to the 10P20 transition at 10.6 $\mu m$ and 10R20 transition at 10.3 $\mu m$. By adjusting the premix pressures in the plasma tubes, we vary the pulse buildup time and hence synchronize the emissions from the two cavities. Collinear propagation of the two beams is obtained by transmitting the beams through an anti-reflection-coated NaCl prism at the appropriate angles of incidence so that the emerging beams overlap and are collinear.

A variable width pulse is sliced from the hybrid oscillator pulse by two 15 mm aperture GaAs Pockel's cells operated in tandem[21]. The GaAs crystals are mounted in impedance-matched transmission type holders of the Los Alamos design[22], and a 3 ns, 23 kV half wave pulse applied from a high pressure (160 psig $N_2$) laser-triggered spark gap (LTSG). The 10 $\mu m$ switching is nearly 100% efficient. Typically the final pulse width after the third germanium polarizer (P3) is just the difference between the electrical delay ($T_E$) and the optical delay ($T_o$) between the Pockel's cells PC1 and PC2. The minimum pulse half width possible from this system is just one half the LTSG risetime.

The approximately 1 mJ pulse from the hybrid oscillator-

Fig.6. The oscillator section of the $CO_2$ laser facility. The two low-pressure sections are visible on the left of the photograph and the two output mirrors mounted on piezoelectric translators, are visible in the foreground.

switchout system is amplified to a 50 mJ level by four passes through a Lumonics 821 HP UV-preionized, 2 atm preamplifier. Spatial filtering is achieved by means of apertures A1, A2 and A3 (Figure 5) and produces a high quality output beam, with a measured power contrast ratio of better than 5 x $10^3$ : 1 after the preamplifier. Figure (7) shows a photograph of the preamplifier system with the associated optical components.

The short pulse is amplified in a 3 atm, large aperture Lumonics 612 high pressure amplifier[21]. In order to extract the stored energy efficiently, the amplifier is used in a triple-pass configuration, with the optical components and the discharge housed in a common enclosure (Figure 8). The 1 cm diameter driver beam is focused before entering the gain medium and expands to approximately 5 cm diameter at the end of the first pass. The beam is then reflected through the discharge to the final pass collimating mirror where it is apertured to 8 cm diameter before completing the third pass. A 1 cm thick saturable absorber cell inserted in front of the second pass return mirror makes the system stable to parasitic oscillations provided the cell contains a partial pressure of at least 3 Torr of $SF_6$. Figure (9) is a photograph of the Lumonics 612 high pressure amplifier system.

The 10 cm diameter beam from the final amplifier is focused into the interaction chamber by an f/15, 150 cm focal length off-axis parabola (Figure 10). Typical parameters of the high pressure pulsed $CO_2$ laser facility are given in Table II. There are two modes in which this facility can be operated. In the static mode, the main interaction chamber, the focussing optics chamber and the injection beamline up to the 25 $\mu$m titanium window, are all filled with background hydrogen gas at a pressure of less than one torr. Tunneling ionization will produce a plasma over a region on the order of twice the Rayleigh length. In the dynamic mode, a puff of gas is released in the interaction region a short time ahead of the main laser pulse. In this case the plasma is produced over a region with dimensions which approximate the size of the puff.

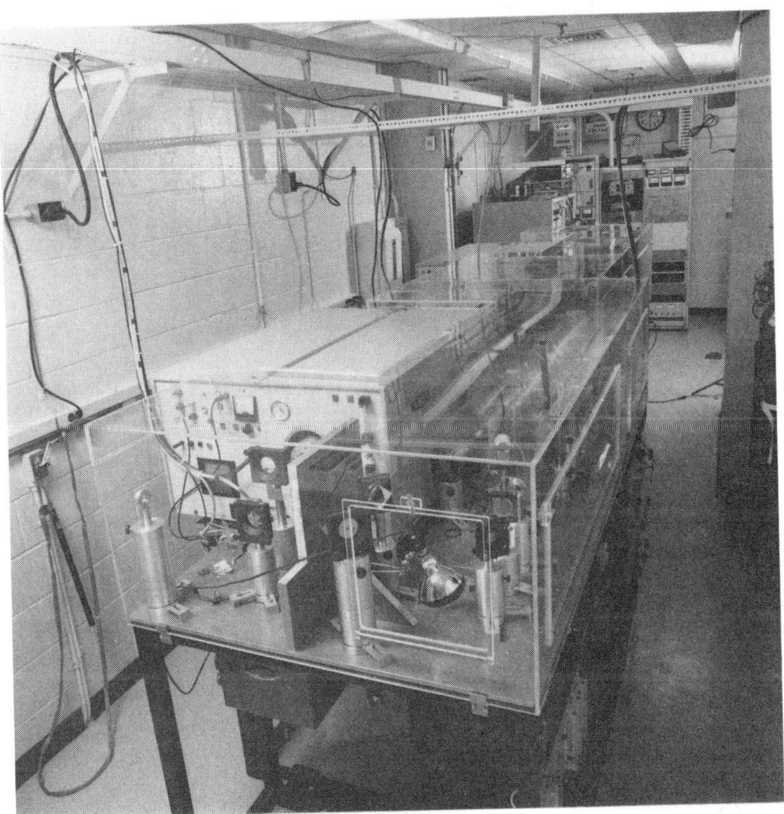

Fig.7. The pre-amplifier system. The short laser pulse from the pre-amplifier system is turned through 90° (in the foreground) and propagates to the input end of the high pressure amplifier located in the adjacent radiation-shielded accelerator vault.

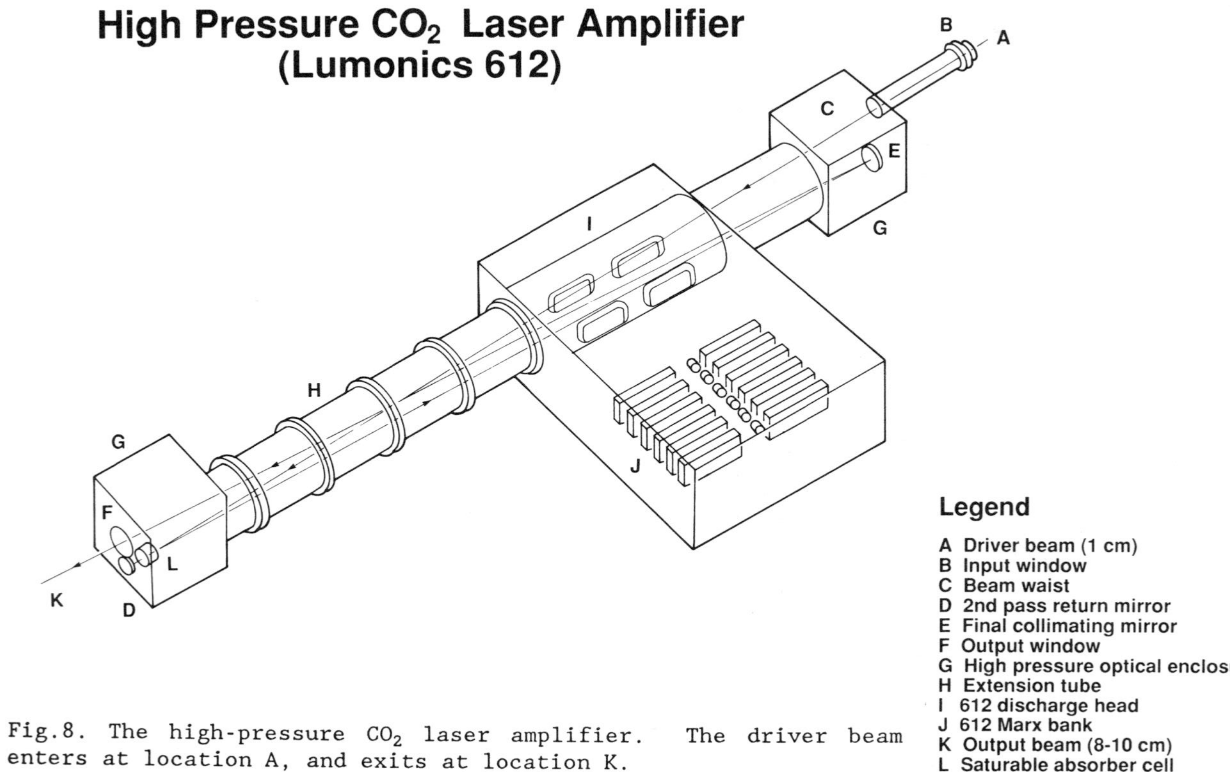


Fig.8. The high-pressure $CO_2$ laser amplifier. The driver beam enters at location A, and exits at location K.

Fig.9. The high-pressure laser amplifier system in the radiation-shielded accelerator vault.

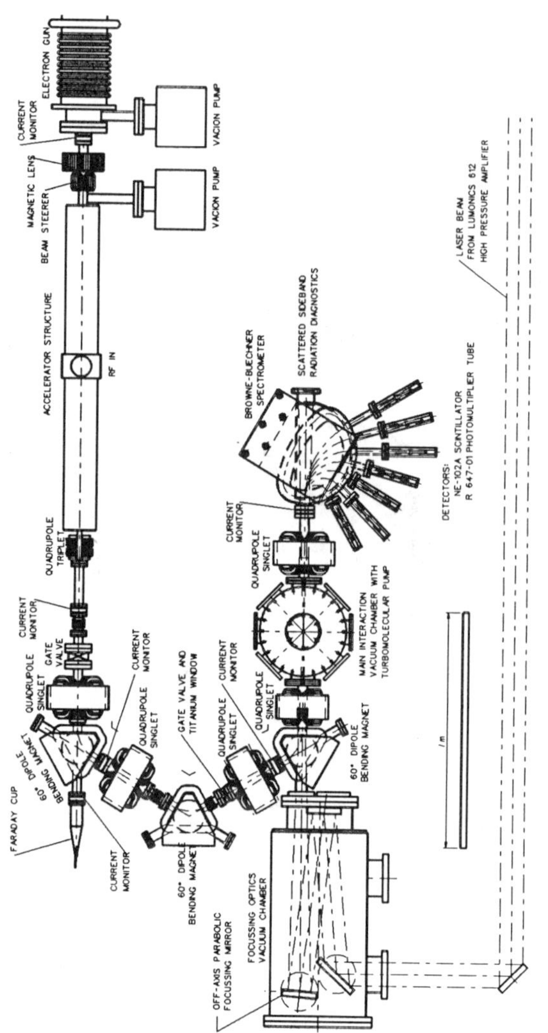


Fig.10. Schematic of the main experimental area of the laser acceleration facility.

Table II. Parameters of the Chalk River High Pressure Pulsed $CO_2$ Laser System

| | |
|---|---|
| Laser Wavelengths | 10.25 μm and 10.59 μm<br>or<br>9.55 μm and 10.55 μm |
| Energy per Wavelength | 30 Joules |
| Laser Pulse Length | 300 - 500 ps |
| Laser Focal Diameter (f/15) | 300 μm |
| Depth of Focus (f/15) | 1.5 cm |
| Focal Power Density | $2 \times 10^{14}$ W/cm$^2$ |

### (ii) Accelerator Facility

The pulsed linear electron accelerator facility is shown schematically in Figure (10). The accelerating structure is mounted on a fixed frame with the beam accelerated horizontally. The output electron beam at the exit of the last accelerating cavity is focused by a quadrupole triplet magnet, turned through 180° in a beamline consisting of three dipole bending magnets and three quadrupole singlets, and brought to a focus by a final focusing quadrupole singlet magnet, in a vacuum interaction chamber. The accelerated electrons from the laser interaction are dispersed in a magnetic electron spectrometer and detected with an array of electron detectors.

The accelerating structure is a 1.6 m S-band (3 GHz) $\pi/2$ standing wave structure. Rf power transmitted by a waveguide from a 2 MW pulsed magnetron is coupled into the central cell through a ceramic window. The width of the driving magnetron pulse is approximately 4 $\mu$s. A thermionic electron gun with a special dispenser-cathode and grid structure, is driven by an avalanche transistor pulser to produce 3 ns duration pulses. With a gap lens (200-300 gauss) and a beam-steering magnet at the exit of the electron gun, the beam is focused approximately 40 cm from the gun exit, which corresponds to the centre of the first accelerating cavity. The vacuum in the electron gun, the accelerating structure and the injection beamline is maintained by 60-litre vac-ion pumps (Figure 10). A gate-valve and a 25 $\mu$m titanium window in the beamline ahead of the third dipole bending magnet isolate the accelerator and the beamline from the main interaction chamber, the focussing optics chamber and the electron spectrometer chamber which are all evacuated with a common turbomolecular pump. A judicious choice of the location of the titanium vacuum window has been made, such that the scattering of the electrons in the window makes the smallest contribution to the emittance growth.

Typical characteristics of the electron linear accelerator are given in Table III. Since the length of a microbunch is 30 ps and the interbunch spacing is 330 ps, we would expect to accelerate single microbunches with laser pulses of 300 - 500 ps duration. Figure (11) shows a photograph of the 10 MeV electron linear accelerator. Figure (12) shows

Table III. Electron Linac Parameters for the Chalk River Laser Acceleration Experiment

| Parameter | Value |
|---|---|
| Electron Energy ($E_0$) | 10 MeV |
| Energy Spread ($\Delta E/E_0$) | ± 2% |
| Input Beam Emittance ($\epsilon_0$) | 1 $\pi$ mm - mrad |
| Acceptance of Accelerating Region | 130 $\pi$ mm - mrad |
| Gun pulse duration | 3 ns |
| Rf frequency | 3 GHz |
| Length of microbunch | 30 ps |
| Microbunch separation | 330 ps |
| Particles in a microbunch | $10^8$ - $10^9$ |
| Expected focal spot diameter | 300 $\mu$m |

Fig.11. The electron linear accelerator and the main interaction vacuum chamber (foreground left) in the accelerator vault. The electron gun is visible on the right.

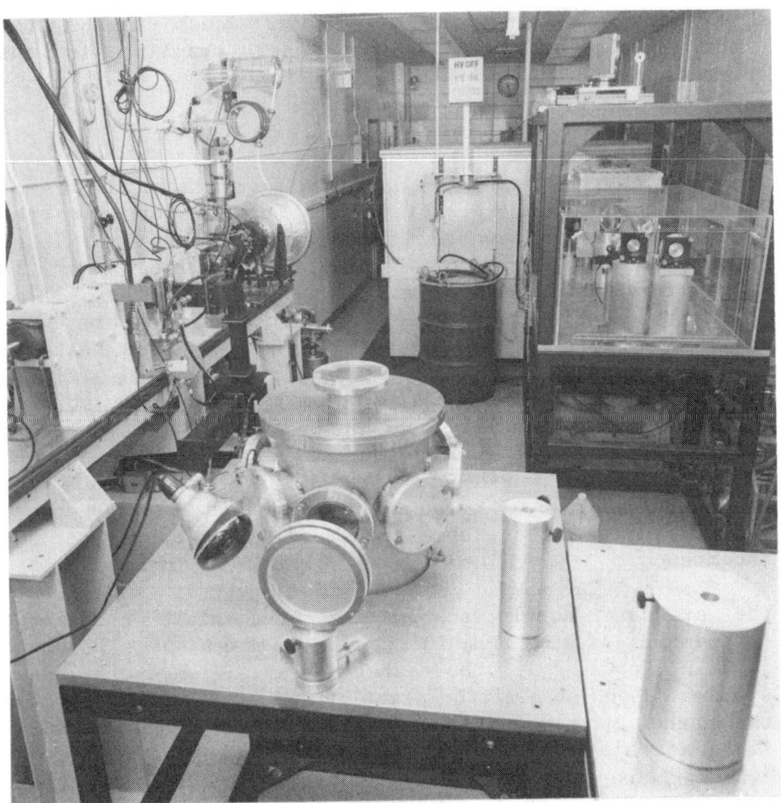

Fig.12. The main interaction vacuum chamber (foreground), the electron linear accelerator (left), the high-pressure amplifier and the optical beamline mirrors in the accelerator vault.

a photograph of the main interaction vacuum chamber, the electron accelerator and the laser beamline in the accelerator vault.

### (iii) Injection Beamline

The high-energy beam of electrons from the exit of the linear electron accelerator has to be transported to the interaction region, where it must overlap with the laser interaction region and be collinear with the laser beam. In order to match the input beam bunch to the plasma interaction region, the input beam emittance should be equal to or smaller than the acceptance of the plasma beatwave region[23]. Furthermore, since the electron beam has a finite energy spread, the transport system must be achromatic in order to ensure a small focal spot. It is also desirable that the system be isochronous to avoid increasing the electron bunch length excessively as it traverses the transport system. A system that would satisfy these requirements must be double-focusing, doubly achromatic and isochronous.

The standard doubly achromatic (beam width and slope independent of momentum in the final image space), double focusing (equality of the radial and axial image distances) systems depend on a mirror plane of symmetry halfway through the magnet system, with an intermediate dispersed radial image at this symmetry plane[24]. The simplest systems that satisfy these requirements consist of either three dipoles, or two dipoles and one quadrupole singlet. In this facility (Figure 10) we have used a variation of the three dipoles system, (with three 60° dipole bending magnets) which reverses the direction of the beam as well as translates it laterally. For added flexibility to the system and ease of mechanical alignment, we have eliminated the fixed edge-focusing at the input and output edges of the dipoles, and replaced it with quadrupole singlets to achieve the same objectives. In addition, we have incorporated a final quadrupole singlet magnet to form the final image. This system still retains enough symmetry features to give good first- and second-order focusing which produces an achromatic image.

The beam transport system for these experiments was designed using the computer programme TRANSOPTR[25], which is a second-order beam-transport design code with automatic internal optimization and general constraints. The code analyses the transport of charged particle beams through a user-defined magnet system, using a matrix formalism to represent the transport elements. Space charge effects (to first order) are included either in two dimensions, treating transverse forces only, or in three dimensions by treating both transverse and longitudinal forces on the beam.

The output emittance of the linac was estimated from the general relationship between the measured transverse emittance and the beam current at the output of linacs as given by the Lawson-Penner empirical scaling law,

$$\varepsilon_n = 5 \times 10^{-5} I^{1/2} (A) \; m.rad \tag{34}$$

For a 10 MeV beam with an output beam current of 100 mA, the normalized emittance $\epsilon_n$ = 2.24 x $10^{-4}$ m.rad. The emittance of the beam is related to the normalized emittance by

$$\varepsilon = \frac{\varepsilon_n}{(\gamma^2 - 1)^{1/2}} \tag{35}$$

where $(\gamma^2 - 1)^{1/2}$ = 20.54 for 10 MeV, and $\epsilon$ = 0.75 $\pi$ mm-mrad. This value of the emittance is expected to be good to within a factor of 3. The beam was modelled using upright phase-space ellipses in the two transverse directions with half widths of x = y = 2 mm and x′ = y′ = 0.75 mrad. The calculated and measured energy spread at the output window of the linac are both $\Delta E/E = \pm 2\%$. The design aim of the TRANSOPTR modelling was to produce an output beam of radius x = y ≈ 150 μm, and a beam waist which coincides with the beam waist of the laser beam. Once a satisfactory system was identified, its properties were studied with the computer programme TRANSPORT[26] which is more suitable for modelling mechanical misalignments and uncertainties in field settings.

Figure (13) shows the radial and axial effective root-mean-square (RMS) beam half-widths as a function of the path length downstream from the linac exit, for a beam with an energy spread of $\Delta E/E = \pm 2\%$. The maximum axial RMS beam full-width is approximately 4 mm compared to the dipole gap

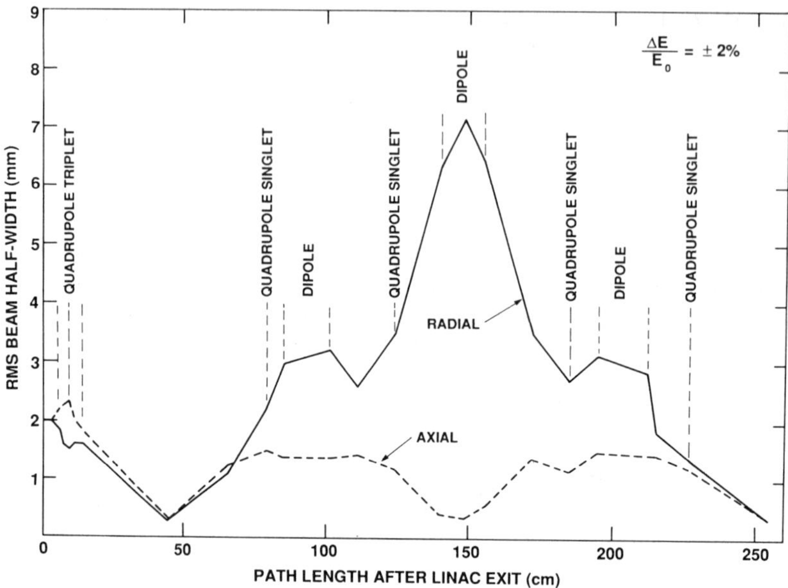


Fig.13. The radial and axial effective root-mean-square (RMS) beam half-widths along the beamline downstream from the linac exit as calculated from TRANSOPTR.

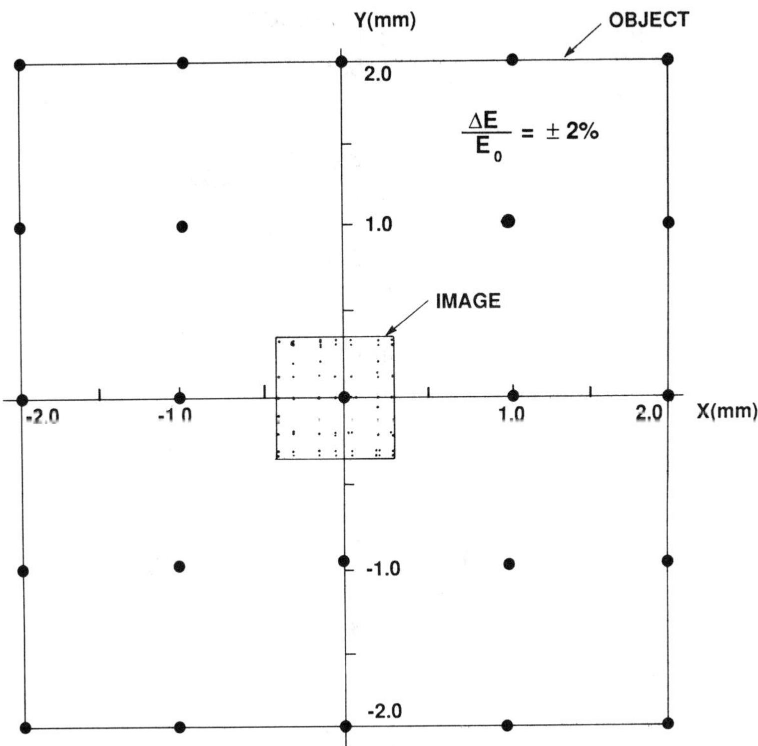


Fig.14. Beam profiles at the object and image planes from TRANSOPTR by ray tracing through the beam optics system. A total of 91 rays were used in this calculation.

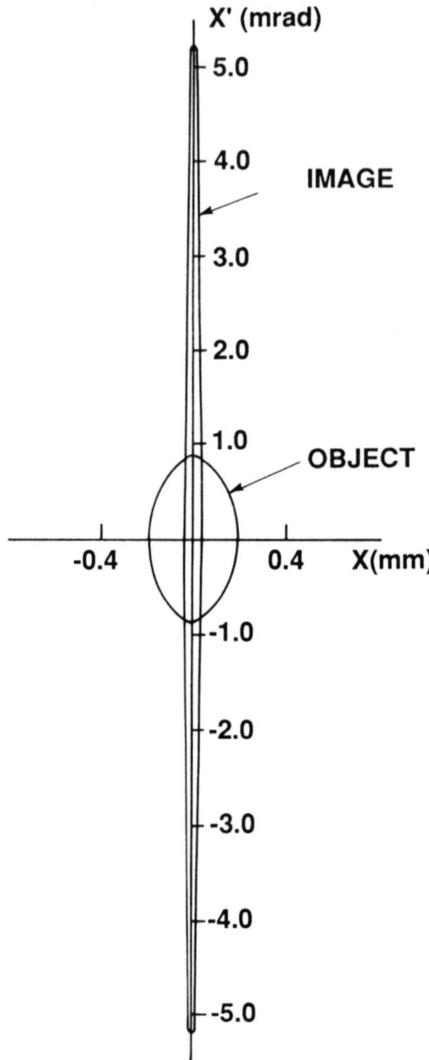


Fig.15. The radial emittance diagram at the object and image planes as calculated from TRANSOPTR.

width of 45 mm. The maximum radial RMS beam full-width is approximately 14 mm and the uniform field region extends over 100 mm. Figure (14) shows the beam profiles at the object and image planes obtained by from TRANSOPTR by ray tracing through the optimized beam optics. The large full dots in Figure (14) show the origin of the individual rays in the object plane and the small full dots give the location of these rays at the image plane. Figure (14) is constructed from second-order calculations of approximately 90 individual rays. Most of the beam is enclosed in a spot diameter of approximately 300 μm although the image is slightly offset from the centreline of the magnet system. Beam-steering magnets are positioned along the entire length of the beamline to steer the beam so that it overlaps with the focused laser beam in the interaction region. A number of non-intercepting current monitors have been also been positioned along the length of the beamline to simplify the setup of the individual beamline components (Figure 10), and maintain the beam alignment during experiments. These current monitors consist of short insulating sections bridged by four 50 ohm resistors which constitute a break in the conduction path of the beam image charge along the beam pipe. Figure (15) shows the radial emittance diagram at the object and image planes. From Figure (15) it is clear that in designing an injection beamline for such a facility, there is essentially a trade-off between the beam size and the beam divergence since emittance is invariant in a conservative system by Liouville's theorem.

The magnetic circuits for the various components in the beamline were modelled using the POISSON[27] group of computer programmes. The calculations give the level of field uniformity for the geometry adopted for the particular component, the required ampere-turns for the driving coils and the fringing fields at the entrance and exit edges of the magnets which are characterized by the field profile coefficient $k_1$ in TRANSOPTR, where

$$k_1 = \int_{-\infty}^{\infty} (B-B_0) \frac{B}{(gB_0^2)} ds \qquad (36)$$

where s is the position along the particle trajectory through the fringing field B, $B_0$ is the magnitude of the magnetic field in the uniform region and g is the pole gap.

The calculated fringing fields were then used to calculate the appropriate field profile coefficients for improved TRANSOPTR calculations. Because of the large values of gap-to-radius ratios (g/R = 0.3) dictated by the dimensions of the laser beam, it was thought prudent to check the calculations for the magnetic circuits against calculations using a fully three-dimensional, finite element computer programme TOSCA[28]. There is good agreement between the two sets of calculations for the dipole magnets as well as the electron spectrometer.

**(iv) Electron Spectrometer**

A broad-range magnetic spectrometer is used to study the accelerated particles from the laser plasma interaction region[29]. In this system a uniform magnetic field with a circular boundary focuses particles from a source outside the field along a focal surface (Figure 10). If $r_e$ is the effective radius of the uniform field region then the bending radius $\rho$ is related to the bending angle $\theta$ by[24]

$$r_e = \rho\frac{\sin\theta}{1+\cos\theta} \qquad (37)$$

The value of $r_e$ is determined by the maximum momentum ($B_0\rho_0 e$) of the particles, the maximum field $B_{max}$ available in the magnet, and the maximum deflection angle $\theta_{max}$. Since $B_{max}\rho = B_0\rho_0$,

$$r_e = \frac{B_0\rho_0}{B_{max}}\tan\frac{\theta_{max}}{2} \qquad (38)$$

If the object and image distances u and v respectively, are measured from the centre of the circular magnet, then

$$\frac{v}{r_e} = \frac{\left(\frac{u}{r_e}\right)}{\left(\frac{u}{r_e}\right)(1-\cos\theta)-1} \qquad (39)$$

and

$$\frac{u}{r_e} = \frac{\left(\frac{v}{r_e}\right)}{\left(\frac{v}{r_e}\right)(1-\cos\theta)-1} \qquad (40)$$

Equations (37) through (40) were conveniently used for the initial design of the spectrometer for uniform magnetic fields of 0.3 T and 0.6 T covering a combined energy range of 5 to 60 MeV. The design was then optimized with the computer design code TRANSOPTR for the seven individual energy channels. The magnetic circuit was modelled using the POISSON group of computer programmes as well as TOSCA and the calculated fringing fields were then used to calculate the appropriate field profile coefficients for improved TRANSOPTR calculations.

In order to provide some vertical focusing, we have incorporated a quadrupole lens in the input end of the spectrometer[30], which not only increases the solid angle over which particles are collected, but also minimizes the number of particles scattered to the walls of the spectrometer vacuum chamber, thereby reducing the background noise levels. The vacuum beamline and the spectrometer vacuum chamber are designed so that the laser beam can propagate through the chamber without obstruction. This is necessary for observing the Thomson scattered laser light in the forward direction which is an important diagnostic for the detection of the plasma waves.

The accelerated electrons are dispersed in the electron spectrometer and detected in a seven-channel array of particle detectors. Each detector consists of a 1 cm aperture NE 102A plastic scintillator and a Hamamatsu Type R647-01 head-on photomultiplier tube enclosed in a mild steel housing which provides shielding from the magnetic fields of the spectrometer as well as the radiation fields within the accelerator vault. A 25 μm titanium vacuum window in front of the housing provides shielding from visible and infrared light within the spectrometer vacuum chamber (Fig. 10). The output from the detector array is fed into a 16 channel variable gain (gain of 2 to 50) Phillips Scientific amplifier followed by a Lecroy 12-channel charge sensitive analog-to-digital converter (ADC) linked to a CAMAC crate and a microcomputer. With this system, single particle detection is possible.

## 4 DISCUSSION

We have designed, built and commissioned an experimental facility to study various aspects of laser and plasma based particle acceleration concepts. The facility consists of a short pulse $CO_2$ laser system and a short pulse high energy electron linear accelerator. Plasma generation is by tunneling ionization of hydrogen gas. A double-focusing, doubly-achromatic injection beamline transports short electron bunches into the laser plasma interaction region so that the laser and electron beams overlap and propagate collinearly through the accelerating region. A magnetic spectrometer with a multichannel detector is used to analyze the accelerated particles from the laser plasma interaction region. The facility has built-in flexibility to study a range of problems in laser and plasma-based acceleration concepts and plasma-based focussing concepts for high energy particle beams.

## ACKNOWLEDGMENTS

We thank R.T.F. Bird and E.B. Selkirk for excellent technical support on all aspects of this work, during the design, construction and commissioning phase of this facility, as well as continuing support. We also thank J.E. Anderchek for major technical contributions to all aspects of this work, D.R. Proulx for mechanical design support, R.J. Kelly, J.F. Weaver, J.D. Balfour, R.J. Bakewell and R.J. Klatt for excellent technical support. We also thank M.S. deJong and R.M. Hutcheon for many useful discussions on electron optics design, and N.H. Burnett of NRC, Canada for many illuminating discussions on multiphoton ionization and laser beam refraction in plasmas.

## 5 REFERENCES

1. T. Tajima and J.M. Dawson, Phys. Rev. Lett. **43**,267 (1979).

2. P. Chen, J.M. Dawson, R.W. Huff and T. Katsouleas, Phys. Rev. Lett. **54**, 693 (1985).

3. P. Sprangle, E. Esarey, A. Ting and G. Joyce, Appl. Phys. Lett. **53**, 2146 (1988).

4. Nizar A. Ebrahim, Physics in Canada **45**, 178 (1989).

5. F.F. Chen, Introduction to Plasma Physics and Controlled Fusion, 2nd ed., Vol. 1, (1984).

6. M.N. Rosenbluth and C.S. Liu, Phys. Rev. Lett. **29**, 701 (1972).

7. S.J. Karttunen and R.R.E. Salomaa, Phys. Rev. Lett. **56**, 604 (1986).

8. P. Gibbon and A.R. Bell, Phys. Rev. Lett. **61**, 1599 (1988).

9. P. Mora, Rev. Phys. Appl. **23**, 1489 (1988).

10. C.M. Tang, P. Sprangle and R.N. Sudan, Phys. Fluids **28**, 1974 (1985).

11. J.P. Matte, F. Martin, N.A. Ebrahim, P. Brodeur and H. Pepin, IEEE Trans. Plasma Sci. **PS-15**, 173 (1987).

12. J.D. Lawson, Rutherford Appleton Laboratory, Chilton, Didcot, U.K., Rep. **RL-83-057**, (1983) Unpublished.

13. R. Fedele, U. de Angelis and T. Katsouleas, Phys. Rev. A **33**, 4412 (1986).

14. L.V. Keldysh, Zh. Eksp. Teor. Fiz. **47**, 1945 (1964) [Sov. Phys. JETP **20**, 1307 (1965)].

15. N.A. Ebrahim, F. Martin, P. Brodeur, E.A. Heighway, J.P. Matte, H. Pepin and P. Lavigne, Proc. 1986 Linear Acc. Conf., Stanford, CA, SLAC Report No. 303, p. 552 (1986).

16. F. Martin, P. Brodeur, J.P. Matte, H. Pepin and N.A. Ebrahim IEEE Trans. Plasma Sci. **PS-15**, 167 (1987).

17. A.E. Dangor, A.K.L. Dymoke-Bradshaw, A. Dyson, T. Garvey, I. Mitchell, A.J. Cole, C.N. Danson, C.B. Edwards and R.G. Evans, IEEE Trans. Plasma Sci. PS-15, 161 (1987).

18. A. Dyson and A.E. Dangor, Laser and Particle Beams, **9**, (1991).

19. R. Rankin, C.E. Capjack, N.H. Burnett and P.B. Corkum, Opt. Lett. (1991).

20. S.C. Mehendale and R.G. Harrison, Opt. Lett. **10**, 603 (1985).

21. K.O. Tan, D.J. James, J.A. Nilson, N.H. Burnett and A.J. Alcock, Rev. Sci. Instrum. **51**, 776 (1980).

22. E.J. McLellan and F.J. Figueira, Technical Digest of Topical Meeting on Inertial Confinement Fusion, Paper Tu C11-1, San Diego (1978).

23. N.A. Ebrahim, Atomic Energy of Canada Limited Report No. **AECL-9389** (1987).

24. J.J. Livingood, The Optics of Dipole Magnets, Academic, New York (1969).

25. E.A. Heighway and M.S. de Jong, Atomic Energy of Canada Limited Report No. **AECL-6975** (1980) (Rev. A).

26. K.L. Brown, F. Rothacker, D.C. Carey and Ch. Iselin, Reports SLAC-91, NAL-91 and CERN 73-16 (1977).

27. M.T. Menzel and H.K. Stokes, LANL Report LA-UR-87-115 (1987); J. Warren, LANL Report LA-UR-87-126 (1987).

28. J. Simkin and C.W. Trowbridge, IEEE Proc. **127**, 368 (1980).

29. C.P. Browne and W.W. Buechner, Rev. Sci. Instrum. **27**, 899 (1956).

30. H.A. Enge, Rev. Sci. Instrum. **29**, 885 (1958).

**Research Trends in Physics: Coherent Radiation Generation and Particle Acceleration**
Editorial Board: J.M. Buzzi, A. Prokhorov (Editor-in-Chief), P. Sprangle, and K. Wille
*La Jolla International School of Physics*, The Institute for Advanced Physics Studies, La Jolla, California

# High Power Particle Accelerators Powered by Intense Relativistic Electron Beams

**M. Friedman and V. Serlin**

Naval Research Laboratory, Plasma Physics Division
Washington, DC 20375-5000

## ABSTRACT

Using novel techniques, electromagnetic energy can be stored, compressed and then used to accelerate high current particle beams to high energy.

## 1. INTRODUCTION

In an on-going research at NRL, a novel acceleration structure was constructed[1]. This accelerator was powered directly by a modulated intense relativistic electron beam (MIREB). The frequency of modulation was 1.3 GHz and the equivalent RF power residing in the MIREB was greater then 3 GW. The average electric field achieved inside a 1 meter-long accelerating structure was 60 MV/m. Higher electric field may be achieved if one uses:

1. MIREBs of higher power.
2. MIREBs of higher frequency.
3. Store and compress the energy of an MIREB and then release it in a very short time duration (power compression).

The first two items were reported by ue earlier[2-5]:

1. A MIREB of power of 30 GW and of a frequency of 1.3 GHz was achieved.
2. A MIREB of power of 3 GW and of a frequency of 3.5 GHz was achieved.
3. A conceptual design of a device that can modulate 30 GW of intense relativistic electron beam up to a frequency of 10 GHz.

In this paper we shall discuss how power compression techniques applied to the MIREBs discussed earlier can achieve high electric field in accelerator structures.

 

## 2. TECHNICAL APPROACH

The power compression scheme utilizes three components: a primary power source of an intense relativistic electron beam, a structure that modulates the electron beam and an energy storage structure. At the end of the storage structure one can substitute as a load an accelerating structure. In the following sections we will describe each of these components.

A. Primary Power Source. The primary power source will be an electron generator. This generator will produce a large diameter thin annular electron beam of current of 70 kA and a voltage of 500 kV. The electron beam will be confined by an external magnetic field to an annular channel. This geometrical configuration can support the generation and propagation of such an intense electron beam. The research issue here will be to demonstrate that the hollow beam can propagate without damage from the diocotron instability(6). We believe that the key to controlling this instability is to minimize the initial disturbance. In previous NRL experiment we propagated a large diameter (13.2 cm) hollow electron beam with current of 30 kA (7) without any evidence of the instability.

B. Modulating Structure. The next step will be to modulate this electron beam. Modulation of intense relativistic electron beams was pioneered by us more than 10 years ago. However, modulation of a large diameter 100 kA coaxial electron beam has only been shown to be possible by computer simulation (5). The objective of this part of the research will be to experimentally verify this theoretical prediction. We believe that we can produce MIREBs with a narrow spectrum up to a frequency of 10 GHz and with a peak RF current >30 kA.

C. Energy Storage System. The RF energy that is contained in the MIREB could be drained with a 30% efficiency(8) and transferred into a storage system. (Note, that extraction efficiency of >50% is possible(4)). The energy storage capability of an RF structure per pulse can be established with the simple formula (for a sufficiently long pulse duration)

$$U = P \cdot Q/\omega \qquad (1)$$

P is the input power, Q is the quality factor of the structure and $\omega/2\pi$ is the frequency of the modulation of the MIREB. Since $P>10$ GW, $Q>1000$ and $\omega/2\pi \sim 1$ GHz the stored energy per pulse $U > 1$ kJ. Of course this energy can only be stored for a short duration, and has to be quickly dumped into a load. An attractive storage system is a backward travelling wave structure with a group velocity $V_g<c$. The energy efficiency will be high if the length of the accelerating structure, L, is such that $L/V_g$ is equal to the duration of the MIREB.

During the loading time of the RF energy into the structure the system is "clean" and no electrons are present in it. Therefore the electric field that can be established inside the structure will be many times the empirical Kilpatrick limit (35 MV/m at 1.3 GHz.)

Once the structure is fully loaded with energy, a single pulse of electrons will traverse and drain this energy. A simple model based upon TEM transmission lines shows that the energy transfer efficiency to electrons should be higher than 50%.

The properties of the RF structure discussed above ensure that the acceleration phase occurs after the original MIREB pulse and the loading of the RF energy have ended. In that case the load is automatically decoupled from the source (in conventional accelerators this decoupling is achieved with high power isolators). Figure 1 displays the RF energy compression system.

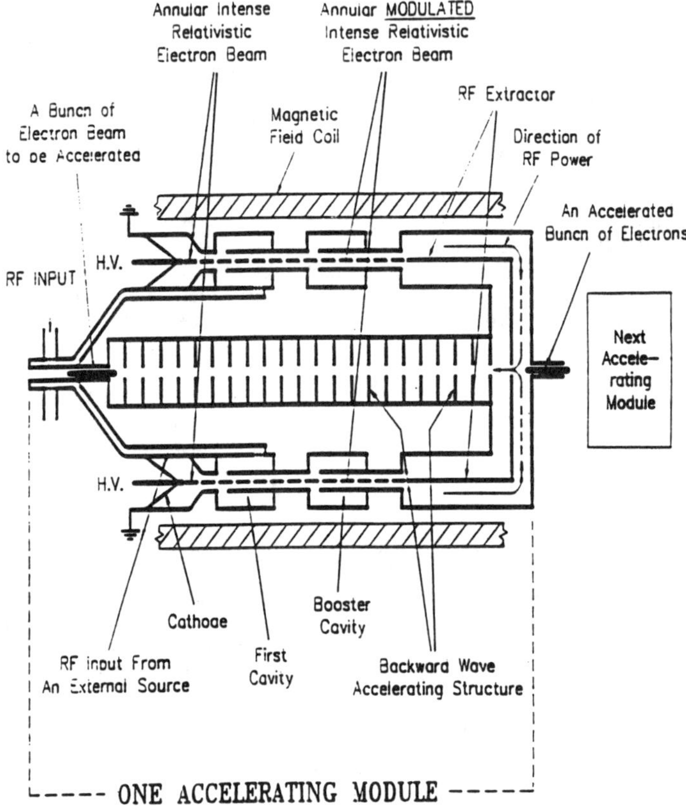


Figure 1. Schematic of a single coaxial accelerator module.

## 3. A LARGE ELECTRON ACCELERATOR

The above device can serve as a single acceleration module powered by a MIREB. Modules can be stacked together. The proper phasing can be accomplished in two different ways:

1. Using a "master" RF generator that locks the phases of the MIREBs to the proper one needed for acceleration of electrons.

2. Using a part of the RF energy from one module to modulate the primary electron beam in the next module. If we use a high power transfer from one module to the next, each accelerating stage will need only one modulating cavity (fig. 2).

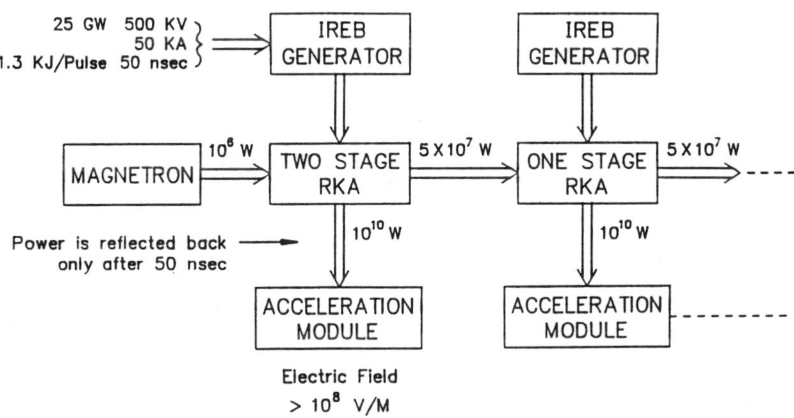


ADVANTAGES

(1) Synchronism is achieved automatically.

(2) No need for isolators, Phase shifters, RF dumps, etc.

(3) High RF power and short duration enable the accelerator to work at high voltage gradient and relax requirements for high vacuum.

(4) Low Q structure relaxes requirement for precise machining.

Figure 2. Block diagram of a large accelerator.

Eventually, with the successful demonstration of the experimental goals at a high power level, the question arises as to the extrapolation of a single shot pulsed power device to an actual accelerator with a reasonable duty cycle. We believe that the rep-rated technology, such as magnetic switches[9], which will make this possible is at an advanced enough stage for a combination of the power holding capabilities and the device we describe to allow for *revolutionary* changes in accelerator design.

## 4. ACKNOWLEDGEMENTS

This work was supported by the Naval Research Laboratory and by the Strategic Defense Initiative Organization, Office of Innovative Science and Technology, Managed by the Harry Diamond Laboratories .

## 5. REFERENCES

1. M. Friedman, J. Krall, Y. Y. Lau, and V. Serlin, "Electron accelerators driven by modulated intense relativistic electron beams", Phys. Rev. Lett. 63, 2468 (1989).

2. M. Friedman, V. Serlin, Y. Y. Lau, and J. Krall, "Relativistic Klystron Amplifier I: high power operation", SPIE Proceedings 2, 1407 (1991).

3. V. Serlin, M. Friedman, Y. Y. Lau, and J. Krall, "Relativistic Klystron Amplifier II: high frequency operation", SPIE Proceedings 8, 1407 (1991)

4. D. G. Colombant, Y. Y. Lau, M. Friedman, J. Krall, and V. Serlin, "Relativistic Klystron Amplifier III: nonlinear beam loading and operating efficiency" SPIE Proceedings 13 , 1407 (1991).

5. J. Krall, M. Friedman, Y. Y. Lau, and V. Serlin, "Relativistic Klystron Amplifier IV: simulation studies of a coaxial-geometry RKA",SPIE Proceedings 23 , 1407 (1991)

6. J. D. Lawson, *The physics of charged-particle beams*, Clarendon Press - Oxford, 1977.

7. M. Friedman, D. Colombant, Y.Y. Lau, V. Serlin and J. Krall, "The Influence of Beam Loading on the Operation of the Relativistic Klystron Amplifier", SPIE Proceedings 2, 1226 (1990).

8. M. Friedman, J. Krall, Y. Y. Lau, and V. Serlin, "Efficient generation of multigigawatt rf power by a klystronlike amplifier", Rev. Sci. Instrum. 61, 171 (1990).

9. D. L. Birx *et al*, UCRL 85738 and in the Third IEEE Int. Pulsed Power Conf., Albuquerque, NM, June 1-3, 1984.

**Research Trends in Physics: Coherent Radiation Generation and Particle Acceleration**
Editorial Board: J.M. Buzzi, A. Prokhorov (Editor-in-Chief), P. Sprangle, and K. Wille
*La Jolla International School of Physics*, The Institute for Advanced Physics Studies, La Jolla, California

# The Problem of Electromagnetic Pulse Self-Focusing in Wake Field Generation

**D.P. Garuchava, N.L. Tsintsadze, and D.D. Tskhakaya**

Institute of Physics
Georgian Academy of Sciences
Tbilisi, Georgia

## Abstract

The case has been considered when nonlinear beam refraction is mainly caused by relativistic effect. It is shown that at nonstationary self-focusing of short pulses the plasma density increases in the range of running focuses.

The phenomena found can turn to be advantageous in laser wake field accelerator. In order to develop the process of acceleration efficiently it is necessary to have a self-focusing electromagnetic pulse, since without it the length of the diffraction broadenting of the beam is much less than the characteristic scales of the particle acceleration.

In the report we discuss some effects, which can take place at self-focusing of relativistically powerful electromagnetic pulses. It is natural to assume, that the role of relativism together with electrostriction increases especially in focal region. That's why it is necessary to solve the selfconsistent problem.

1.Let's consider the propagation of intensive HF electromagnetic waves circularly polarized in the transparent plasma.

$$\vec{E} = (\vec{e}_x + i\vec{e}_y)\frac{1}{2}E(r,z,t)\exp\left\{-i\omega_o t + ik_0 z\right\} + c.c \qquad (1)$$

$\vec{e}_x$ and $\vec{e}_y$ are the unit vectors along the corresponding axis, r is the transverse coordinate. For the slowly changing amplitude we have equation:

 

$$2ik_o \frac{\partial E}{\partial \xi} + \Delta_\perp E + \frac{\omega_0^2 - k_0^2 c^2}{c^2} E - \frac{\omega_{pe}^2}{c^2} \frac{1}{\gamma} \frac{\langle n_e \rangle}{n_0} E = 0 \quad (2)$$

$$\tau = t - z/V_g, \qquad \xi = z \quad (3)$$

$$V_g = \frac{\partial \omega_0}{\partial k_0}$$

$$\gamma = \sqrt{1 + I} \quad (4)$$

$$I = \frac{e^2 |\vec{E}|^2}{m_e^2 c^2 \omega_0^2} = \frac{|\vec{p}_e|^2}{m_e^2 c^2} \quad (5)$$

$\gamma$ is the relativistic factor .The other notations are clear.

This equation is valid if the characteristic lengths of the field variations along z axis and in the transvers direction are larger, than the length of the electromagnetic wave:

$$L_z \gg L_\perp \gg \lambda_o \qquad \lambda_0 = \frac{2\pi}{k_0} \quad (6)$$

Plasma is considered to be transparent and the length of the HF field packet is much larger than that of the wave:

$$V_g^2 \tau_0^2 \gg \lambda_0^2, \quad (7)$$

$\tau_0$ is the duration of the electromagnetic pulse.

Plasma "coldness" conditions in the most experiments are fulfilled with great reserve. It is more difficult to neglect the inertia of electrons. The necessary conditions for this purpose,

$$\frac{\omega_{pe}^2 \tau_0^2}{\gamma} \gg 1, \tag{8}$$

for ultrarelativistic beams, $\gamma \gg 1$, is rigid. (Here $\omega_{pe}$ is the electron plasma frequency). But in the most experiments using picosecond and even femtosecond lasers in the transparent plasma this condition is fulfilled for not too high intensities.

Then from the electron motion equation for slowly changed electric field one can write

$$\langle\vec{E}\rangle = -\frac{m_e c^2}{e} \vec{\nabla} \gamma \;, \tag{9}$$

Let's ignore the ion motion ,

$$\omega_{pi}^2 \tau_0^2 \ll 1 \,. \tag{10}$$

Under these conditions if the width of the packet is less than its length $V_g^2 \tau_0^2 \gg a^2$ we obtain the following expresssion from Puasson equaticn for the concentration perturbation:

$$\frac{\langle n_e \rangle}{n_0} = 1 + \frac{\delta n_e}{n_0} = 1 + \frac{c^2}{\omega_{pe}^2} \Delta_\perp \gamma \,. \tag{11}$$

So we have quasistationary system consisting in both local and non-local nonlinearities. From the form of the nonlinear term we see that in the general case it is impossible to separate relativistic and strictional effects. The division is possible only in the case of weak relativistic effect and small concentration perturbation.

From the eq.(11) it is evident that for the convex distribution of the beam intensity, the concentration pertúrbation on the beam axis is always negative,

$$\Delta_\perp \gamma < 0, \qquad \delta n_e < 0. \tag{12}$$

As the particle concentration can not be negative,

$$\langle n_e \rangle \geq 0, \tag{13}$$

the condition $\frac{\delta n_e}{n_0} \geq -1$ must be fulfilled. It should be noticed that $\frac{\delta n_e}{n_0}$ is arbitrary.

At the points, where the condition (13) is violated and $\langle n_e \rangle = n_0 + \delta n_e = 0$ ,

$$\frac{\delta n_e}{n_0} = -1 ,$$

the following system of equations should be solved:

$$2ik_0 \frac{\partial E}{\partial \xi} + \Delta_\perp E + \frac{\omega_0^2 - k_0^2 c^2}{c^2} E = 0 , \tag{14}$$

$$1 + \frac{\delta n_e}{n_0} = 0 \tag{15}$$

In this region the nonlinearity is saturated. All the electrons are forced out the axial region of a beam. This phenomenon is called "electron cavitation".

For the sake of simplisity let's consider the self-focusing of a Gaussian beam in the abberationless approximation [1].

Let's consider the electric field in the form:

$$E = E_0(r,\xi,\tau) \exp (ik_0 S(r,\xi,\tau)), \tag{16}$$

$$I = \frac{e^2 E^2}{m_e^2 c^2 \omega_0^2} ,$$

$$I = \frac{I_0(\tau)}{f^2(\xi,\tau)} \exp\left\{ - \frac{r^2}{a^2 f^2(\xi,\tau)} \right\} , \tag{17}$$

$$S = \frac{r^2}{2} \beta(\xi,\tau) + \psi(\xi,\tau), \tag{18}$$

$$\beta = \frac{1}{f} \frac{\partial f}{\partial \xi} , \tag{19}$$

where S is eikonal of the wave, $\beta$ is the front curvature and $\psi$ is the phase climb on the beam axis. It is assumed that wave front of the beam in the nonlinear medium is remained spherical but the front curvature is changed. The negative $\beta < 0$ corresponds to the convergent front, while positive $\beta > 0$ - to the divergent one, and $\beta = 0$ - to the flat one. The following conditions are carried out on the boundary:

$$\xi = 0 , \qquad \beta = \beta_0, \qquad \psi = 0 ,$$

$$f(0,\tau) = 1, \qquad \frac{\partial f}{\partial \xi} = 0. \tag{20}$$

For the dimensionless width of the beam in the paraxial approximation[1],

$$r^2 \ll a^2 f^2, \tag{21}$$

we obtain the following equation

$$\frac{\partial^2 f}{\partial \xi^2} = - \frac{1}{k_0^2 a^4} \frac{1}{f^3} \left\{ \frac{\omega_{pe}^2 a^2 f^2}{2c^2} \frac{I_0/f^2}{(1+I_0/f^2)^{3/2}} + \right.$$

$$\left. + \frac{I_0/f^2(4 + I_0/f^2)}{(1 + I_0/f^2)^2} - 1 \right\} \tag{22}$$

The first, second and third terms are specified by relativistic effect, electron striction and diffraction, respectively. From this equation we see that the beam is focused, when the expression in the brackets is positive. It means that the following condition must be fulfilled:

$$\frac{\omega_{pe}^2 a^2}{2c^2} I_0 > \frac{1 - 2I_0}{\sqrt{1 + I_0}} \tag{23}$$

If $I_0 \geq 0.5$ the beam of arbitrary width begins focusing. The critical intensity of self-focusing is determined from the equation:

$$I_{cr} \left| \frac{\sqrt{1 + I_{cr}}}{1 - 2I_{cr}} \right| = \frac{2c^2}{\omega_{pe}^2 a^2} \tag{24}$$

As the expression in the right-hand side of the inequality (23) is less than unity, then the condition of self-focusing has the form:

$$I_0 > \frac{2c^2}{\omega_{pe}^2 a^2} \qquad \text{if} \qquad I_0 < 0.5,$$

or $$\tag{25}$$

$$I_0 \geq 0.5$$

From the condition of the concentration positivity (13) on the beam axis (r=0) we have

$$\frac{\omega_{pe}^2 a^2 f^2}{2c^2} \geq \frac{I_0/f^2}{\sqrt{1 + I_0/f^2}} \tag{26}$$

For those $f$ at which this inequality (26) is not fulfilled, instead of (22) one should solve equation:

$$\frac{\partial^2 f}{\partial \xi^2} = \frac{1}{k_0^2 a^4} \frac{1}{f^3} . \tag{27}$$

This equation follows from eqs. (14), (15), written for the case of "cavitation".

This means that when dimensionless width of the beam at self-focusing reaches the value $f_n$, determined from the equation:

$$f_n^4 \left(1 + I_0/f_n^2\right)^{1/2} = \frac{2c^2}{\omega_{pe}^2 a^2} I_0 , \tag{28}$$

all electrons are forced out from the axial region, and the saturation of nonlinearity takes place. And one should solve the equation (27) for the cavitation. As we see from the solution of the equation, the beam begins to diverge:

$$f^2 = 1 + \frac{\xi^2}{k_0^2 a^4} . \tag{29}$$

The offective potential for the eqs.(22), (27) has - the form:

$$\frac{\partial^2 f}{\partial \xi^2} = - \frac{\partial}{\partial f} V(f) , \tag{30}$$

$$V(f) = \frac{1}{k_0^2 a^4} \left\{ \frac{3}{2} \frac{1}{f^2} \frac{1}{1 + I_0/f^2} + \frac{\omega_{pe}^2 a^2}{2c^2} \frac{1}{\sqrt{1 + I_0/f^2}} - \right.$$

$$\left. - \frac{1}{I_0} \ln \left(1 + I_0/f^2\right) \right\} \qquad f > f_n , \tag{31}$$

$$V_2(f) = \frac{1}{k_0^2 a^4}\left(\frac{1}{2f^2} + c\right) \qquad f < f_n. \qquad (32)$$

The constant c is determined from the continuity condition at the "cavitation" boundary point $f = f_n$:

$$V_1(f_n) = V_2(f_n) \qquad (33)$$

From expressions (31), (32), (33) follows that the potential hole is asymmetric:

$$V(f) = \frac{\omega_{pe}^2}{2k_0^2 a^2 c^2} = H_0 \qquad f \to \infty ,$$

$$V(f) \to \infty \qquad f \to 0. \qquad (34)$$

The regime of beam propagation depends on the intensity $I_0$ and front curvature $\beta_0$ on the boundary. They determine the energy level, at which the dimensionless width of the beam $f$ changes:

$$V_0 = \frac{1}{2}\beta_0^2 + \frac{1}{k_0^2 a^4}\left\{\frac{3}{2}\frac{1}{1 + I_0} + \frac{\omega_{pe}^2 a^2}{2c^2}(1 + I_0)^{-1/2} - \frac{1}{I_0}\ln(1 + I_0)\right\} \qquad (35)$$

For sufficiently intensive beams with small curvature on the boundary, when the energy level is smaller than the height $H_0$,

$$V_0 < H_0, \qquad (36)$$

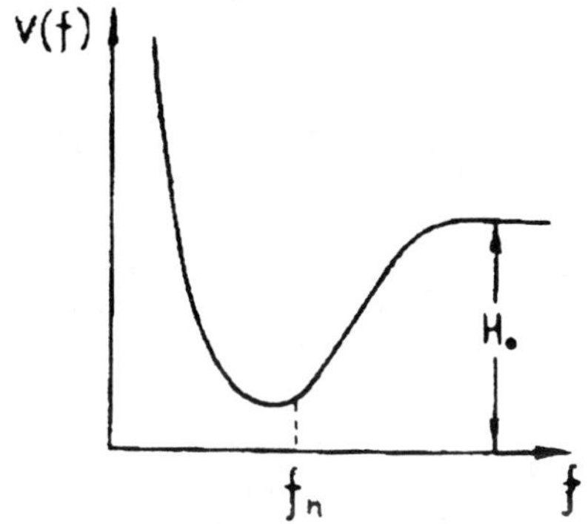


Fig.1. The qualitative dependence $V = V(f)$

the self-sustaining channel is formed. The beam width changes periodically, but these oscillations are not sinusoidal.

For the beams withovercrytical intensities and large front curvature on the plasma boundary ,

$$V_0 > H_0 , \tag{37}$$

convergent beam finally is always disfocused after self-focusing.

Of course above mentioned is valid for the paraxial abberationless approximation. However, after "the cavitation" the axial rays begin to diverge while the rays at periphery, where the electron concentration does not fall to zero, continue to converge. Abberation is sufficient and it can not be neglected any more. But the computations show that, taking into account the abberation, above described general dynamics of self-focusing remains valid.

The results of numerical calculations of eqs. (2), (11) are shown in Fig.2.

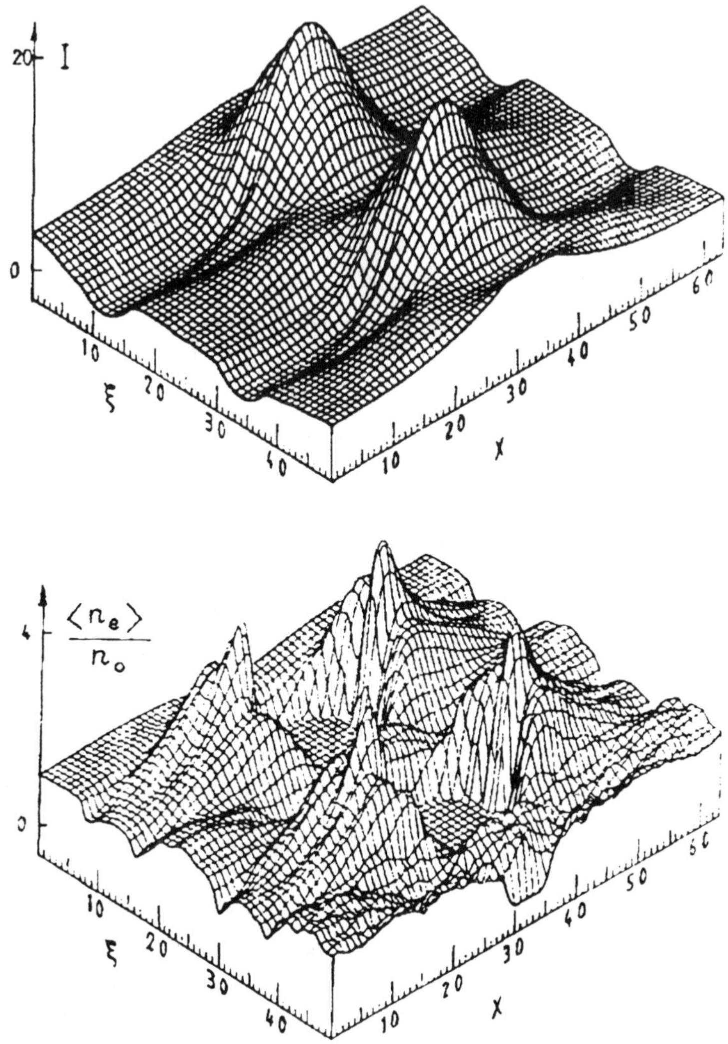

Fig.2. Distribution of the intensity I and concentration $\frac{\langle n_e \rangle}{n_0}$ in the case of $\frac{\omega_{pe}^2 a^2}{c^2} = 40$, $\frac{\omega_{pe}^2}{\omega_0^2} = 0.1$, $I_0 = 0.5$, $\frac{\xi}{2k_0 a^2} \equiv \xi$, $\frac{x}{a} \equiv x$.

For simplicity we have considered the two-dimensional beam ($\Delta_{\perp}=\frac{\partial^2}{\partial x^2}$, i.e. the beam width along the axis y is much bigger than the beam width $a$ along axis x). The distribution of the intensity on the boundary is cosine-like. The small perturbations on higher harmonics are superposed on the main harmonic. As it is seen from Fig.2 the electron concentration in the focus areas goes to zero. After self-focusing the beam begins to broaden. In those points, where the central rays, diverging from the axis, intersect the cross converging lateral rays, the so called "lateral" focuses are formed. (Intensity in them is less than the intensity in the focus-area on the axis). The beam intensity on the axis begins to decrease and becomes less than that at peripheries. After this the electrons will be forced out of the lateral focuses towards the axis. Electron concentration on the axis will dominate over diffraction again. So, the self-focusing channel is formed. It is found that the more boundary intensity exceeds the critical intensity, the longer the channel is, where the electron concentration goes to zero.

2. Now we'll investigate a strange phenomenon, which can take place at the self-focusing of powerful short pulses of electromagnetic radiation. The effect is strange indeed, since the self-focusing can be accompanied not by decrease of particles concentration [2-4], but by compression of the plasma. It will be shown below, that this effect can be observed only for the relativistic intensity of radiation. Unlike above-mentioned self-focusing of the long pulse we'll consider below the short pulses. The pulse length is less than its width,

$$V_g^2\tau_0^2 \leq a^2 . \tag{38}$$

The physical essence of the compression effect is the following. The relativistic increase of the electron mass in HF field leads to the decrease of the electron plasma frequency and increase of the plasma refraction coefficient[5]. In the weak relativistic case it is possible to separate strictional and relativistic effects from each other, and to represent the refractive index N in the following form:

$$N^2 = \varepsilon_0 + \delta\varepsilon , \tag{39}$$

where $\varepsilon_0 = 1 - \frac{\omega_{pe}^2}{\omega_0^2}$ is the linear part of dielectric permeability and $\delta\varepsilon$ is the nonlinear addition caused by the concentration perturbation and the relativistic weighting of the electron mass:

$$\delta\varepsilon = \frac{\omega_{pe}^2}{\omega_c^2}\left(\frac{\delta m_e}{m_e} - \frac{\delta n_e}{n_0}\right), \tag{40}$$

$$\frac{\delta n_e}{n_0}, \quad \frac{\delta m_e}{m_e} \ll 1,$$

$$\frac{\delta m_e}{m_e} = \frac{1}{2} I = \frac{1}{2} \frac{e^2 |E|^2}{m_e^2 \omega_0^2 c^2} .$$

The necessary requirement for self-focusing is positivess of the perturbation of dielectric permeability $\delta\varepsilon > 0$. In the nonrelativistic case, when we completely neglect the variation of electron mass, $\delta m_e = 0$, the necessary concition for self-focusing is $\delta n_e < 0$. It means that we have the impoverishment of electrons in focus area. If the weighting of electron mass dominates over the relative perturbation of concentration,

$$\frac{\delta m_e}{m_e} > \frac{\delta n_e}{n_0} , \tag{41}$$

then the self-focusing can be accompanied by plasma compression , $\delta n_e > 0$.

The equation for the slowly changing amplitude of the HF electric field (see (1), (2)) with weak relativistic intensity and for the small concentration perturbation has the following form:

$$2ik_0 \frac{\partial}{\partial \xi} E + \Delta_\perp E + \frac{\omega_{pe}^2}{c^2}\left(\frac{1}{2} I - \frac{\delta n_e}{n_o}\right) E = 0 . \qquad (42)$$

It is impossible to solve analytically equation (42) together with equations of gas dynamics, i.e. with the equation for $\delta n_e$. We'll confine ourselves to the investigation of the simplest abberationless case without taking into account a backaction of the density perturbation on the behaviour of HF amplitude. It means, that the electrostriction can be neglected in comparison with the relativistic effect:

$$\frac{\delta n_e}{n_0} \ll I. \qquad (43)$$

In spite of defects of such approximations, only in this case it is possible to represent the solutions in the form of elementary functions [1, 6-8], and then determine the tendency of the development of the process.

Eq. (42) we can rewrite as·

$$2ik_0 E + \Delta_\perp E + \frac{\omega_{pe}^2}{c^2} \frac{1}{2} IE = 0, \qquad (44)$$

$$I = \frac{e^2 |E|^2}{m_e^2 \omega_0^2 c^2} \ll 1 .$$

For further analysis we'll consider the intensity profile parabolic in space, Gaussian in time and with the power of a beam much higher than the critical one.

The corresponding solution of this equation we can represent in the form [1]:

$$I = \frac{I_{00}}{f^2}\left(1 - \frac{r^2}{4a^2f^2}\right) \exp\left\{- \frac{\tau^2}{2\tau_0^2}\right\}, \tag{45}$$

$$f^2 = 1 - \frac{\xi^2}{R^2} \exp\left(-\frac{\tau^2}{2\tau_0^2}\right),$$

$$R = 2\sqrt{\frac{2}{I_{00}}}\ \frac{k_0 c}{\omega_{pe}}\, a, \qquad r \le 2af,$$

where $2a$ is the width of the beam at the point $\xi = 0$ and the dimensionless beam width $f$ satisfies the boundary conditions (20).

The focusing length $Z_f$, determined by the condition $f = 0$, follows the change of the pulse in time. Ceratinly, at the point of the focus ($f = 0$) expression (45) is not valid, since $I \to \infty$ and the condition $I \ll 1$ is violated. Bellow we'll take small values of the difference:

$$1 - \frac{\xi^2}{R^2}, \tag{46}$$

but it will not violate the weak relativistic condition $I \ll 1$.

Let us consider hydrodynamicval approximation,

$$\omega_{pe}^2 a^2 \gg V_{Te}^2, \tag{47}$$

($V_{Te}$ is the thermal velocity of electrons) and assume, that the characteristic freauency of the pulse is smaller than the ion plasma frequency:

$$\omega_{pi}^2 \tau_0^2 \gg 1. \tag{48}$$

Then from the motion and contitnuity equations we obtain following equations for the perturbation of particles concentrations

$$\left(\frac{\partial^2}{\partial\tau^2} - c_s^2\Delta_\perp\right)\frac{\delta n_e}{n_0} = \frac{1}{2}\frac{c^2}{V_g^2}\frac{m_e}{m_i}\left(\frac{\partial^2}{\partial\tau^2} + V_g^2\Delta_\perp\right)I(\xi,\ r,\ \tau), \quad (49)$$

$$\frac{\delta n_e}{n_0} \approx \frac{\delta n_i}{n_0}\ . \quad (50)$$

As a rule, in most laboratory devices condition,

$$a^2 \gg c_s^2\tau_0^2, \quad (51)$$

is fulfilled. In this case the second term in the left-side of equation (49) is small in comparison with the first one, but we'll keep it to prevent emergence of the terms proportional to $\tau$ and to describe properly the distribution of the concentration along the axis z:at $z \to +\infty$ (or $\tau \to -\infty$) the perturbation of concentrations must vanish.

From the quite general consideration we can show the possibility of the positive values of density perturbation. Let us assume that radiation intensity is localized in space and time and $a$ and $\tau_0$ are the characteristic spatial and temporal scales of these localizations:

$$I(\xi,\ r,\ \tau) = I(\xi, \rho,\ \tau'), \qquad \rho = \frac{r}{a},\ \tau' = \frac{\tau}{\tau_0}\ . \quad (52)$$

The characteristic scale of the localization along the axis $z \equiv \xi$ is the focusing length.

The solution of eq.(49) we can represent in the form:

$$\frac{\delta n_e}{n_0} = \frac{1}{2}\frac{m_e}{m_i}\frac{c^2}{V_g^2}\Bigg\{I + \frac{1}{2\pi}\frac{V_g^2\tau_0^2}{a^2}\Delta_\perp\int_0^{2\pi}d\varphi\int_0^{\infty}ds\int_0^{s}dx\ I\left(\vec{\rho} + \vec{i}\,\frac{c_s\tau_0}{a}\sqrt{s^2-x^2},\ \tau'-s\right), \quad (53)$$

where $\vec{i}$ is the unit vector, $\vec{i} \equiv (\cos\varphi,\ \sin\varphi, 0)$.

From (53) it follows, that in the first front of the pulse, when $\tau = t - \frac{z}{V_g}$ is negative, there is no density perturbation. Behind the pulse the perturbation decreases as $\left(\frac{c_s\tau}{a}\right)^{-3}$ $(c_s\tau \gg a)$. In the transversal direction perturbation $\delta n_e$ is localized in the area of the size:

$$r \le c_s\tau \tag{54}$$

If the beam width satisfies the condition ,

$$a^2 > V_g^2\tau_0^2 , \tag{55}$$

the second negative term in (53) can be less than the first one. (This second term is negative since the intensity I has the maximum on the axis z, $\rho = 0$). It means that in the area of braod pulse localization the particles concentration increases, $\delta n_e > 0$.

These general conclusions can be illustrated with help of the model pulse (45). For such a pulse the general solution of (53) gives the following expression for the perturbation on the axis z at the maximum point of the pulse $\tau = 0$ $(z = V_g t)$:

$$\left.\frac{\delta n_e}{n_0}\right|_{r=0} = \frac{1}{2}\frac{m_e}{m_i}\frac{c^2}{V_g^2}\frac{I_{00}}{1-\xi^2/R^2}\left(1 - \frac{V_g^2\tau_0^2}{a^2}\right) \tag{56}$$

Consequently, under the condition:

$$\frac{V_g^2\tau_0^2}{a^2} < 1 \tag{57}$$

the perturbation of the particles concentration is positive, $\delta n_e > 0$ (see (50)).

In the area of the pulse, in its first front $z > V_g t$, the perturbation follows the pulse profile

$$\left.\frac{\delta n_e}{n_0}\right|_{r=0} = \frac{1}{2}\frac{m_e}{m_i}\frac{c^2}{V_g^2} I_0 \exp\left\{- \frac{\tau^2}{2\tau_0^2}\right\} \tag{58}$$

Behind the pulse outside its localization $z < V_g t$,

$$(V_g t - z) \gg V_g \tau_0 \frac{a}{c_s \tau_0} ,$$

the positive perturbation decreases in a degree way :

$$\left.\frac{\delta n_e}{n_0}\right|_{r=0} = \sqrt{2\pi}\, \frac{m_e}{m_i} \frac{c^2}{V_g^2} I_o \frac{V_g}{c_s} \frac{V_g \tau_o}{a} \left(\frac{a}{c_s \tau}\right)^3 . \tag{59}$$

We ought to pay attention to the parameter $\frac{V_g \tau_0}{a} < 1$, which, according to the results, plays an important role in the phenomenon of plasma compression. It is known, that ponderpmotive force is proportional to the gradient of the radiation intensity. It means, that ratio of the transversal and longitudinal components of the ponderomotive force is just equal to this parameter:

$$\frac{\nabla_r I}{\nabla_z I} \approx \frac{V_g \tau_0}{a} < 1 \tag{60}$$

This ratio helps us to explain the physical picture of the plasma compression effect described above.

The ponderomotive force sweeps the plasma particles from the area of HF field localization both in the transversal and longitudinal directions (the direction, in which the pulse moves; see Fig.3). But the hydrodynamical velocity of particles $v$ is much less than the group velocity of the pulse, $v \ll V_g$. Thus the pulse overlaps the particles [9].

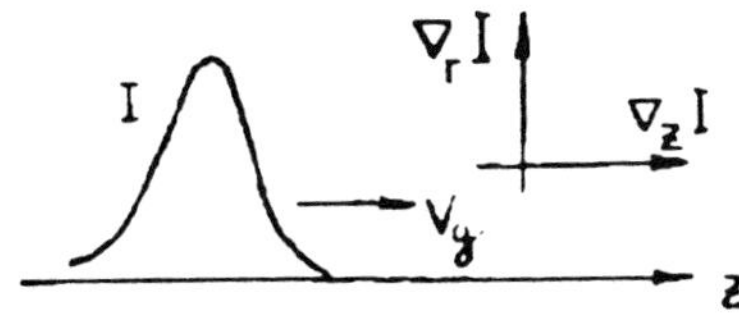


Fig.3. The qualitative radial distribution of the beam intensity and the directions of ponderomotive force components.

Under the condition (60) the rate of sweeping in the longitudinal direction is higher than in the transversal one. Consequently, the ponderomotive force of high frequency field groups the particles in its localization area.

## REFERENCES


1. Akhmanov S.A., Sukhorukov A.p., Khokhlov R.V. Uspekhi Phys.Nauk, 1979, v.93, 19.
2. Askar'yan G.A. Uspekhi Phys.Nauk, 1973, v.111, p.249.
3. Lee Y.C., Liu C.S., Chen U.H., Nishikawa K. Plasma Phys. and Controlled Nuclear Fusion Research. 1074, v.3, p.207.
4. Max C.E. Phys.Fluids, 1976, v.19, p.74.
5. Litvak A.G. ZETP, 1969, v.57, p.629.
6. Lugovoi V.I. Prokhorov A.M. Uspekhi Phys.Nauk, 1973, v.111, p.204.
7. Zakharov V.E., Sobolev V.V., Sinakh V.S. ZETP, 1971, v.60, p,136.
8. Konno K., Suzuki H. Preprint NUP-A-7810, Nihon University,1987.
9. Tsintsadze N.L., Tskhakaya D.D. ZETP, 1977, v.72, p.480.

**Research Trends in Physics: Coherent Radiation Generation and Particle Acceleration**
Editorial Board: J.M. Buzzi, A. Prokhorov (Editor-in-Chief), P. Sprangle, and K. Wille
*La Jolla International School of Physics*, The Institute for Advanced Physics Studies, La Jolla, California

# Effects of Plasma Inhomogeneity in Laser Accelerators

**L.M. Gorbunov and A.N. Moskalev**

P.N. Lebedev Physics Institute
Russian Academy of Sciences
Moscow, Russia

Two types of laser accelerators attract most attention in last years - plasma beat-wave accelerator (PBWA) /1-3/ and laser wake-field accelerator (LWFA) /1,4,5/. In both cases the quick wave of electron charge density (plasma wave) is used for electrons acceleration and only method of wave excitation differs one accelerator from another. In the case of PBWA the plasma wave is excited by means of two-frequency laser pulse with moderate intensity. The effective excitation is reached by carring out the resonant condition $\omega_1 - \omega_2 \simeq \omega_p$, which connects the difference of laser wave frequencies $\omega_{1,2}$ with plasma frequency $\omega_p = \sqrt{4\pi e^2 N/m}$ determined by plasma electron concentration $N$. Hence the strict limitation on plasma density follows and even small variations in plasma density influence essentialy on the process of acceleration. In the another case LWFA the charge-density wave is excited by a short intensive one-frequency laser pulse and any strict limitations on plasma density are absent. However in this case plasma density variations have essential influence on the acceleration process.

 

This report reviews some results of theoretical and numerical investigations of plasma inhomogeneity influence on plasma wave excitation and electron acceleration in both types of accelerators. The results were obtained in Plasma Theory Department of Lebedev Physics Institute.

## Part I. PLASMA BEAT-WAVE ACCELERATOR

### 1. Basic equations

The equation for complex slowly varying amplitude of plasma wave excited by two-frequency electromagnetic pulse has been discussed in many papers (e.g. Refs. 6-12). It is

$$-2i\frac{\partial q}{\partial \xi} - q\cdot\delta + \frac{3}{2}q|q|^2 = a_1 a_2^* \qquad (1)$$

where $q = eE^{l}(\xi, x)/2mc\omega$ is dimensionless amplitude of plasma wave; the variable $\xi = \frac{\omega}{v_g}(x - v_g t)$ determines a distance from leading edge of laser pulse in moving with pulse frame; dependence upon variable $x$ is connected with a density variation $\delta = (\omega_p^2(x) - \omega^2)/\omega^2 = \delta N(x)/N_0$ ; $v_g$ is the group velocity of pulse, $N_0 = m\omega^2/4\pi e^2$ is the resonant plasma concentration in linear approximation, $a_{1,2} = eE^{tr}_{1,2}/2mc\omega_{1,2}$ are dimensionless amplitudes of electromagnetic waves. The assumptions $|q| < 1$ , $|a_{1,2}| < 1$ and $|\delta| < 1$ were used by derivation of Eq. (1) . Besides dissipation, thermal motion of electrons, ion dynamivs, electromagnetic cascading, generation of plasma wave harmonics, transverse effects are omitted. All thise problems has been discussed in literature and conditions of validity Eq.(1) were formulated in Ref. 13.

Transforming to real quantities in Eq.(1) ( $q = -iQ\exp(i\varphi)$, $a_{1,2} = A_{1,2}\exp(i\varphi_{1,2})$) we obtain :

$$2(dQ/d\xi) = -A_1A_2\cos\theta \tag{2}$$

$$2(d\theta/d\xi)Q = Q\delta(x) - \frac{3}{2}Q^3 + A_1A_2\sin\theta \tag{3}$$

where $\theta = \varphi - \varphi_1 - \varphi_2$. The real electric field of plasma wave is expresses through values $Q$ and $\theta$ by means of relation

$$E(\xi, x) = 2Q(\xi, x)\cos[\xi + \theta(\xi, x)] \tag{4}$$

The trajectory of accelerated electron is descrobed in the form $x = v_g(t - t_0) + x_0 + x_1(t)$, where $x_0$ and $v_g$ are position and velocity of particle in initial time $t_0$ respectively. From equation of motion of relativistic particle for the value $y_1 = \frac{x_1\omega}{c}$ we obtain

$$\frac{d^2y_1}{d\tau^2} = 2\left[\gamma_0^{-2} - 2\frac{dy_1}{d\tau}\sqrt{1-\gamma_0^{-2}} - \left(\frac{dy_1}{d\tau}\right)^2\right]^{3/2} Q(y,\eta)\cos[\eta + \theta(y,\eta)] \tag{5}$$

where $\tau = \omega t$ is dimensionless **time**, $y = \frac{\omega}{c}x_0 + \tau - \tau_0 + y_1$; $\eta = y - \tau$. The electron energy in initial time $t_0$ is equal $\varepsilon_0 = mc^2\gamma_0$. In the next time the energy is calculated by means of formula

$$\gamma(\tau) = \frac{\varepsilon}{mc^2} = \left[\gamma_0^{-2} - 2\frac{dy_1}{d\tau}\sqrt{1-\gamma_0^{-2}} - \left(\frac{dy_1}{d\tau}\right)^2\right]^{-\frac{1}{2}} \tag{6}$$

We prepouse that the pulse form, its initial position and density profile of plasma are given. Also we believe to be known initial position of a accelerated electron (place of injection). The electric field accelerating of the particle is calculated from Eq. (4). Then Eq. (5) is used to obtain the values $y_1$, $(dy_1/d\tau)$ and energy (6) over small time interval $\Delta\tau$. After that by means of Eq. (4) we determine the field acting on the particle by new position of pulse, coordinate and velocity of the particle. From Eq. (5) we obtain the values $y_1$, $(dy_1/d\tau)$ and energy in time $2\Delta\tau$ and so on. Thus we investigate the plasma wave excitation and electron acceleration in inhomogeneous plasmas. In our concideration the electric field of plasma wave is calculated not only in place of particle but in all space between the electron and leading edge of laser pulse.

## 2. Plasma wave excitation and electron acceleration in plasma with linear density profile

For square form laser pulse the analytic implicit solution of Eqs. (2) and (3) can be obtained in inhomogeneous plasmas by consideration the detuning $\delta$ as parameter depending from coordinate $X$. The analysis of this solution is presented in Ref.14. However, the results of numerical investigations have more obvious feature. The evolution of amplitude $Q$ (envelope) and phase $\theta$ are shown in Fig. 1 for pulse propogating in direction of increasing concentration. The sloping dotted lines in Fig.1 schow the region of nonlinear phase stabilization of plasma wave to be discussed in Ref. 14. Also in Fig.1 the corresponding intervals of density variations are shown.

The influence of plasma inhomogeneity on the electron acceleration depends essentialy on the relation between two scales.

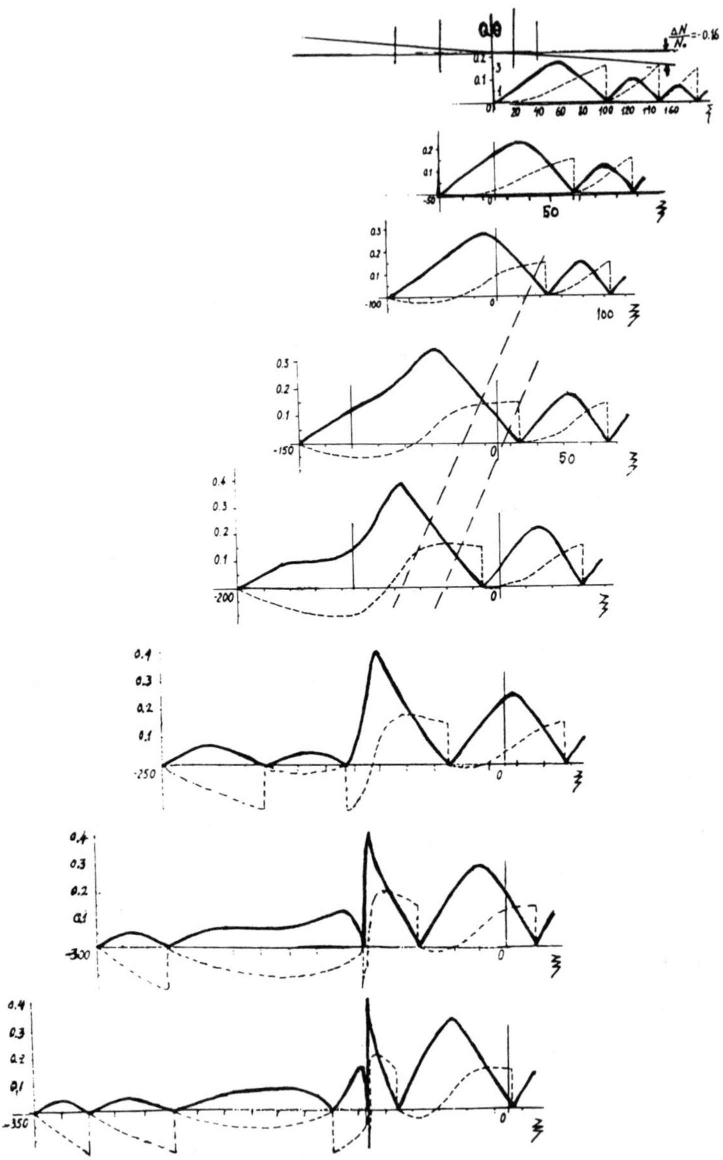

Fig.1. Variation of amplitude $Q$ (solid line) and phase $\theta$ (dashed line) of plasma wave excited by propogation of square laser pulse in inhomogeneous plasma in direction of increasing density ( $A_1A_2 = 7,3\cdot10^{-3}$, $\delta_c \simeq 0,1$ ).

The first one is connected with inhomogeneity and determines the length of resonant excitation region of plasma waves $\ell_N$ . The second scale characterises the distance of dephasing on $\pi$ the wave and particle in homogeneous plasma by electron acceleration (so-colled length of acceleration which is nearly equal to $\ell_a \simeq \gamma_0^2 (c/\omega)$ ). We shall speak that inhomogeneity has small scale if the inequality $\ell_N \ll \ell_a$ is fulfilld. In the case of opposite inequality ( $\ell_N > \ell_a$ )it is usefull to speak about the long scale inhomogeneity.

The time variation of electron energy $\gamma(\tau)$ is shown in Fig.2. The particles were injected in space where plasma density was 20 % lower the linear resonant density $N_0$ . In moment of injection the leading edge of pulse is places in different points. In Fig. 2 (a,b,c) it, is shown three groups of the curves with various distances between the leading edge of the laser pulse and particle injection point. The exact value of distance is lebeled near every curve. Small difference in distancen corresponds to injection of particle in various periods of plasma wave. The electric field was nearly equal in all initial positions of particles.

According to Fig. 2 in initial stage the electron is accelerated and then it begins to brake. This result is immediate consequence of plasma inhomogeneity which introduces the variation of plasma wave phase and leads to change the sign of electric field acting on particle. Then the electric field changes the sign once more and electron begins the acceleration again. The number

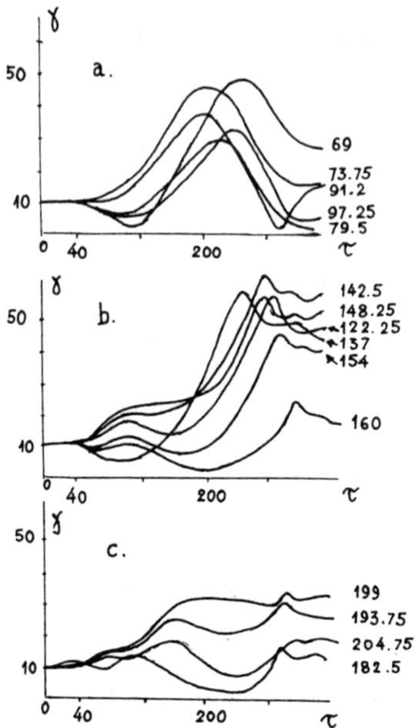


Fig.2. Variation with time $\tau = \omega t$ the energy of electrons to be injected in plasma wave by $\delta$ = -0,2 in different points with respect to leading edge of laser pulse.

of such stages of acceleration and deacceleration due to distance between the particle and leading edge of laser pulse in time of injection. This number in that larger that distance is longer.

The electron obtains main energy by crossing the region of nonlinear resonance. If electron is injected on short distance from leading edge of laser pulse then the accelerating field is sufficiently low and gain of energy relatively small (Fig. 2a).

For electron to be injected on long distance from leading edge of laser pulse the amplitude of plasma wave is great but in narrow region. Out of this region the phase of wave has jumps and acting on the particle field many times changes the sign (Fig. 2 c). There is the certain interval of initial distances between particle and leading edge for which the phase of plasma wave is practicaly constant in the vicinity of resonant density (nonlinear phase stabilisation /14/ ) and amplitude of plasma wave is sufficiently high (Fig. 2 b). From presented in Ref. 14 equations we obtain for considered conditions that interval of phase stabilization begins from $\xi$ =125 and cames to end by $\xi \simeq$ 170. Inside of this interval for every initial particle position there is corresponding range of density variation. But realy the permissible density variations are not wide and equal nearly 4 % . This interval of nonlinear phase stabilization is shown in Fig. 1 by means of sloping dotted lines.

For laser pulse of square form propagating in direction of decreasing density the main difference consists in more slow phase variation in region $\delta < \delta_c$. The interval of nonlinear phase stabilization is wider but amplitude of wave is lower. This result due to difference in excitation of nonlinear plasma oscillations with frequencies more higher ( $\delta > \delta_c$) and more lower ( $\delta < \delta_c$ ) than resonant frequency ( $\delta \simeq \delta_c$). More smooth phase variation leads to increasing of acceleration time. But efficienty of acceleration is not so high because of smaller wave amplitude and decreases with time by going the particle out the resonant region.

These examples show that in inhomogeneous plasmas the electron acceleration depends on the sign of density gradient and efficienty of acceleration is higher by motion of particle in direction of more dense plasma.

We investigate also excitation of plasma wave of laser pulse with linearly increasing intensity on leading edge. The jumpe of phase vanish expect the region of nonlinear resonance ( $\delta \simeq \delta_c$). Therefore the transitions from a acceleration stage to deaccele-ration one are more smooth. The regime of nonlinear phase stabi-lization is safed for considered form of laser pulse.

The most considered runs correspond to inequality $\ell_N < \ell_a$ . In the limit $\ell_N \ll \ell_a$ the distance between accelerated particle and leading edge of laser pulse is practicaly constant. For given retardet time which is proportional to this distance the electric field of plasma wave in inhomogeneous plasma (various values detuning ) was investigated in Ref. /22/.

## 3. Electron acceleration in plasma with hump-shaped form of concentration

Considered above the linear form of density variation is very convenient for investigation the influence of plasma inhomo-geneity on electron acceleration as more simple example of density variation. However, nonmonotonic density variation is more naturaly. In this part we present some results of numerical inves-tigation the influence on process of electron acceleration a region with higher plasma density (hump-shaped form of density).

We have considered in this section only laser pulses of square form. In Fig. 3 the energy of electrons as function of time are shown. The ground density was nearly 2 % lower and maximum density was nearly 9 % higher than linear resonant density.

In Fig. 3a the results are shown for relatively narrow hump the length of which is more smaller than length of acceleration $l_a$ . In all calculations the injection place was in the same point indicated in Fig. 3a, but the leading edge of laser pulse has different initial positions. Corresponding distances are shown by means of figures by curves. It is clear that narrow hump has a week influence on process of acceleration.

For hump with longer scale the acceleration increases visible in the region of higher density, however, sufficiently narrow region arises with reduced rate of acceleration and in this region the electron is even deaccelerated (Fig. 3b). The dotted line in Fig. 3b shows the energy of electron in homogeneous plasma with density ~ 2 % lower the linear resonant density $N_0$ . Thus the density splash increases the rate of acceleration for some positions of electrons relatively the leading edge of laser pulse.

Figure 3a shows the results of calculation for more wide hump. In this case the scale length of hump approaches to the acceleration scale length. The change of wave phase because of plasma inhomogeneity leads in fact to absence of acceleration in the region with more higher density. As result this region is losted for acceleration and energy of accelerated particles is reduced.

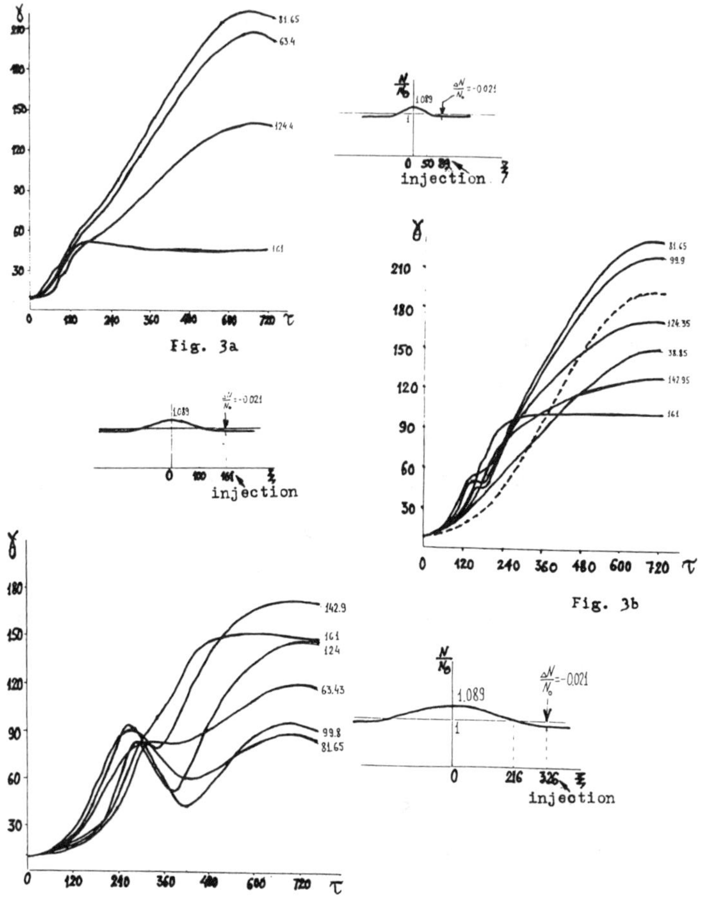

Fig. 3a

Fig. 3b

Fig. 3c

Fig.3. Electron energy $\gamma$ as a function of time $\tau = \omega t$ for hump-shaped form of density

a) scale length of hump $\xi_0$ =100

b) scale length of hump $\xi_0$ = 200

c) scale length of hump $\xi_0$ = 432.

The deshed line in Fig. 3b shows the electron energy in homogeneous plasma.

This conclusion is confirmed of calculations with more higher hump of concentration (Fig. 4 ).

In summary, the density splash with very small scale length has week influence on process of acceleration. For more dimension of hump it is possible to incress of electron energy if density in splash is approached to nonlinear resonant value ( $\delta \simeq \delta_c$ ). The long splash reduces essentialy the acceleration efficiency. The change of electron energy is irregular and acceleration is practicaly absence by motion of particle across density hump.

As one example else we present the numerical results modeling the experiment /16/ where it is measured the dependense of plasma wave excitation and electron acceleration from plasma density.

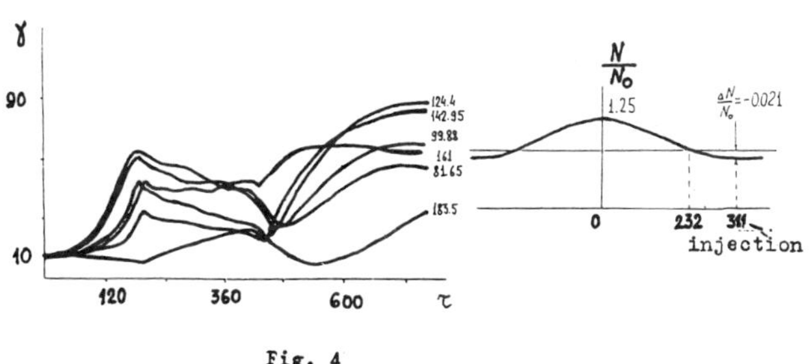


Fig. 4

Fig.4. Energy of electrons $\gamma$ as a function of time $\tau$ crossing the region of dense plasma with large scale length ( $(N_{max}/N_0)=1,25$).

The laser pulse was more longer than the length of plasma. The plasma wave begins detected when the maximum value of density exceeds the linear resonant value $N_0$ . The accelerated electrons were detected in more narrow interval of maximum values of density near $N_0$.

In corresponding to experiment of Ref. 16 conditions (long laser pulce, small plasma dimension) the amplitude of plasma wave is shown in Fig. 5 (a,b,c) in two times (two laser pulse positions) and for three values maxima of density. It is clear that by exceeding density maximum the form of function $Q$ becomes compli-

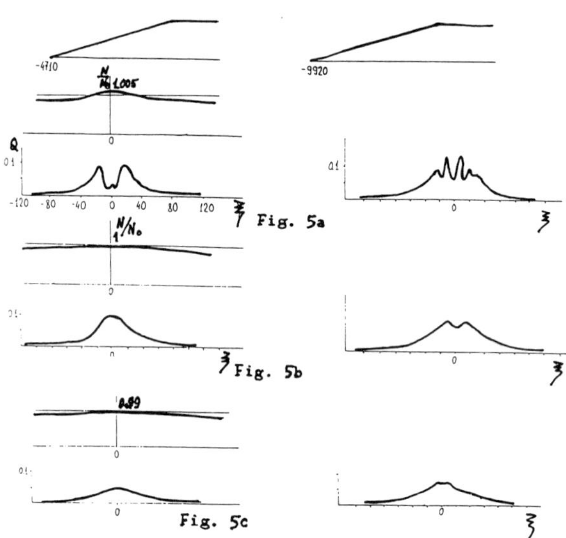


Fig.5.Plasma wave amplitude $Q$ as a function of distance at two moments of time and for three maximum values of density a) $(N_{max}/N_0)=1{,}005$ ; b) $(N_{max}/N_0)=1$ ; c) $(N_{max}/N_0)=0{,}99$.

cated however always there is the resonant region where amplitude of plasma wave is sufficiently high. By increasing of density maximum still more the regions of plasma wave excitation move apart and become more narrow (Fig. 6). As shown above the efficiency of electron acceleration in this case reduces (Fig. 4).

## 4. The influence of longitudinal magnetic field on excitation of charge density waves

A sufficiently weak longitudinal magnetic field increases the scale length of plasma waves excitation region in inhomogeneous plasma for pulses with transverse dimensions of the order of $c/\omega$, where $\omega = \omega_1 - \omega_2$ /17/. This result is connected with the opportunity to satisfy the resonant conditions for wave excitation by simultaneous change both plasma density and direction of wave propagation.

It was considered two overlap axial simmetric wave beams being produced of two waves with linear polarization along OX

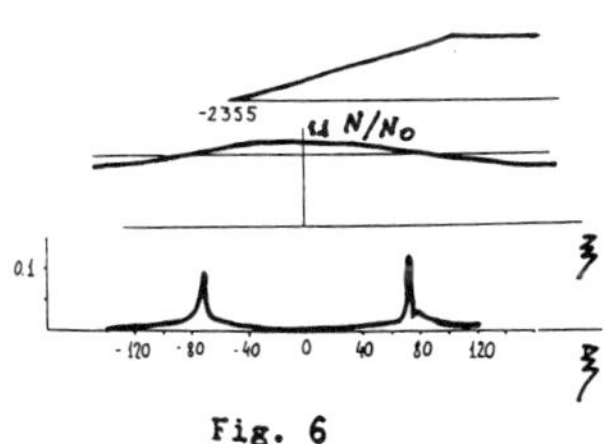


Fig. 6

Fig. 6. Plasma wave amplitude $Q$ excited by laser pulse with slow increasing leading edge in a inhomogeneous plasma ( $\xi_f$ =9420).

axis. The beams propogate along $Oz$ axis in inhomogeneous plasma. The plasma density depends only on variable $z$ (Fig. 7). For the Gaussian form of beams in linear approximation the density perturbations $\delta n$ on the frequency $\omega$ are given by

$$\delta n = \frac{E_{10} E_{20} \omega_p^2 K^2}{m \omega_1 \omega_2 8\pi\omega^2} \mathrm{Re} \left[ e^{iKz - i\omega t} F(\rho, z) \right] \tag{7}$$

$$F(\rho, z) = \frac{d^2}{8\pi K^2} \int_0^\infty dK_\perp \, K_\perp J_0(K_\perp \rho) e^{-\frac{d^2 K_\perp^2}{8}} \frac{K^2 + K_\perp^2}{\left(i\frac{\nu}{\omega} - \frac{z}{L}\right) - \frac{\Omega^2}{\omega^2} \frac{K_\perp^2}{K^2 + K_\perp^2}} \tag{8}$$

where $\rho$ is transverse coordinate, $K = K_1 - K_2$ ; $E_{10}, E_{20}$ are amplitudes of waves with frequencies $\omega_1$ and $\omega_2$ , $\nu$ is collision frequency of electrons, $L$ is scale length of the density variation in point where take place equality $\omega = \omega_p(z)$. The cyclotron electron fre uency $\Omega$ is assumed to be small then $\omega$ .

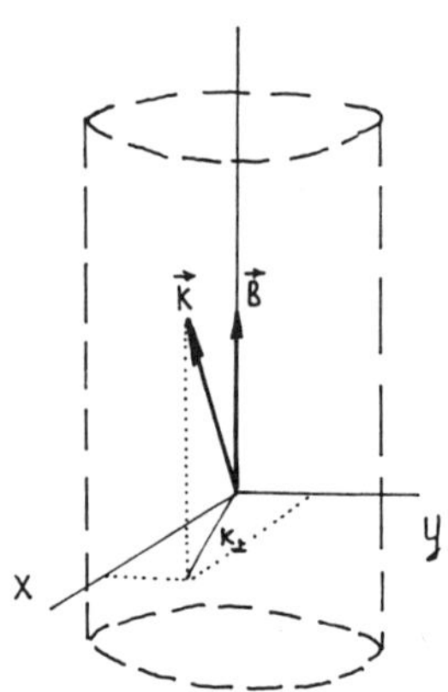


Fig.7. Scheme of considered model. The dotted lines show wave beams.

On a beam axis ( $\rho$ =0) the equation (8) is expressed through the tabulated functions. Fig. 8 shows the value $|F|$ characterising the charge density wave amplitude as a function of variable $z/L$. It is clear that increase of magnetic field leads to grow of excitation region dimension, however, amplitude becomes lower. When the unequality $(\Omega/\omega)^2 > (\nu/\omega)$ is fulfild the width of excitation region is equal approximately $\ell_H \simeq L(\Omega/\omega)^2$.

For conditions near to experimental situation /18/ ( $\omega_p \simeq 10^{13}$ $c^{-1}$, $d \simeq 3.10^{-3}$ cm ) the magnetic field 100 kG gives rise of the excitation region up to scale length of plasma inhomogeneity. However in plasma beat-wave accelerator not dissipation but nonlinearity determines the amplitude of a plasma wave. The problem in this case is not considered now.

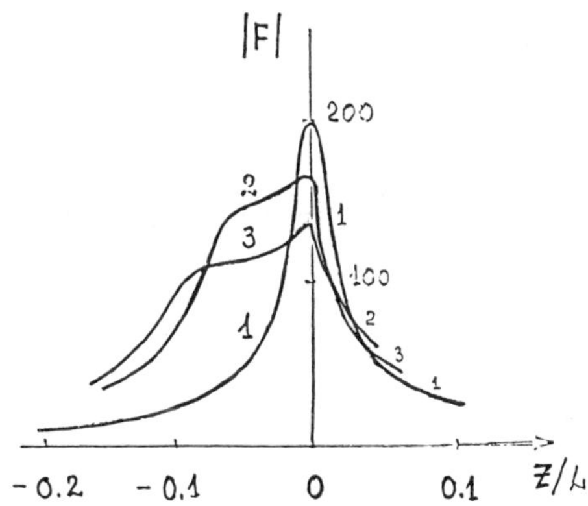


Fig.8. Variation dimensionless plasma wave amplitudes with coordinate. The curves 1,2,3 are correspond to parameter $(\Omega/\omega)^2 = 0;\ 0,1;\ 0,14.$ $\left[ (\nu/\omega) = 10^{-2}, \frac{(Kd)^2}{8} = 1 \right]$.

## PART II. LASER WAKE FIELD ACCELERATOR

### 5. Optimal plasma density

In laser wake field accelerator the charge density wave is excited by means of short one-frequency laser pulse and unlike the plasma beat wave accelerator the plasma density is not determined any sharp conditions. However, in this case there is some optimal plasma density producing the maximum gain of electron energy. To determine the optimal plasma density we consider shortly the main parameters of a wake field accelerator.

Let us consider the electron injected in maximum of potential $\varphi_m'$ with velocity $v_0'$ in the frame moving with the plasma wave. After/wards of acceleration the maximum electron energy in laboratory frame is given by

$$W_m = \gamma_p W_m' \left(1 + \beta_p \sqrt{1 - (mc^2)^2/W_m'^2}\right)$$

where $v_p$ is the phase velocity of wave, $\beta_p = v_p/c$ , $\gamma_p = (1-\beta_p^2)^{-\frac{1}{2}}$ , $W_m' = mc^2\gamma_0' + e\varphi_m'$ , $\gamma_0' = (1 - v_0'^2/c^2)^{-\frac{1}{2}}$ . By condition $v_0' \ll c$ the maximum energy gain of accelerated electrons is

$$\Delta W_m = W_m - W_0 = mc^2\gamma_p \left(a + \beta_p\left[\sqrt{a(2+a) + (v_0'/c)^2} - (v_0'/c)\right]\right) \quad (9)$$

where $a = e\varphi_m\gamma_p/mc^2$ , $\varphi_m$ is the maximum value of potential in laboratory frame. For harmonic wave $\varphi' = \varphi_m' \cos k_p' x'$ the parameter $a$ is expressed through the electric field amplitude $E_m$ and the wave number $k_p$ in laboratory frame : $a = eE_m\gamma_p/mc^2 k_p$ .

By the laser pulse generation of a charge density wave in plasmas we have $\gamma_p = \omega_o/\omega_p$ and $k_p \simeq \omega_p/c$ , where $\omega_o$ is the laser frequency, $\omega_p = \sqrt{4\pi e^2 N/m}$ is the plasma frequency which is determined of the electron concentration $N$ .

In the limit $a \gg 1$ from the Eq. (9) the well known result follows /1-3/

$$W_m \simeq 2mc^2\gamma_p a = 2eE_m\gamma_p^2 k_p^{-1}$$

This expression is valid if the energy of a accelerated electron is more higher than its initial energy.

The time of electron passage through the accelerating phase, determines the length of acceleration

$$L_a \simeq \gamma_p \int_0^{x_o'} dx' \left\{ 1 - \left[\gamma_o' + \frac{e}{mc^2}\left(\varphi_m' - \varphi'(x')\right)\right]^{-2} \right\}^{-\frac{1}{2}}$$

where $x_o'$ is the point of zeroth potential ( $\varphi'(x_o')$ =0 ). For the harmonic potential this expressen reduced to elliptic integrals and has in some limiting cases the next form

$$L_a \simeq \begin{cases} \dfrac{\pi\gamma_p^2 c}{k_p v_o'} , & a \ll \frac{1}{4}\left(v_o'/c\right)^2 \\ \dfrac{\gamma_p^2}{k_p\sqrt{a}} \ln\left(\dfrac{8c\sqrt{a}}{v_o'}\right), & 1 \gg a \gg \frac{1}{4}\left(v_o'/c\right)^2 \\ \dfrac{\pi\gamma_p^2}{k_p}\left[1 + \dfrac{1}{\pi\sqrt{a}}\ln\left(8c/v_o'\right)\right] , & a \gg 1 \end{cases} \qquad (10)$$

The length of acceleration is infinite by zero value $v_o'$ . The used as a rule expression for acceleration length $L_a = \pi\gamma_p^2/k_p$ is valid not only by condition $a \gg 1$ but by unequality $(v_o'/c\gamma) \gg \exp(-\pi\sqrt{a})$ which is connected with initial velocity of particle.

Let's consider the Gaussian form of a laser pulse. In linear approximation that pulse excites the plasma wave with the amplitude /4/

$$E_m = \frac{a_o^2 mc^2}{eL} \sqrt{\pi}\, \nu^2 e^{-\nu^2} \tag{11}$$

where $a_o = V_E/c$, $V_E$ is the amplitude of quiver electron velocity in laser field, $L$ is the pulse dimension, $\nu = (k_p L)/2$.

The curves for the dimension-less energy $W_m = 4\,\Delta W_m / mc^2 a_o^2 (k_o L)^2$, the acceleration length $l_a = 8 L_a / \pi L (k_o L)^2$, the amplitude of plasma wave $e_m = e E_m L / a_o^2 mc^2$ and the gain of acceleration $p = \frac{2}{\pi} (W_m / l_a)$ are shown in Fig. 9. The calculations were

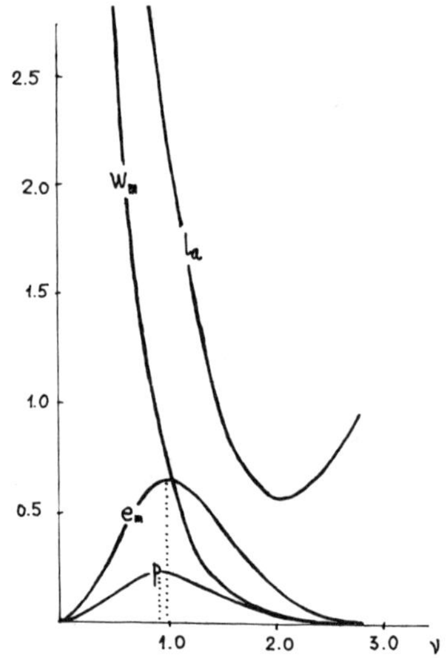


Fig.9. The maximum energy $W_m$, the length of acceleration $l_a$, the gain in energy $p$ and the amplitude of plasma wave with respect to $\nu = k_p L/2$

produced by $(v_0'/c)=10^{-2}$, $a_0^2=10^{-2}$, $(\omega_0 L/c)=2.10^3$, which correspond to pulse duration 1 ps for neodim glass laser by the intensity $3{,}7.10^{17}$ W/cm$^2$.

The most higher gain of acceleration arises by $\nu \simeq 1$, when the charge density wave length $\lambda_p = 2\pi/k_p$ is equal $\pi L$. This conclusion is suffciently clear because of by $\nu=1$ there is the maximum of plasma wave amplitude. However, that coincidence of maxima for the functions $\mathcal{E}_m$ and $P$ takes place only by condition $a_0^2 k_0 L \gg 0.1 \left[\ln(8c/v_0')\right]^2$.

6. The electron acceleration in inhomogeneous plasmas

The nonlinear equation /19, 20 / used for consideration of charge density wave potential $\phi_0$ generated by given laser pulse

$$\frac{d^2\phi_0}{d\xi^2} + \frac{k_p^2(x)}{2}\left(1 - \frac{1+|a_E|^2/2}{\phi_0}\right) = 0 \tag{12}$$

where $\xi = x - v_g t$, $k_p \simeq \omega_p(x)/c$ is the wave number due to variable plasma concentration, $a_E(\xi) = eE(\xi)/m\omega_0 c$ is the dimensionless amplitude of a laser pulse. The motion equation of particle has the form analogous to Eq. (5)

$$\frac{d^2y}{d\tau^2} = \ell_0 \frac{d\phi_0}{d\xi}\left[1 - \frac{v_p^2}{c^2} - 2\frac{v_p}{c}\frac{dy}{d\tau} - \left(\frac{dy}{d\tau}\right)^2\right]^{3/2} \tag{13}$$

where $y = \frac{x_1}{\ell_0}$, $\tau = \frac{ct}{\ell_0}$, $\ell_0$ is the scale of laser pulse length. The scheme of calculation is the same as indicated above.

The calculations were performed for the Gaussian form of pulse

$$|a_E|^2 = a_0^2 \exp\left[-(2\xi/\ell_0)^2\right]$$

with the feature dimension $\ell_0=3.10^{-3}$ cm which corresponds to the pulse duration 0,1 ps . The value $a_0^2$ was equsl to 1,83 and it can be realized by mean of Nd -glass laser with the energy 4 J by the transverse size $3.10^{-3}$ cm.

As in the case of plasma beat wave accelerator we shall consider firstly the linear profile of plasma density. The energy of electrons $\gamma(\tau)$ accelerated in a plasma with increasing density are shown in Fig.10. The injection of electrons was produced in fixed point on different distances from the laser pulse. The accelerating field in injection place was the same in all cases. It is clear that electrons are accelerated most effective by some initial position. Beginning acceleration from this position the electrons have the gain in energy more higher than in homogeneous plasmas (dotted line in Fig. 10).

The calculations of electron acceleration in a plasma with decreasing density shown the reduction of gain in energy by any distances between the laser pulse and the injection points (Fig.11). The negative influence of plasma inhomogeneity is the more the particle is on the longer distance from the laser pulse.

The reason of such dependence the gain in energy from the sign of plasma concentration gradient can be explained the next manner. The phase of a charge density wave has the form $\Phi = \omega_p(x)\left(t - \frac{x}{v_g}\right)$ and a phase velocity determining propagation of point with constant phase is $v_\phi = v_g\left[1 - \frac{d \ln \omega_p(x)}{dx}(tv_g - x)\right]^{-1}$.
This velocity increases by motion of a wave in direction of more dense plasma and it decreases by a wave motion in more rarefied plasma. In process of acceleration the particle obtains the

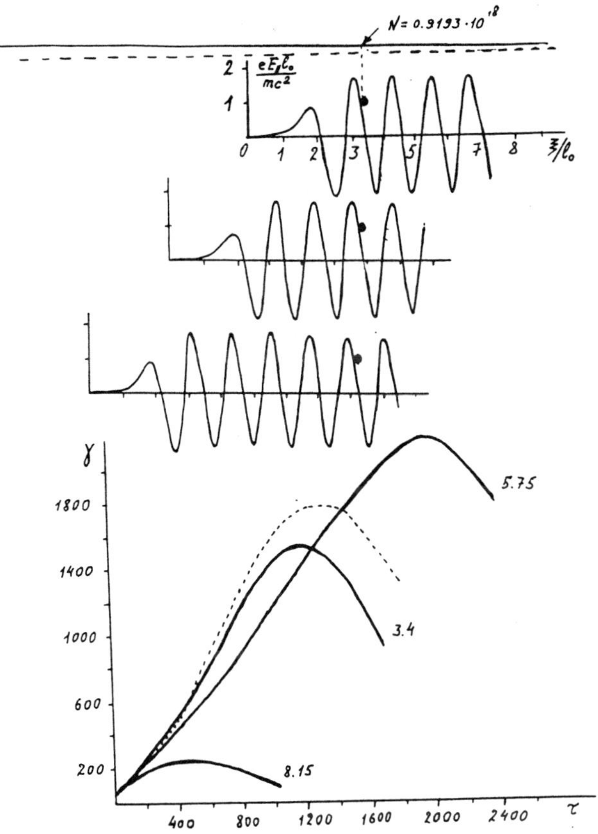


Fig. 10. Electron energy $\gamma$ as a function of time by acceleration in plasma with linearly increasing density. The three curves correspond different distances between laser pulse and point injection of particle. The dotted line shows $\gamma(\tau)$ in homogeneous plasma.

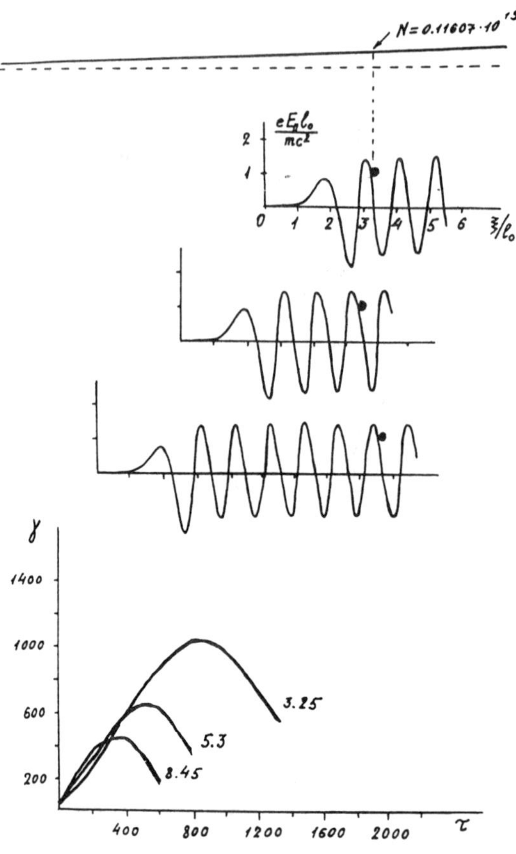


Fig.11. Electron energy $\gamma$ by acceleration in plasma with lineary decreasing density. The distances between the laser pulse and point of injection are indicated near the curves.

energy and its velocity becomes higher than that for wave. In inhomogeneous plasma with increasing density the wave catch up a particle and can produce more effective acceleration. The particle moves more longer time in accelerating phase of wave.

The electric field and potential generated by short laser pulse in a plasma with a strong density gradient are shown in Fig. 12. It is clear seen the wave length decreasing because of plasma inhomogeneity.

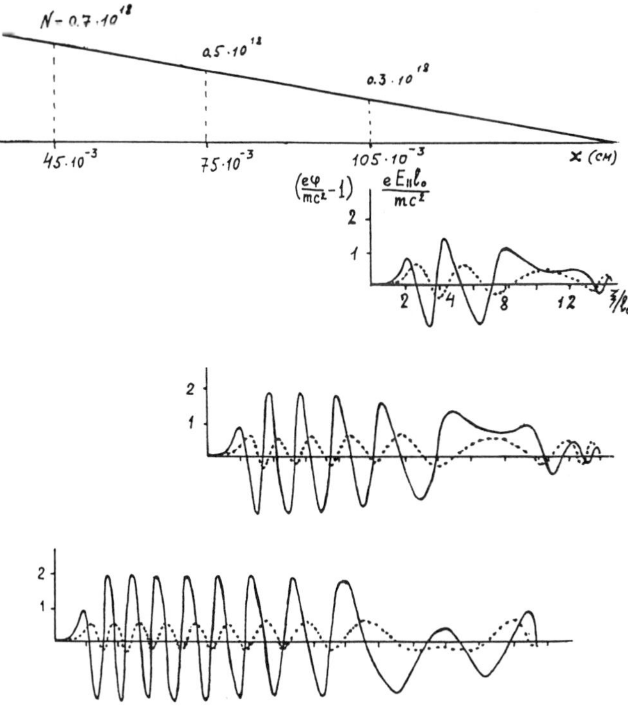


Fig.12. The field (solid line) and potential (dotted line) produced of short laser pulse in inhomogeneous plasma.

With aim to consider more realistic situation we investigated the transmission of a laser pulse and electron acceleration in plasma layer with variable density. The Fig. 13 shows the plasma density variation in space, the place of particle injection and their initial position relatively the laser pulse, the time evolution of accelerated particles energy. In the rare plasma both in increasing density region and decreasing density region any effective acceleration is absent. In this regions the phase of wave changes very fast and particles are not trapped in the wave. The plasma density varies most smaller near the maximum and regular acceleration produces only in this region. The result of

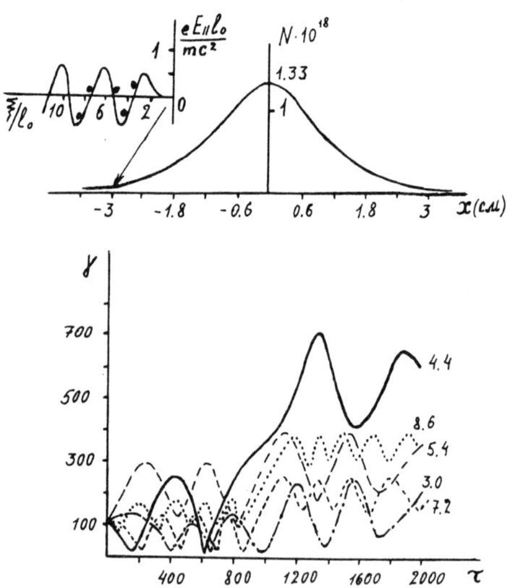


Fig.13. The plasma density (a) and the electron energy for different initial position of laser pulse.

acceleration depends on a distance between the particle and laser pulse by injection. For every plasma density profile there is some convenient particle positions with the maximum gain in energy.

## 7. Increase of acceleration efficiency by means of special plasma density profile

In laser wake field accelerator it is possible by means of a special form of plasma density to hold the particle in accelerating phase more long time and essentialy to rise the efficiency of the method /21/.

From condition of constant wave phase in particle place and by supposition of constant acting force on accelerated electron we obtain the required plasma density profile

$$\frac{N(x)}{N(x_0)} = \left[ \frac{x_0/\ell}{\frac{x}{\ell} - \beta_b \left[ \sqrt{\left( \left[ (x-x_0)/\ell \right] + \gamma_p \right)^2 - 1} + \sqrt{\gamma_p^2 - 1} \right]} \right]^2 \tag{14}$$

where $\ell = mc^2/|e|E$ , $E$ is the plasma wave amplitude which is assumed to be constant, $\beta_b = v_g/c$ , $\gamma_p = (1 - v_p^2/c^2)^{-\frac{1}{2}}$, $v_p$ is the initial electron velocity which is not equal in general case to velocity of wave. The plasma density profiles corresponding to a constant force acting on electron are shown in Fig. 14. For considered above (§6) situation $\gamma_p = \gamma_b$ the plasma density increases. This conclusion is in agreement with numerical calculations.

The infinity of plasma density (Fig. 14) corresponds to approach of electron to laser pulse in process of acceleration. The stationary phase can be obtained in this situation only by

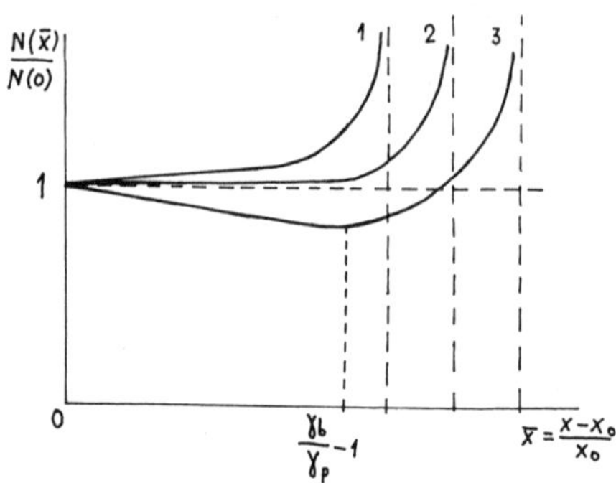


Fig.14. The plasma density profiles producing the constant force acting on electron by $\gamma_p > \gamma_b$ (1), $\gamma_p = \gamma_b$ (2), $\gamma_p < \gamma_b$ (3).

means of more rapid increasing density.

## Conclusion

As it follows from our analysis the influence of plasma inhomogeneity on the electron acceleration differs for PBWA and LWFA schemes.

In PBWA scheme the plasma wave is excited by means of resonant force and the determined value of plasma density must be fulfilled with accuracy of percent or even less. In opposite case the plasma wave amplitude and phase variations destroy the process of acceleration. The longitudinal magnetic field for laser beams with diameter of order the plasma wave length can reduce the requairements on level of plasma homogeneity.

In LWFA scheme the requirements are lower as on value of plasma density as the permissible variations of density. The optimum plasma density corresponds to equality of plasma wave length and dimension of laser pulse. The influence of plasma inhomogeneity depends essentialy from distance between a laser pulse and a accelerated particle. In general case the level of permissible density variations is some tens of percents. By increasing of distance between a laser pulse and a accelerated particle the requirement on plasma homogeneity are more sharp. For every determined distance it is possible indicate the corresponding plasma density profile giving the gain in energy more higher than in homogeneous plasma.

## References

1.Tajima T., Dawson J.M. Phys.Rev.Lett.,1979, 43, 267.

2.Chen F.F. Laser Accelerators, Preprint PPG-1107, UCLA, 1987.

3.Fainberg Ja.B. J.Plasma Phys (USSR), 1987, 13, 607.

4.Gorbunov L.M., Kirsanov V.I. Sov.Phys.JETP, 1987, 66, 290.

5.Sprangle P., Esarey F., Ting A., Joyse G. Appl.Phys.Lett., 1988, 53, 2146.

6.Rosenbluth M.N., Liu C.S. Phys.Rev.Lett., 1972, 29, 701.

7.Tang C.M., Sprangle P., Sudan R.N. Phys.Fluids, 1985, 28, 1974.

8.McKinstrie C.J., Forslund D.W. Phys.Fluids, 1987, 30, 904.

9.Mori W.B. IEEE Trans., Plasma Sci., 1987, 15, 88.

10.Karttanen S.J., Salomaa R.R. Phys.Rev.Lett., 1986, 56, 604.

11.Noble R.J. Phys.Rev.A., 1985, 32, 460.

12.Mendonca J.T., J.Plasma Phys., 1985, 34, 115.

13.Gibbon P., Bell A.R. Phys.Rev.Lett., 1988, 61, 1599.

14.Gorbunov L.M., Kirsanov V.I. Sov.Phys.JETP, 1989, 69, 329.

15.Gorbunov L.M., Moskalev A.N. Proc. of International Workshop on Strong Microwaves in Plasmas, Suzdal, 1990 (in press).

16.Kitagawa Y., et al. Preprint Inst.of Laser Engineering, Osaka, Japan, N90018, 1989.

17.Gorbunov L.M., Romanov A.B. Sov.Phys.,Lebedev Inst.Reports, 1991 (in press).

18.Clayton C.E., Joshi C., Darrow C., Umstadter D. Phys.Rev.Lett., 1985, 54, 2343.

19.Bulanov S.V., Kirsanov V.I.,Sakharov A.S. JRTP Lett., 1989, v.50, p. 176.

20.Berezhiani V.I., Murusidze I.G. Proc.IY Intern.Workshop on Nonlinear and Turbulent Processes in Phys. Kiev, ND, 1989, v.1, p. 235.

21.Gorbunov L.M., Kirsanov V.I., Mtingwa S., Ramazashvili R.R. Sov.Phys., Lebedev Inst. Reports, 1989, N10, 27.

22.Karttunen S.J., Salomaa R.R.E. Phys.Scripta, 1989, 39, 741.

**Research Trends in Physics: Coherent Radiation Generation and Particle Acceleration**
Editorial Board: J.M. Buzzi, A. Prokhorov (Editor-in-Chief), P. Sprangle, and K. Wille
*La Jolla International School of Physics*, The Institute for Advanced Physics Studies, La Jolla, California

# Beatwave Excitation of Plasma Wave and Electron Acceleration

**Y. Kitagawa, K. Sawai, K. Mima, K. Nishihara,**
**H. Azechi, K.A. Tanaka, H. Takabe, and S. Nakai**

Institute of Laser Engineering, Osaka University
Suita, Osaka, 565, Japan

## ABSTRACT

Reported is the simultaneous observations of a beatwave excited plasma wave and energetic electrons. A plasma wave is excited when the beatwave frequency of a laser with 10.6 and 9.57 μm equals the plasma frequency. The Stokes sideband measurement gives the wave amplitude $dn/n_0$ of 5 %. Plasma electrons with the energies more than 10 MeV are observed at the resonant density. The experiment offers a possibility for the plasma beatwave accelerator.

## INTRODUCTION

Since Tajima and Dawson have proposed the laser beatwave accelerator as a promising collective acceleration scheme,[1] there have been many theoretical studies[2-5] because of a number of potentially important applications in the fields of plasma physics, astrophysics and particle accelerator physics, but only few experimental studies. Clayton et al. have first demonstrated the plasma wave excitation by a resonant laser beatwave,[6] followed by the work of Ebrahim et al.[7] The other work of Ebrahim et al. suggests the beatwave accelerated electrons.[8] In this letter, we present the simultaneous observations of the 10.6 and 9.57 μm beat wave excited plasma wave and electrons with the energies more than 10 MeV. The electric field of the wave is more than 1 GV/m. The results offer us a possibility for the beatwave accelerator.

What we did for the beatwave acceleration are: 1) Double line oscillation of $CO_2$ laser, 2) Resonant plasma production, 3) Plasma wave excitation, 4) Observation of the energetic electrons, 5) Computer simulations and 6) Summary.

## 1) DOUBLE LINE OSCILLATION OF $CO_2$ LASER

A schematic of the apparatus is shown in Fig. 1. A $CO_2$ laser, the LEKKO VIII electron-beam controlled laser,[9] produces a 0.4 ns rise and 1 ns full width at half maximum pulse containing both 150 J in the 10.591 μm line (10p(20)) and 150 J in the 9.569 μm (9P(22)). The latter was injected from a CW oscillator to the LEKKO VIII oscillator cavity. An f/10 NaCl lens focuses the laser beam of a 23 cm diameter to less than a 1 mm diameter spot size, resulting in an intensity $I_0 \geq 2\times10^{13}$ W/cm$^2$ for each line. The absolute intensity[10] will be between $2\times10^{13}$ and $2\times10^{14}$ W/cm$^2$.

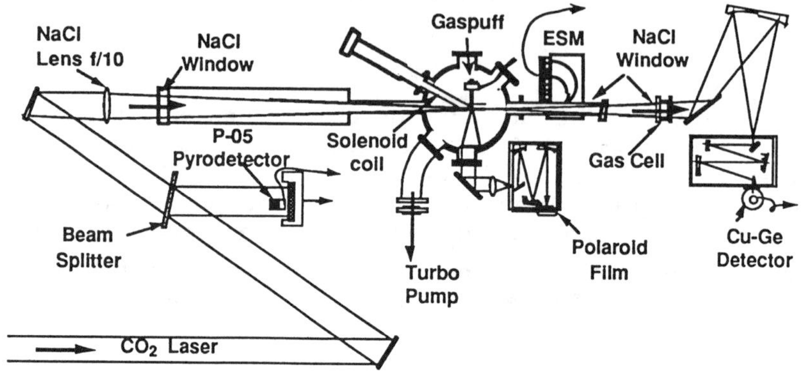


Fig.1. Experimental setup for the beatwave acceleration.

## 2) RESONANT PLASMA PRODUCTION

At the focal point in a vacuum chamber, an electromagnetically driven gas-puff injects 1-cc hydrogen gas 300 μs prior to the laser irradiation, which produces almost fully ionized hydrogen plasmas. Charging voltage for the gas-puff from 6 kV to 9 kV as well as the filling pressure from 1 to 10 atm easily controls electron densities $n_0$ from $3\times10^{16}$ to $7\times10^{17}$/cm$^3$ at the focal point. The beatwave frequency between 10.6 and 9.57 μm equals the plasma frequency for $1.1\times10^{17}$/cm$^3$. From the Stark broadening measurement of the hydrogen Balmer α line (6563 Å) emission we estimated the electron density at the focal point. 8 Å full width at the half-power corresponds to $1.0\times10^{17}$/cm$^3$ for the electron temperature of larger than 10 eV.[11]

The spectral resolution is less than 1 Å, i.e., $1\times10^{16}$/cm$^3$, but the reconstruction error from the Polaroid film is about 1 Å. The measurement of the Balmer α line emission area shows that the plasma is cylindrical with a diameter of 1.8 mm and an axial length of 7 mm. Figure 2 shows thus obtained densities as a function of the voltage applied to the gas-puff. Plasma wave excitation and electron acceleration are very sensitive to the electron density. Since the data from shot to shot vary by a factor of 2-3 around the averaged densities, we monitored the density for almost all shots. In this experiment we injected no electrons so that we did not use the solenoid coil in the chamber.

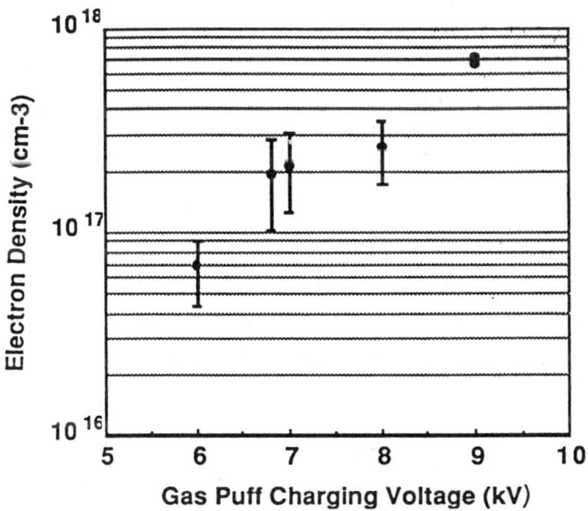


Fig. 2. Electron density as a function of the voltage charging to the gas-puff.

## 3) PLASMA WAVE EXCITATION

The double-line laser, as soon as producing resonant density plasma, excited the plasma wave in it. The ratio of the laser frequency to the plasma frequency $\gamma_p$ is 10. The excited plasma wave in turn forward-scatters the pump waves to Stokes (11.857 μm) and anti-Stokes (8.7 μm) sidebands and also to their harmonic series. The Stokes line intensity $P_s$ is expressed by the Bragg scattering formula as[12]

$$\frac{P_s}{P_0} = \left[ \frac{\pi}{2} \frac{\delta n}{n_0} \frac{n_0}{n_c} \frac{L}{\lambda_0} \right]^2, \qquad (1)$$

where $P_0$ is the 10.6 μm pump intensity, δn the electron density perturbation, $n_c$ the cutoff density, L and $\lambda_0$ are the length of the excited plasma wave and the laser wavelength. We measured $P_s$ and $P_0$ using a monochrometer with a liquid He-cooled Cu-Ge detector to confirm the plasma wave excitation and to estimate its amplitude $\delta n/n_0$. The detector sensitivities to 10.6 μm and 11.86 μm are almost same. The signal ratio $P_s/P_0$ was expected to be very small and indeed was from 0.001 to 0.08, so that to detect $P_s$ we had to put a gas cell on the light path, as shown in Fig. 1. The cell, containing a 100 Torr mixture of $SF_6$ and $C_2F_3Cl$, absorbed 99.9% of 10.6 μm and 94% of 9.57 μm pumps, preventing damages on the monochrometer optics as well as suppressing the stray signal level to less than 0.001 $P_0$. To detect $P_0$ and thus to obtain an absolute ratio of $P_s/P_0$ , we replaced the cell with teflon foil attenuators. The attenuation includes ± 25 % error. Note that the pulse rise and width of the Stokes line $P_s$ and thus of the plasma wave agreed with those of $P_0$ within the detector-scope (Tektronix 7104) response time of <1 ns. $P_0$ was also measured by a P-05 pyroelectric detector for each shot.

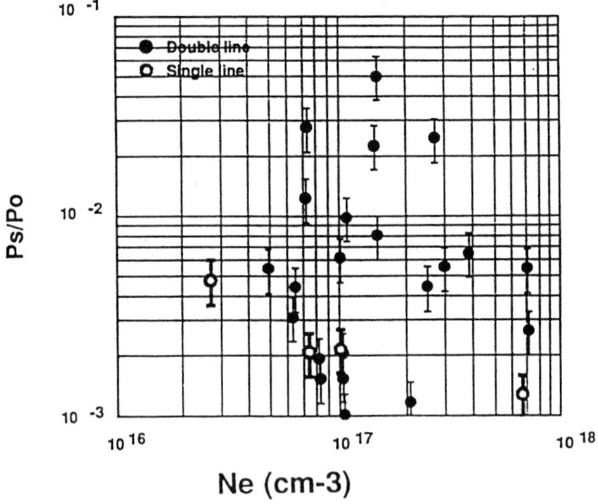


Fig. 3. Forward scattered Stokes sideband intensity $P_s$ at 11.857 μm divided by 10.6 μm pump intensity $P_0$ is plotted as a function of the electron density: Solid circles for double line and white circle for single line irradiation. Vertical bar is due to $P_0$ calibration error of 25%. The detection limit is 0.001.

Figure 3 shows Stokes signal $P_s/P_0$ as a function of $n_0$. They show a peak of 0.05 at $1.4\times10^{17}/cm^3$ and a broad resonance feature from 0.7 to $3\times10^{17}/cm^3$. The stray light (noise) level is about 0.001. The reason why it is so broad will be that, whenever $n_0$ is higher than the resonant density, either the front or the rear of the focal point should satisfy the resonant condition to excite the wave. The vertical bar in the figure is due to $P_0$ calibration error of 25 %. From Eq. (1) the amplitude $\varepsilon = \delta n/n_0$ giving $P_s/P_0 = 0.05$ (0.007 - 0.06) is 5 % (2 - 6 %) supposing L = 3 mm. Since we have not directly measured L, we assumed L to equal the electron-ion collision mean free path. The mean free path is 3 mm or the mean free time is 10 ps for the electron temperature of 100 eV. The value is roughly the homogeneous length out of the 7 mm plasma. If the plasma wave duration really equals the laser duration (= 1 ns), this may suggest that the finite-length (~3 mm) waves are excited not once but more times in tandem within 1 ns. $\varepsilon = 5$ % corresponds to the wave field of 1.5 GV/m much smaller than the Rosenbluth and Liu saturation value of 20 % or 6 GV/m.[13)] Even for the single line irradiation, we detected small but not stray signals at 11.857 μm around $10^{17}/cm^3$. They are shown by white circles in the figure, corresponding to $\varepsilon \sim 1$ %. They may be due to instabilities such as a stimulated Raman scattering (SRS), since, when the intensity is $10^{14}$ W/cm², it is on the threshold level for the forward Raman scattering (FRS).[14,15)]

The plasma wave with $\varepsilon$ and $\gamma_p$ can trap and accelerate either plasma electrons or injected electrons having the initial kinetic energies not less than

$$W_{min} = mc^2[\gamma_p(\gamma_p\varepsilon + 1)(1 - \beta_p\beta') - 1] \tag{2}$$

to final accelerated energies of

$$W_{max} = mc^2[\gamma_p(\gamma_p\varepsilon + 1)(1 + \beta_p\beta') - 1], \tag{3}$$

where $\beta_p = \sqrt{1 - 1/\gamma_p{}^2}$ and $\beta' = \sqrt{1 - 1/(\gamma_p\varepsilon + 1)^2}$.

The plasma wave of $\varepsilon = 5$ % accelerates 1.4 MeV electrons to 12 MeV in 7 mm distance. The energy gain is more than 10 MeV.

## 4) OBSERVATION OF THE ENERGETIC ELECTRONS

An 8 channel electron spectrometer ESM placed 85 cm from the focal point covers the energy of from 2 to 22 MeV. The entrance aperture is 8 mm in diameter, i.e.,$7\times10^{-5}$ sterad. Each channel, whose window width is 0.6 MeV, has a scintillator (Pilot-U) - photomultiplier coupling linked to the CAMAC. By using a 3 MeV electron beam from an induction linac, we have confirmed that the

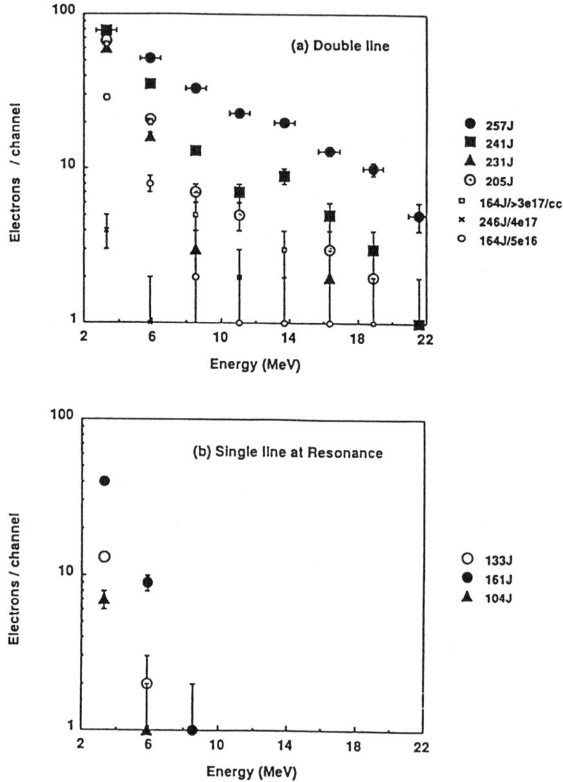


Fig. 4. Energy spectra of the electrons forward emitted from the hydrogen plasma. Number of Electrons per channel (a) with the double line both at and out of resonance and (b) with the single line irradiation at the reso-nance($1\text{-}2\times10^{17}/cm^3$). Same symbols are from the same shot, of which the parameters are as laser energies / $n_0$. Vertical error is one electron (detection limit). Horizontal bar is channel width.

ESM can correctly analyze the electron energy. Figure 4 (a) shows number of electrons emitted into the ESM from the plasma with the double line irradiation. Same symbols correspond to the same shot. Large symbols are at the resonance ($1\text{-}2\times10^{17}/cm^3$) and small ones are out of resonance. The parameters on the right-hand side show the incident laser energies/the electron density for each shot. We estimated the noise level by comparing the data with and without plasmas, which gave us ±one electron as the vertical error bar in the figures. The horizontal bar shows the channel width. The electron spectra seem to have two components, i.e., low energy (<10 MeV) and high energy (>10 MeV) components, specially for the resonant density shots. With the single line irradiation at resonance, as seen in Fig. 4 (b), the spectra have only low energy components (<8 MeV). They are like those with double line but out of resonance.

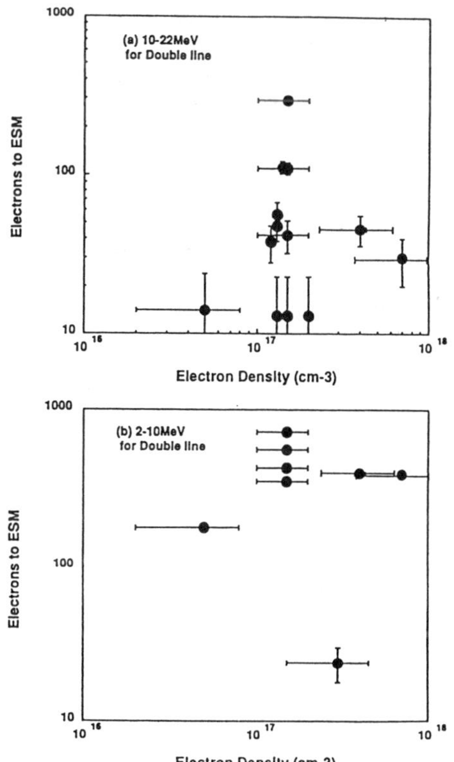


Fig. 5. (a) Number of the high energy electrons (10-22MeV) emitted into the ESM aperture versus the electron density with the double line irradiation. (b) Number of the low energy electrons(2-10MeV). Vertical error bar is 10 electrons. Horizontal error bar is due to the density measurement error.

We integrated the area of the low and high energy components separately and plotted them as a function of the electron density. Thus we found that even with the double line irradiation, the high energy components show a resonant feature to the density, but the low energy components do not show such a feature(see Figs. 5(a) and (b)). In Figs. 5, the vertical axes correspond to the integrated numbers of the electrons emitted into the ESM aperture.

The vertical error bar is ±10 electrons and the horizontal bar is due to density measurement error. It seems that the high energy (>10 MeV) electrons are due to the beatwave-excited plasma wave. It is, however, not consistent with ε of 5 %, because, although the low energy components are plausible to be trapped by the wave potential, the most of the high energy components are too high to be accelerated by one wave. 1.5 GV/m (5%) times 3 mm only provides 4.5 MV. Suppose the amplitude of 6% and the acceleration distance of 7 mm, which is the plasma length, the gain becomes 14 MeV, but the distance is twice the length of the plasma wave. It is not yet understood. It may be explained by the multiple stage acceleration, for instance.[16] However even this does not account for the data between 14MeV and 22MeV.

The causes of the low energy components emitted both for single and double line irradiations are as well not clear. With the double line irradiation, the low energy components show not a sharp but a broad resonance, as in Fig. 5 (b). The beatwave would also affect to these electrons.

In Figs. 3 and 5, the resonant density seems to be not $1.1\times10^{17}$ but $1.4\times10^{17}/cm^3$. So the absolute density may be corrected by 1.2 times.

## 5) COMPUTER SIMULATION

We simulated how the injected 0.6MeV electrons are accelerated by the beatwave excited fields using a fluid-particle code

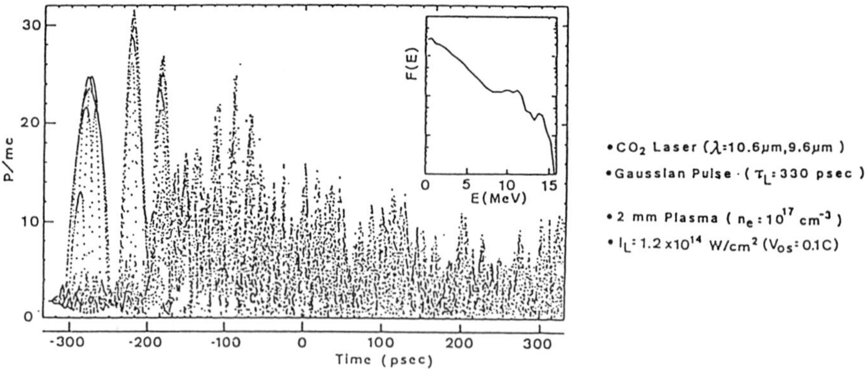


Fig. 6. Beatwave acceralation of electrons by Fluid-Particle Code.

ILESTA(1-D). The laser pulse is Gaussian with 330ps HWHM at the intensity of $1.2 \times 10^{14}/cm^2$. The plasma length is 2mm with $n_e = 10^{17}/cm^3$. The results are shown in Fig. 6. The vertical axis is the electron momentum. 0 on the horizontal axis is the laser peak timing. The small window shows the time integrated electron spectrum, which shows a two componemt feature (lower and larger than 10MeV).

## 6) SUMMARY

As a summary we simultaneously observed the beatwave excited plasma wave and energetic electrons. The 10.6 and 9.57 μm beat-wave excited the plasma wave of $\varepsilon = 5$ % at the resonant density. The field is 1.5 GV/m. We observed, then, plasma electrons with the energies more than 10 MeV. The experiment not only encourages us to inject electron beams into the system and to accelerate them, but also offers a possibility for the plasma beatwave accelerator in near future.

We thank C. Yamanaka for his encouragement and S. Nakayama for preparing the ESM. One of the author, Y. Kitagawa, acknowledges F. F. Chen, C. Joshi and their UCLA group for useful discussions.

**References**

1) T. Tajima and J. M. Dawson, Phys. Rev. Letters **56**, 267 (1979).
2) T. Katsouleas, J. M. Dawson, Phys. Rev. Letters **51**, 392 (1983).
3) D. W. Forslund and J. M. Kindel, Phys. Rev. Letters **54**, 558 (1985).
4) C. Darrow et al., Phys. Rev. Letters **56**, 2629 (1986).
5) K. Mima et al., Phys. Rev. Letters **57**, 1421 (1986).
6) C. E. Clayton et al., Phys. Rev. Letters **54**, 2343 (1985).
7) N. A. Ebrahim et al., Proc. 1986 linear Accelerator Conf. (Stanford, CA), p.552, 1986. SLAC Report-303.
8) N. A. Ebrahim et al., IEEE Trans. Nuclear Science, **NS-32**, 3539 (1985).
9) C. Yamanaka et al., IEEE J. Quantum Electronics, **QE-17**, 1678 (1981).
10) K. Terai et al, Review Laser Engineering, **12**, 37 (1984) in Japanese.
11) G. Bekefi, "*Principle of Laser Plasmas*", Ch. 13. John Wiley & Sons, N. Y. 1976.
12) R. E. Slusher and C. M. Surko, Phys. Fluids **23**, 472 (1980).
13) M. N. Rosenbluth and C. S. Liu, Phys. Rev. Letters **29**, 701

(1972).
14) K. Estabrook and W. L. Kruer, Phys. Fluids **26**, 1892 (1983).
15) C. Joshi et al., Phys. Rev. Letters **47**, 1285 (1981).
16) T. Tajima, laser and Particle Beams, **3**, 351 (1985).

**Research Trends in Physics: Coherent Radiation Generation and Particle Acceleration**
Editorial Board: J.M. Buzzi, A. Prokhorov (Editor-in-Chief), P. Sprangle, and K. Wille
*La Jolla International School of Physics*, The Institute for Advanced Physics Studies, La Jolla, California

# The Last Results from the UA2 Experiment at the CERN Collider

**The UA2 Collaboration**

Presented by
Stefano Lami

## ABSTRACT

After three runs between 1988 and 1990, large samples of W and Z boson events were accumulated by the upgraded UA2 detector at the CERN SPS Collider. New improved results for the W and Z production cross-sections $\sigma_W$ , $\sigma_Z$ and for the mass ratio $m_W/m_Z$ have been obtained. The measurement of the cross-sections ratio has been used to extract a new value for the W boson width, $\Gamma_W = 2.07 \pm 0.13$ (stat) $\pm$ 0.07 (syst) GeV. The bosons mass ratio provides a new value for the weak mixing parameter $\sin^2\theta_W$. By combining this result with recent measurements of the Z mass from $e^+e^-$ colliders we obtain a new precise value of the W mass : $m_W = 80.61 \pm 0.34$ (stat) $\pm$ 0.20 (syst) GeV.

## 1. Introduction

Since the first observation of W and Z bosons at the CERN SPS Collider, their properties have been the subject of intensive studies. Today, the properties of Z bosons can be studied to a high level of accuracy at $e^+e^-$ machines, with the advantages of high statistics, virtually zero background and an intrinsic energy calibration derived from the machine beam optics. The study of W bosons however is still in the domain of proton-antiproton colliders. The upgraded UA2 non-magnetic detector has recently completed its data taking at the CERN proton-antiproton Collider at a center of mass energy of 630 GeV. In the present paper I will describe briefly the data analysis, and final results from the 88-89 runs as well as very preliminary results including also the more recent data are here presented. Successful operation of the Antiproton Accumulator Complex and the SPS provided peak luminosities up to 5 $10^{30}$ $cm^2$ $s^{-1}$. The total integrated luminosity amounted to 13.5 $pb^{-1}$, yielding large samples of $W \rightarrow e\nu$ and $Z \rightarrow e^+e^-$ decays. These have been analyzed with the aim of significantly improving our knowledge of the ratio of the W and Z masses. And the measurement of the bosons production cross sections is useful to reach a new precise value for the W width. Full details concerning these analyses for the previous samples can be found in Ref. [1,2].

## 2. The UA2 Detector

The UA2 detector consists of a central and forward tracking system equipped with preshower detectors for electron identification and surrounded by electromagnetic and hadronic calorimeters. The central calorimeter covers the pseudorapidity region $|\eta| < 1.0$. Each of the 240 cells is longitudinally segmented into an electromagnetic and two hadronic compartments. The endcap calorimeters, consisting of 384 cells in total, are longitudinally segmented into an electromagnetic and a hadronic compartment. The granularity of both systems is approximately $\Delta\phi \times \Delta\eta = 15^o \times 0.2$ . The calorimeter is hermetic up to $|\eta| = 3$ ($\theta = 5^o$) with electromagnetic calorimetry extending to $|\eta| = 2.5$. Details of the construction and performance of the various detectors can be found in Ref. [1,2] and references therein. Fig.1 shows a longitudinal view of one quarter of the new UA2 detector.

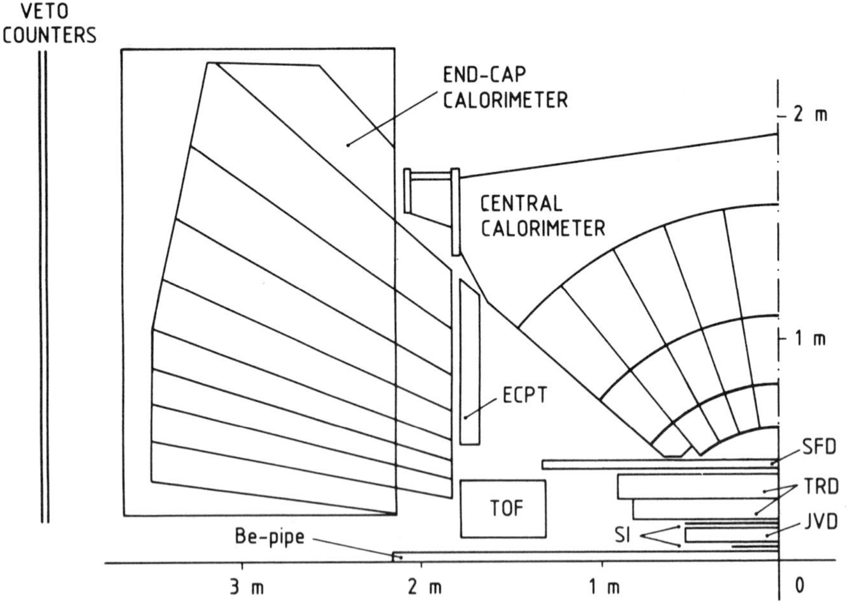


**Figure 1 :** Schematic longitudinal view of the UA2 detector showing one quadrant.

Precise initial calibration of the calorimeter and accurate tracking of changes over a period of years are mandatory to minimize the errors in measuring the energy. The calorimeters have been calibrated on test beams of electrons, pions and muons of several energies and the response of all cells has been studied. The energy scale was tracked by periodic calibrations using a $^{60}Co$ source. Each year from 1986 to 1990 a portion of the calorimeter was removed from the detector and re-calibrated on a test beam, leading to a 1% estimated error on the energy scale for the electromagnetic calorimeter. Changes in the response (up to $\pm$ 10 %) due to impact point and direction of the incident particle have been accurately parametrized. Because of the different construction, the energy scale uncertainties of the central and the endcap cells are not fully correlated. Also, the so called "edge" cells of the central calorimeter ($0.8 < |\eta| < 1.0$) have a substantially poorer response than the remaining part due to a shorter electromagnetic compartment (this modification was required to allow more radial space for the central tracking system).

## 3. Electron Measurement

For both W and Z samples considered here, the trigger was based on the information of the calorimeter, basically requiring the presence of at least one electron with $P_T > 10$ GeV (W trigger) or of at least two electrons with $P_T > 5$ GeV , $\Delta\phi > 60^o$ ( Z trigger ). The trigger had 3 levels, the first one using analog sums of the signals from the cells with $|\eta| < 2$. At the second level, electron and jets clusters were reconstructed in a dedicated processor using a fast digitization of the calorimeter signals. At the third level FASTBUS processors performed a complete calorimeter analysis using the final digitization and the full set of calibration constants. The W trigger was fully efficient for $P_T > 12$ GeV, the Z one for $P_T > 6$ GeV.

In the offline analysis the directions of the charged particles were reconstructed using the tracking detectors. The event vertex was required to be within 25 cm from the center of the detector. Candidate electron tracks were defined as pointing to calorimeter e.m. clusters and the preshower detectors were used to match the trajectories of candidate electron tracks with the position of the e.m. shower. Also, the lateral and longitudinal profiles of each shower were required to be consistent with those expected for a single isolated electron incident along the track direction as determined from test beam data. These cuts and their efficiencies are described in detail in Ref. [2].

The electron energy was corrected for the impact point and track direction dependence of the calorimeter response and for the average energy loss in the preshower detectors. To minimize the influence of particles produced in association with the W or Z (underlying event) the cells containing the largest fraction of the cluster energy (typically two cells) were used to determine the electron energy.

In order to obtain a high-quality energy measurement, for the masses measurement we applied fiducial cuts by rejecting those W candidates where the electron hit one of the edge cells ( $0.8 < |\eta| < 1.0$) in the central calorimeter, those where the electron was in the forward calorimeters (about 22 % of the total), and those where the electron hit was too close to a cell boundary, since in this region the corrections were large and the shape of the response was not Gaussian. For Z candidates, at least one of the electrons had to satisfy these same requirements.

## 4. Neutrino Measurement

Transverse momentum balance was used to measure the energy of the undetected neutrino:

$$\mathbf{P}_T(\nu) \approx - [\, \mathbf{P}_T(e) + \mathbf{P}_T \text{ (hadrons)} \,] \qquad (1)$$

The equality is only approximate, since the angular coverage is not 100% and non linearities in the response of the hadron calorimeter for low energy particles affect the measurement. In this approximation $\mathbf{P}_T(W) = - \mathbf{P}_T(\text{hadrons})$. Since the resolution on $P_T(\nu)$ depends on the resolution on $P_T(e)$ and $P_T(\text{hadrons})$ and the latter deteriorates when the hadron activity in the event increases, for the mass measurement the W candidates with $P_T(W) > 20$ GeV (about 5% of the total) were removed from the sample. The validity of Eq. (1) was checked by using the $Z \rightarrow e^+e^-$ events, where $P_T(Z)$ could be measured directly from the electron momenta. We found an excellent agreement between the momentum balance of the Z and the detected hadrons projected along a direction which maximizes the sensitivity to the energy scale and a calculation that includes a full modeling of the calorimeter response.

## 5. The Data Samples and the cross sections

A kinematical selection was used to reduce the QCD background on the W sample to less than 1 % :

$$20 < P_T(e) < 60 \quad \text{GeV}$$
$$20 < P_T(\nu) < 60 \quad \text{GeV}$$
$$40 < m_T < 120 \quad \text{GeV}$$

where the transverse mass, $m_T$, was defined as usually from

$$m_T^2 = 2\, P_T(e) P_T(\nu)\, (1 - \cos\phi_{e\nu}).$$

With these cuts the 1988-89 W sample contained 1676 events with the electron in the central calorimeter, of which 3.8 % were due to the processes $W \rightarrow \tau\nu$ followed by $\tau \rightarrow e\nu$. In the forward region the sample contained 365 events.

The measured cross section is determined from the equation

$$\sigma_W = (N_W - N_\tau) / \varepsilon\eta L$$

where $N_W$ is the observed number of W events, $N_\tau$ is the contribution from $W \rightarrow \tau\nu$, $\eta$ is the acceptance of the geometrical and kinematic selections, $\varepsilon$ is the overall electron detection efficiency, and L is the integrated luminosity, determined with a 7% uncertainty.

The results for the different calorimeter regions were combined to give a best estimate of the W cross section using weigths proportional to the product of efficiency and acceptance for each region; the published value for the 1988-89 sample is ( see Ref.[2]) :

$$\sigma_W = 660 \pm 15 \text{ (stat)} \pm 37 \text{ (syst)} \quad \text{pb.}$$

This procedure is also used to evaluate $\sigma_Z$ and is more reliable for combining results from small event samples than the use of weights proportional to the number of observed events.

For the Z sample, events were required to lie in the range

$$70 < m_{ee} < 120 \quad \text{GeV}$$

where $m_{ee}$ is the invariant mass of the $e^+e^-$ pair. To increase the detection efficiency, at most one of the electron tracks was allowed to satisfy slightly looser selection criteria (see Ref. [2]), leading to a sample with an estimated background of < 1%. If the event contained a third e.m. cluster with $E_T > 5$ GeV, a three body mass was computed as appropriate for $Z \rightarrow e^+e^-\gamma$ events. With the 1988-89 Z sample we measured a cross section of

$$\sigma_Z = 70.4 \pm 5.5 \text{ (stat)} \pm 4.0 \text{ (syst)} \quad \text{pb.}$$

### 6. The W boson width

Most of the theoretical and experimental systematic uncertainties cancel partially or totally in the ratio of the cross sections. The result is ( Ref. [2]) :

$$R = 9.38\,^{+0.82}_{-0.72}\ \text{(stat)} \pm 0.35\ \text{(syst)}\ .$$

After combining the statistical and systematic errors the 90% confidence interval for R is

$$8.2 < R < 10.9.$$

The measurement of the cross sections ratio R can be used to extract a value for the W width, since all other quantities are either measured at $e^+e^-$ machines or can be reliably calculated in the framework of the Standard Model. By using a weighted average for the Z width, $\Gamma_Z = 2.496 \pm 0.016$ GeV, the result is

$$\Gamma_W = 2.25 \pm 0.18\ \text{(stat)} \pm 0.07\ \text{(syst)}\ \text{GeV}\ .$$

By including also the 1990 run data we obtain the following preliminary results for the bosons cross sections

$$\sigma_W = 702 \pm 12\ \text{(stat)} \pm 20\ \text{(syst)}\ \ \text{pb}$$
$$\sigma_Z = 68.5 \pm 4.3\ \text{(stat)} \pm 2.4\ \text{(syst)}\ \ \text{pb},$$

and a value $R = 10.25\,^{+0.70}_{-0.63}$ (stat) $\pm$ 0.35 (syst) for the cross sections ratio, which finally gives a W width

$$\Gamma_W = 2.07 \pm 0.13\ \text{(stat)} \pm 0.07\ \text{(syst)}\ \text{GeV},$$

in good agreement with the value of 2.10 GeV expected in the case of a heavy top quark.

## 7. The mass fits and the results

Our goal is to measure the mass ratio with the smallest possible errors. The endcap and central calorimeters do not have identical systematic errors and the relative populations of W and Z events in the calorimeters are different, so the scale factor does not cancel in taking the mass ratio. The 1988-89 Z selection resulted in 54 events with both electrons satisfying the fiducial cuts for the central calorimeter (see sect. 3). The events in this sample (the C-C sample) and the W events (1203 central events for the 1988-89 runs) had their masses measured in the same calorimeter regions, and therefore the energy scale uncertainty did not affect the mass ratio $m_W / m_Z$ . However, the C-C sample contained only 32 % of the total Z events. To increase the sample size, we used also 94 $Z \rightarrow e^+e^-$ events with one electron outside the fiducial region of the central calorimeter. For these events (the so-called $P_T$-constrained sample) we determined the energy of the electron outside the fiducial region using the $P_T$ balance rather than by a direct measurement. This method is similar to the one used for the W neutrino, with one important difference: since the electron direction was measured precisely by the tracking system, the full momentum (and not only the transverse component) could be reconstructed. In this way the mass $m_{ee}$ was essentially derived from the energy measured in the central fiducial region, and was affected by the corresponding scale uncertainty. The price one had to pay was of course a worse resolution ( 4 GeV against 2 GeV on the average) due to the fact that the energy of only one electron was really measured and that the resolution on $P_T$ (hadrons) affected the energy determination of the other electron.

The masses of the W and Z have been determined using a maximum likelihood fitting procedure. The Kolmogorov-Smirnov test was used to provide a confidence level for the results. A detailed Monte Carlo simulation of the W and Z production and decays, followed by a carefully tuned model of the detector response to the decay products, was used to evaluate the effect of systematic uncertainties and to compute small corrections.

For the $Z \rightarrow e^+e^-$ decay the invariant mass of the electron pair , $m_{ee}$ (or, eventually, $m_{ee\gamma}$) was measured directly. In this case we used an analytical likelihood function which approximates well the expected line shape . The effects neglected by the simplified function were introduced as small corrections to the final results.Separate fits were performed for the C-C and the $P_T$ -constrained samples (see fig. 2). The results are summarized in Table 1. Before combining these results, the corrections and systematic errors were evaluated. We considered the following effects:

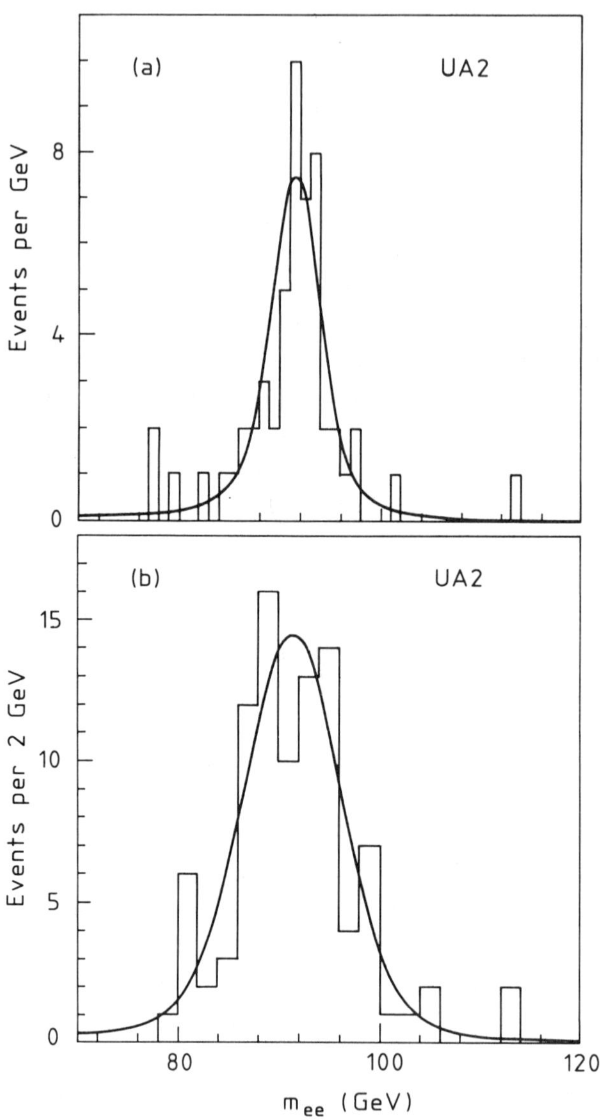


**Figure 2 :** The fit to the invariant mass distributions for the Z samples with the fitted curves superimposed. (a) the central Z sample, (b) the $P_T$-constrained Z sample.

- variations in the underlying event contribution to the electron energy scale;
- the effect of radiative decays , i.e. unresolved $\gamma$ from $Z \rightarrow e^+e^- \gamma$;
- variations in the parameters of the model for measuring $P_T$ (hadrons) and their effect on the $P_T$ -constrained fit.

| Sample | Parameter | 1 Parameter fit | 2 Parameter fit |
|---|---|---|---|
| Central | $m_Z$ (GeV) | 91.69 ± 0.43 | 91.70 ± 0.45 |
| | $\Gamma_Z$ (GeV) | 2.5 | $2.96^{+0.98}_{-0.78}$ |
| | Conf. Level | 96 % | 99 % |
| $P_T$ (constr) | $m_Z$ (GeV) | 91.51± 0.57 | 91.53 ± 0.59 |
| | $\Gamma_Z$ (GeV) | 2.5 | $2.94^{+1.18}_{-0.94}$ |
| | Conf. Level | 96 % | 97 % |

**Table 1** : A summary of the fits to different Z samples (1988-89 data). The error shown is the statistical error from the fit.

These effects lead to an overall correction of -140 MeV, an overall systematic error of 120 MeV, and an additional systematic error of 100 MeV for the $P_T$-constrained fit. This additional systematic error was added in quadrature to the statistical error of the $P_T$-constrained fit, and then the final result was obtained by taking the weighted average of the two one-parameter fit results of Table 1 and adding the -140 MeV correction ( a 1% scale error is included) :

$$m_Z = 91.49 \pm 0.35 \text{ (stat)} \pm 0.12 \text{ (syst)} \pm 0.92 \text{ (scale) GeV} \quad \text{(1988-89 data)}$$

For the $W \rightarrow e\,\nu$ decay, the lack of knowledge of the longitudinal momentum of the neutrino forced the use of transverse variables in the fitting procedure, namely $P_T(e)$, $P_T(\nu)$ and transverse mass, $m_T$. In this case, likelihood functions were obtained numerically for the three variables using the Monte Carlo simulation. The results are summarized in Table 2, where it appears that the smallest statistical error belongs to the transverse mass fit. Fig. 3 shows the fit to the transverse mass assuming a fixed W width, for the 1988-89 sample.

| Distribution | Parameter | 1 Parameter fit | 2 Parameter fit |
|---|---|---|---|
| $m_T$ | $m_W$ (GeV) | 80.75 ± 0.31 | 80.78 ± 0.31 |
| | $\Gamma_W$ (GeV) | 2.1 | $1.89\,^{+0.47}_{-0.40}$ |
| | Conf. Level | 84 % | 89 % |
| $P_T$ ( e ) | $m_W$ (GeV) | 80.79 ± 0.38 | 80.83 ± 0.39 |
| | $\Gamma_W$ (GeV) | 2.1 | $1.60\,^{+0.78}_{-0.68}$ |
| | Conf. Level | 95 % | 97 % |
| $P_T$ ( ν ) | $m_W$ (GeV) | 80.32 ± 0.41 | 80.33 ± 0.42 |
| | $\Gamma_W$ (GeV) | 2.1 | $2.03\,^{+0.82}_{-0.72}$ |
| | Conf. Level | 83 % | 88 % |

**Table 2** : A summary of the fits to different W distributions (1988-89 data). The error shown is the statistical error from the fit.

The analysis of systematic effects is more complex than in the case of the Z. We considered the following effects:

- variations of the $P_T$ (W) distribution and variations in the parameters of the model which describes the measurement of $P_T$ (hadrons) ;
- different parton distributions used in the Monte Carlo generating the likelihood function;
- possible scale errors in the measurement of $P_T(\nu)$;
- uncertainties in the electron resolution;
- variations in the underlying event contribution to the electron energy scale;
- effects of finite Monte Carlo statistics used to generate the likelihood functions;
- the effect of radiative decays , i.e. unresolved $\gamma$ from $W \rightarrow e\,\nu\,\gamma$.

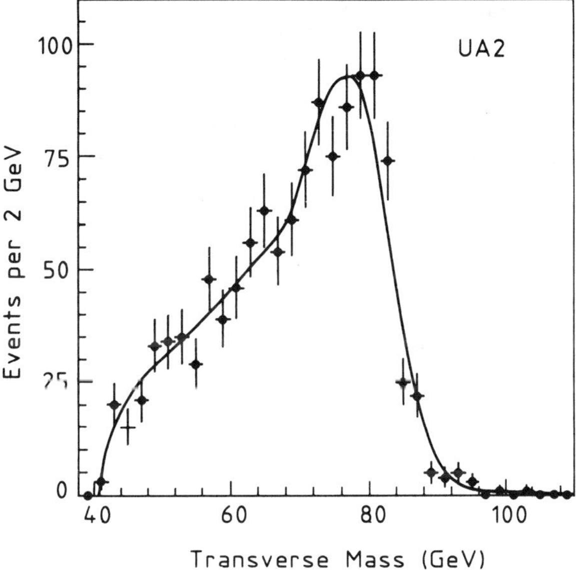


**Figure 3 :** The fit to the transverse mass distribution for the W sample with the fitted curve superimposed.

These effects lead to an overall correction of +40 MeV, +60 MeV and +160 MeV, and an overall systematic error of 210 MeV, 300 MeV and 470 MeV for the $m_T$, $P_T(e)$ and $P_T(\nu)$ fits respectively. The fit to the transverse mass gives the smallest errors, both statistical and systematic, and is therefore quoted as our final value (see Ref.[1]):

$$m_W = 80.79 \pm 0.31 \text{ (stat)} \pm 0.21 \text{ (syst)} \pm 0.81 \text{ (scale) GeV} \quad \text{(1988-89 data)}$$

where a 1% energy scale error has been included.

A preliminary study including also the 1990 data sample results in 2052 W events in the central calorimeter and 244 Z events ( a sample of 94 central events and 150 events for the $P_T$constrained sample ), and the preliminary UA2 results for the boson masses are

$$m_Z = 91.34 \pm 0.28 \text{ (stat)} \pm 0.08 \text{ (syst)} \pm 0.91 \text{ (scale) GeV}$$

$$m_W = 80.75 \pm 0.23 \text{ (stat)} \pm 0.21 \text{ (syst)} \pm 0.81 \text{ (scale) GeV}$$

## 8. Comparison with the Standard Model

As already stated , the mass ratio $m_W/m_Z$ is virtually free from errors due to the energy scale: any possible effect due to non-linearities in the calorimeter response was minimized by choosing an energy of 40 GeV for the calibration, which is close to the mean value expected for W and Z decays. A preliminary result, including also the 1990 data, is :

$$m_W/m_Z = 0.8841 \pm 0.0037 \text{ (stat)} \pm 0.0022 \text{ (syst)}$$

This can be combined with the most recent results from LEP and SLC [3] for the Z mass : $m_Z = 91.177 \pm 0.031$ GeV, to give a rescaled W mass:

$$m_W = 80.61 \pm 0.34 \text{ (stat)} \pm 0.20 \text{ (syst) GeV.}$$

We can relate the mass ratio to the electroweak mixing parameter $\sin^2\theta_W$ using the renormalization scheme of Sirlin [4] which, in the case of a single Higgs doublet ($\rho = 1$), reads :

$$\sin^2\theta_W = 1 - (m_W/m_Z)^2$$

Our result is :

$$\sin^2\theta_W = 0.2184 \pm 0.0065 \text{ (stat)} \pm 0.0039 \text{ (syst)}$$

which agrees well with the values obtained from a world average of ν-nucleon Deep Inelastic Scattering experiments [5]:

$\rho = 1$ (fixed) $\quad \sin^2\theta_W = 0.2314 \pm 0.0029$ (stat) $\pm 0.0049$(syst)

$\rho = 1.0036 \pm 0.0118 \quad \sin^2\theta_W = 0.2346 \pm 0.0105$ (stat) $\pm 0.0079$(syst)

From our data we can also put limits on the value of $m_{top}$ using the relationship between $m_Z$ (which is precisely measured from LEP and SLC) and $\sin^2\theta_W = 1 - (m_W/m_Z)^2$ and exploiting the dependence of the radiative corrections $\Delta r$ on $m_{top}$ [6]:

$$m_Z = A / [\sqrt{(1 - \Delta r)} \sin\theta_W \cos\theta_W]$$

where $\quad A = (\pi\alpha/\sqrt{2}\, G_\mu)^{1/2} = 37.2805 \pm 0.0003$ GeV

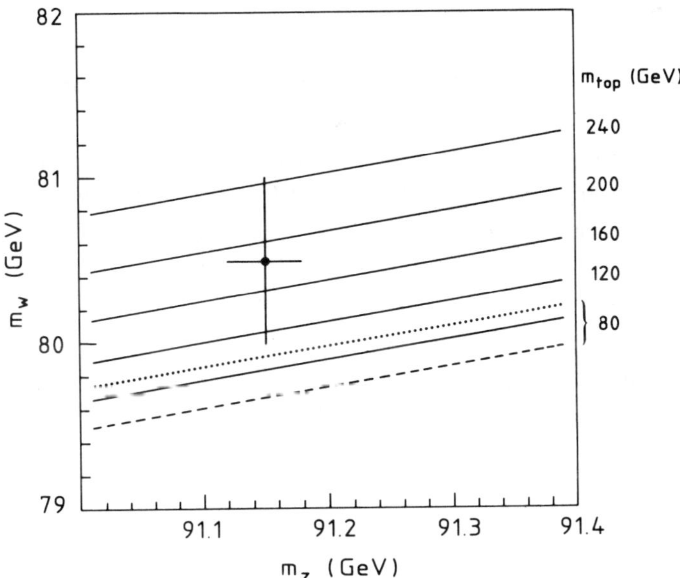


**Figure 4 :** The comparison with the Minimal Standard Model predictions. The solid lines indicate the allowed values for $m_W$ and $m_Z$ for a given $m_{top}$ with $m_{Higgs}$ = 100 GeV. The dotted (dashed) line indicates the prediction for $m_{top}$ = 80 GeV with $m_{Higgs}$ = 10 (1000) GeV. The data point is defined in the text.

The relationship between $m_Z$ and $m_W$ is plotted for several values of $m_{top}$ in Fig. 4. The UA2 result, in combination with that of LEP and SLC, has been marked by a data point whose errors reflect the combined statistical and systematic errors on the measurements. Computing the limits on $m_{top}$ we found :

$m_{top} > 90$ GeV

$m_{top} < 280$ GeV at 90 % Confidence level.

## 9. Conclusions

Precise values for the W and Z masses have been measured from large samples of $W \to e\nu$ and $Z \to e^+e^-$ events accumulated by the UA2 detector.

After a careful analysis of systematic effects, an improved preliminary result is obtained for the mass ratio $m_W/m_Z = 0.8841 \pm 0.0037$ (stat) $\pm$ 0.0022 (syst). This has been combined with recent measurements of the Z mass from LEP and SLC to give an absolute measurement of the W mass $m_W = 80.61 \pm 0.34$ (stat) $\pm$ 0.20 (syst) GeV. A new value for the electroweak mixing parameter, $\sin^2\theta_W = 0.2184 \pm 0.0065$ (stat) $\pm$ 0.0039 (syst), has also been reported. The results of these measurements are in good agreement with the Minimal Standard Model, and give further support to the hypothesis that the top quark is heavier than the W.

## References


[1] J. Alitti et al. (UA2 Collaboration) : Phys. Lett. **241B** (1990) 150.

[2] J. Alitti et al. (UA2 Collaboration) : Z. Phys. C - Particles and fields **47** (1990) 11.

[3] G. Abrams et al. (Mark II Collaboration) : Phys. Rev. Lett. **63** (1989) 2173;
D. Decamp et al. (ALEPH Collaboration) : Phys. Lett. **231B** (1989) 519 and CERN preprint CERN-EP/89-169;
P. Aarnio et al. (DELPHI Collaboration) : Phys. Lett. **231B** (1989) 539;
B. Adeva et al. (L3 Collaboration) : Phys. Lett. **231B** (1989) 509 and L3 preprint #004;
M.Z. Akrawy et al. (OPAL Collaboration) : Phys. Lett. **231B** (1989) 530.

[4] A. Sirlin : Phys. Rev. **D22** (1980) 971.

[5] G.L. Fogli and D. Haidt : Z. Phys. C - Particles and fields **40** (1988) 379.

[6] W.J. Marciano and A. Sirlin : Phys. Rev. **D29** (1984) 945.

**Research Trends in Physics: Coherent Radiation Generation and Particle Acceleration**
Editorial Board: J.M. Buzzi, A. Prokhorov (Editor-in-Chief), P. Sprangle, and K. Wille
*La Jolla International School of Physics*, The Institute for Advanced Physics Studies, La Jolla, California

# On the Generation of Large-Amplitude Plasma Wakefields with Low Phase Velocities by an Intense Short Laser Pulse

**I.G. Murusidze and L.N. Tsintsadze**

Institute of Physics, Georgian Academy of Sciences
Tbilisi, Georgia

The excitation of large-amplitude plasma waves is the main problem in the development of new plasma based high-energy particle accelerators. The idea of generating plasma waves having a phase velocity close to the speed of light using a single laser pulse was originally proposed by Tajima and Dawson /1/. Soon this idea was lost and only after a decade it has been developed by Sprangle et al. /2/ as the laser wake-field accelerator scheme (LWFA). Following the papers /3,4/ it is getting clear that the real success of the LWFA scheme is associated only with the relativistically intense ($E_L > mc\omega/e$) short ($\tau_L < 2\pi/\omega_p$) laser pulses.

The maximum amplitude attainable by a relativistic plasma wave is one of the fumdamental questions. The formula by Akhiezer and Polovin /5/ define the maximum or the so-called wave-breaking amplitude of a fully relativistic plasma space-change waves in a cold plasma:

$$E_{max} = \frac{mc\omega_p}{e}\sqrt{2(\gamma_{ph}-1)} \tag{1}$$

where $\gamma_{ph} = (1 - v_{ph}^2/c^2)^{-\frac{1}{2}}$ and $v_{ph}$ is a phase velocity of the waves.

This result predicts that the closer the phase velocity is to the speed of light, the larger value of $E_{max}$ is to be expected, and the amplitude approaches infinity as $v_{ph}$ approaches C. It has been shown that for plasma waves driven by a laser pulse these predictions fail.

The problem of propagating large-amplitude laser pulses through plasmas has many vague points. One of them is a group velocity of such pulses. For the pulses with a given amplitude and unchanged shape (these assumptions restrict the validity of our results by a short time period) the phase velocity of driven plasma waves is definitively related to the group velocity of a driving pulse.

Previous papers have considered the generation of plasma wakefields by the laser pulses moving with group velocities close (or equal) to the speed of light C, and hence, they could not answer the question: How does the maximum amplitude of driven waves depend on the group velocity of the laser pulse (or on the phase velocity of excited wakefields)?

We start from Max well's equations and the equations for cold relativistic electron fluid (ions are stationary) assuming that all quantities depend on $t$ and $z$ being the direction of propagation of the intense laser pulse. Describing the fields associated with the laser pump and plasma response by the vector and scalar potentials $\vec{A}_{\perp}(t,z)$ , $\varphi(t,z)$ we have the following field equations:

$$\left(\frac{\partial^2}{\partial z^2}-\frac{\partial^2}{\partial t^2}\right)\vec{A}_{\perp} = \frac{n}{\gamma}\vec{A}_{\perp} \qquad (2)$$

$$\frac{\partial^2 \varphi}{\partial z^2} = n-1 \qquad (3)$$

here the Conlomb gauge is used, $div\vec{A}_{\perp}=0$, hence $A_z=0$.

The relativistic fluid equations can be written in the form:

$$\frac{\partial}{\partial t}P_{\parallel}+\frac{\partial}{\partial z}\gamma=\frac{\partial}{\partial z}\varphi \qquad (4)$$

$$\frac{\partial}{\partial t} n + \frac{\partial}{\partial z} \frac{n p_{\parallel}}{\gamma} = 0 \tag{5}$$

where $\gamma = \sqrt{1 + p_{\parallel}^2 + p_{\perp}^2}$, $\vec{p}_{\perp} = \vec{A}_{\perp}$

**Note that in Eqs.(2-5) the following dimensionless quantities have been introduced:**

$$t \to \omega_p t \;;\quad z \to K_p z \;;\quad K_p = \omega_p / c \;;\quad \omega_p = (4\pi e^2 n_0 / m)^{1/2};$$

$$\vec{p} \to \vec{p}/(mc);\quad \vec{A}_{\perp} \to \frac{e\vec{A}_{\perp}}{mc^2};\quad \varphi \to \frac{e\varphi}{mc^2};\quad n \to \frac{n}{n_0}.$$

Now we assume that all quantities are the functions of the new independent variables $\xi$, $\tau$, where $\xi = z - v_p t$ and $\tau = t$. Then Eqs.(2-5) become:

$$\left( \frac{1}{\gamma_p^2} \frac{\partial^2}{\partial \xi^2} + 2v_p \frac{\partial^2}{\partial \xi \partial \tau} - \frac{\partial^2}{\partial \tau^2} \right) \vec{A}_{\perp} = \frac{n}{\gamma} \vec{A}_{\perp} \tag{6}$$

$$\frac{\partial^2}{\partial \xi^2} \varphi = n - 1 \tag{7}$$

$$\frac{\partial}{\partial \tau} p_{\parallel} + \frac{\partial}{\partial \xi} (\gamma - v_p p_{\parallel} - \varphi) = 0 \tag{8}$$

$$\frac{\partial n}{\partial \tau} + \frac{\partial}{\partial \xi} \left[ n \left( \frac{p_{\parallel}}{\gamma} - v_p \right) \right] = 0 \tag{9}$$

here $\gamma_p \equiv 1/\sqrt{1 - v_p^2}$, $v_p$ $(\to v_p / c)$ is the dimensionless phase velocity which equals to the group velocity of the laser pulse.

Thus, this closed set of equations describes the self-consistant interaction of intense laser pulses with cold plasma.

If the laser-pulse duration, $\tau_L$, is small compared to $\tau_e$, where $\tau_e$ is the characteristic time of changing the envelope of $\vec{A}_{\perp}$, then the so-called quasistatic approximation is valid /4/. Formally we set $\partial/\partial\tau \Rightarrow 0$ in the fluid equations. In this case it can be integrated.

Now we can write the coupled set of field equations which describes the ID nonlinear laser-plasma interaction within the quasustatic approximation for laser pulses of arbitrary polarization and arbitrary intensities:

$$\left(\frac{1}{\gamma_p^2}\frac{\partial^2}{\partial\xi^2}+2v_p\frac{\partial^2}{\partial\xi\partial\tau}-\frac{\partial^2}{\partial\tau^2}\right)\vec{A}_\perp=\frac{v_p\vec{A}_\perp}{\sqrt{(1+\varphi)^2-\frac{1}{\gamma_p^2}(1+\vec{A}_\perp^2)}} \quad (10)$$

$$\frac{\partial^2}{\partial\xi^2}\varphi=\gamma_p^2\left\{v_p\frac{1+\varphi}{\sqrt{(1+\varphi)^2-\frac{1}{\gamma_p^2}(1+\vec{A}_\perp^2)}}-1\right\} \quad (11)$$

Let's first examine the case of a square-shaped laser radiation pulse having the circular polarization which propagates in the $z$ direction with unchanged shape and the group velocity $v_g=v_p$.

Integrating Eq.(11) we obtain:

$$\left(\frac{d\varphi}{d\xi}\right)^2+2\gamma_p^2\left\{1+\varphi-v_p\left[(1+\varphi)^2-\frac{\gamma_{0\perp}^2}{\gamma_p^2}\right]^{1/2}\right\}=$$

$$2\gamma_p^2\left\{1-v_p\left[1-\frac{\gamma_{0\perp}^2}{\gamma_p^2}\right]^{1/2}\right\} \quad (12)$$

where $\gamma_{0\perp}^2=1+A_{0\perp}^2$

with the help of Eq.(12) we find:

$$E_{max}^2(v_{p_1})-E_{max}^2(v_{p_2})=\left(\sqrt{\gamma_{p_1}^2-1}-\sqrt{\gamma_{p_1}^2-\gamma_{0\perp}^2}\right)^2-$$

$$\left(\sqrt{\gamma_{p_2}^2-1}-\sqrt{\gamma_{p_2}^2-\gamma_{0\perp}^2}\right)^2 \quad (13)$$

where $E_{max}(v_{p_{1,2}})$ denotes the electric field amplitude at the phase velocities $v_{p_1}$ and $v_{p_2}$, respectively.

Equation (13) shows that when $v_{p_1} < v_{p_2}$ (i.e. $\gamma_{p_1}^2 < \gamma_{p_2}^2$) this difference is positive, i.e. if $v_{p_1} < v_{p_2}$, $E_{max}(v_{p_1})$ will be always greater than $E_{max}(v_{p_2})$.

The numerical analysis of Eq.(11) in the case of the Gaussian pulses confirm that the lower is the phase velocity of the wakefield, the greater is its maximum amplitude. Fig.(1) shows the dependence of the wakefield amplitude on the phase velocity for gaussian 1 psec pulses ( $\tau_{FWHM}$=1 psec - full width at half maximum) of various amplitudes.

In Fig.(2) the numerical solutions of wakefields produced by two Gaussian pulses with $A_{\perp max}$=1.4 $mc^2/e$ (solid line) and $A_{\perp max}$=2.8 $mc^2/e$ (dashed line) are presented.

Note that the both pulses have identical duration, $\tau_{FWHM}$ = 0.1 psec, but in the first case (solid line) $v_{p_1}$ =0.95 C ( $\gamma_{p_1}$ = 3.16), whereas in the second one (dashed line) $v_{p_2}$ =0.99 C ( $\gamma_{p_2}$ =10).

Thus, we can conclude that Gaussian pulses with lower

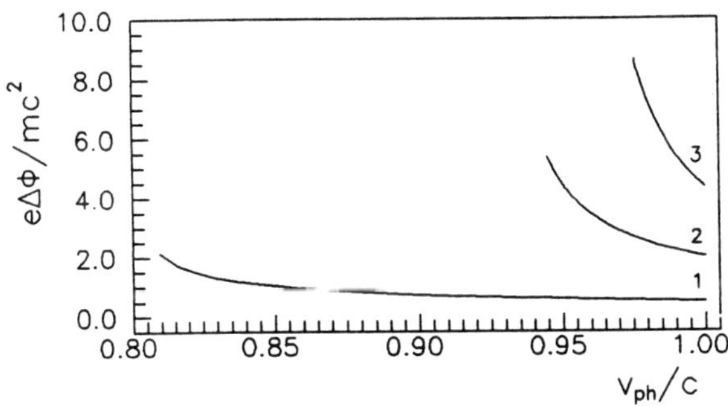


Fig. 1. Dependence of the wakefield amplitude on the phase velocity for Gaussian pulses ($\tau_{FWHM}$=1.psec) of various amplitudes. Curve 1 corresponds to $A_{\perp max}$=1.4mc$^2$/e; curve 2 - $A_{\perp max}$= 2.8mc$^2$/e; curve 3 - $A_{\perp max}$=4mc$^2$/e.

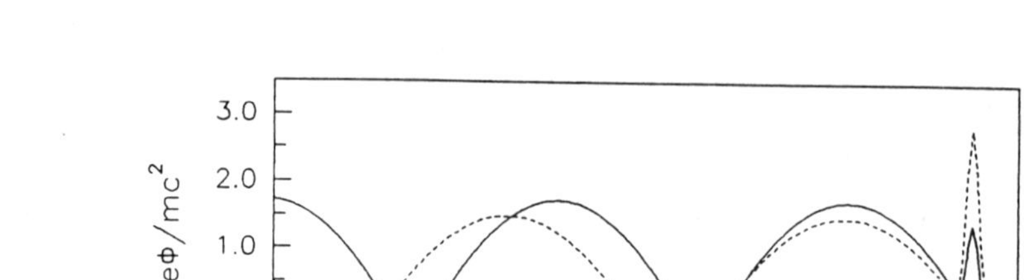


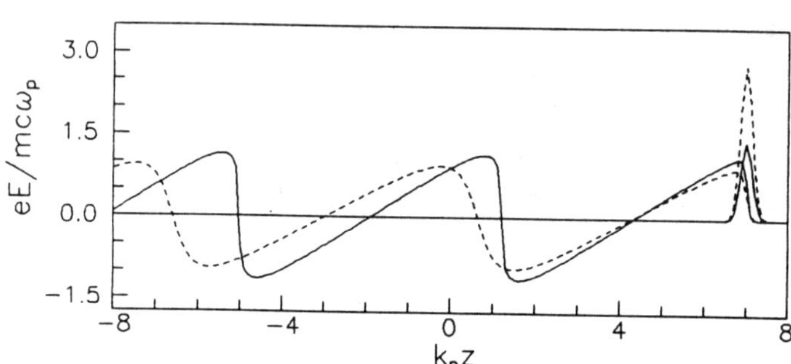


Fig. 2. The electric fields and electrostatic potentials of the wakefields excited by two Gaussian pulses with identical duration. $\tau_{FWHM}$=0.1psec, and different amplitudes. Solid curve corresponds to the case $A_{1max}=1.4mc^2/e$ and $v_{ph}=0.95c$. Dashed curve corresponds to $A_{1max}=2.8mc^2/e$ and $v_{ph}=0.99c$.

**amplitudes and lower group velocities are able to generate the wakefields having greater amplitudes than wakefields produced by Gaussian pulses with higher amplitudes but phase velocities close to the speed of light.**

## REFERENCES

1. Tajima T., Dawson J.M., Phys.Rev.Lett., 43, (1979) 263.
2. Sprangle P., Esarey E., et al., Appl.Phys.Lett., 53, (1988) 2146.
3. Berezhiani V.I., Murusidze I.G. Phys.Lett. 148, (1990), 338.
4. Sprangle P. et al. Phys.Rev.A, 41, (1990) 4463.
5. Akhiezer A.I., Polovin R.V. Sov.Phys. JETP 3, (1956) 696.

**Research Trends in Physics: Coherent Radiation Generation and Particle Acceleration**
Editorial Board: J.M. Buzzi, A. Prokhorov (Editor-in-Chief), P. Sprangle, and K. Wille
*La Jolla International School of Physics*, The Institute for Advanced Physics Studies, La Jolla, California

# Excitation of a Large Amplitude Plasma Wave by a Short Laser Pulse Using Resonator Scheme

**K. Nakajima**

National Laboratory for High Energy Physics
Tsukuba, Ibaraki, Japan

## ABSTRACT

It is shown that a large amplitude plasma wave is resonantly excited by a round-trip propagation of a short laser pulse in a plasma. The scheme results in a resonator of plasma wave that is useful for acceleration of a charged particle beam. The method is presented to avoid wave saturation due to detuning resulting from large amplitude nonlinear oscillation.

## INTRODUCTION

A plasma based accelerator concept, which would make use of the large amplitude plasma wave, has recently received a great deal of attention because in principle plasma can support orders of magnitude higher accelerating field gradients than conventional accelerators can produce. This makes the plasma based accelerator a breakthrough for the next generation of particle accelerators required to accelerate charged particles to ultrahigh energies in a compact space.

Excitation of plasma waves with large amplitude and highly relativistic phase velocity can be acomplished by a number of schemes, known as the beat wave accelerator[1] (BWA), the plasma wake-field accelerator[2] (PWFA) and the laser wake-field accelerator[3] (LWFA). In the BWA scheme the plasma wave is resonantly excited by beating of two electromagnetic waves with a frequency difference equal to the plasma frequency. The PWFA utilizes a high current bunched electron beam as a driver traveling through a plasma in which a plasma wave is left behind the electron bunch. In the LWFA, suggested together with the BWA and recently receiving renewed attentions, a short, high power, single frequency laser pulse produces accelerating wake-fields as it propagates through a plasma. The excitation of plasma waves by a single electromagnetic wave packet is regarded as the plasma wake-field mechanism with a laser pulse acting as a negative charged macro particle moving through the plasma at the group velocity of a wave packet.

These concepts raise some problems that have to be resolved. The BWA requires fine tuning of two laser frequencies and plasma density within less than 1 % to satisfy the resonance conditions. Although the plasma wave is efficiently excited in the BWA scheme, the growth of the wave amplitude saturates due to detuning the resonance condition between the beating frequency of electromagnetic waves and the plasma frequency which reduces as the amplitude increases because of the relativistic plasma oscillation. Main problem in the PWFA scheme is technological difficulties for producing a short, high-current, optimally shaped bunched relativistic beam. Since the PWFA is regarded as the energy transformer in which the driving beam is decelerated to generate the wake-field for accelerating the trailing beam, an essential requirement is a high transformer ratio of the accelerating wake-field to the decelarating field within the driving bunch. Although both the PWFA and the LWFA concepts require no fine adjustment of the plasma density, excitation of the plasma wave due to a non-resonant single pump is rather inefficient than the BWA mechanism due to two pumps. In the LWFA it is essential to have a sufficiently intense laser pulse with a shorter length than the plasma wavelength. The problems with regard to the particle acceleration arise in common with all plasma-based accelerators. The phase detuning between the plasma waves and the accelerated particles, the energy depletion of the driving beam and the stably transported distance of both laser and particle beams in the plasma without diffraction and instabilities strictly limit the acceleration length and the energy gain. Some of these problems may be resolved by novel schemes extensively investigated. However, most of such schemes are usually accompanied with technological difficulties.

The resonant excitation of the plasma wave in the PWFA scheme has been proposed and its first proof-of-principle experiment has been examined in the KEK linac[4] by using a high-intensity electron beam consisting of multiple bunches. It is straightforwardly conceivable that a resonant scheme for the LWFA can be constructed by use of muti-traversal propagation of a short laser pulse. In this scheme the efficiency of wave excitation is remarkably raised due to the buildup of plasma waves. Problems arising from the resonance condition will be relaxed because phase matching for the laser frequency is not necessary in the resonator. The amplitude saturation due to nonlinear detuning will be also approximately avoided. Therefore the plasma wave can continuously grow to a large ampltude limited by a so-called wave breaking. From an accelerator point of view, no problem related to the acceleration length occurs because the resonator length should be chosen to be smaller than the diffraction length or the phase detuning distance over which the accelerated particle overruns a correct acceleration phase of the plasma wave. For an actual application to the accelerator, the particle will be accelerated to an ultra high energy through multi-stage of the plasma wave resonator.

## EXCITATION OF PLASMA WAVE

We consider electron density oscillations in a plasma excited by an impulse of the intense short laser pulse. For an unmagnetized, cold plasma of classical electrons and immobile ions, the linearized equations describing the motion of the plasma electron fluid are

$$m_e \frac{\partial \boldsymbol{v}}{\partial t} + m_e \nu_{ei} \boldsymbol{v} = \nabla(e\phi + \phi_{NL}), \tag{1}$$

$$\frac{\partial n}{\partial t} + n_0 \nabla \boldsymbol{v} = 0, \tag{2}$$

$$\nabla^2 \phi = 4\pi e n, \tag{3}$$

where $\boldsymbol{v}$ and n are the (first order) electron velocity and density perturbation, $n_0$ the unperturbed density, $\nu_{ei}$ the electron ion collision frequency, $\phi$ the electrostatic potential of the plasma, $\boldsymbol{E} = -\nabla\phi$, and $\phi_{NL}$ is the ponderomotive potential defined by averaging the nonlinear force over $2\pi/\omega_0$, exerted on plasma electrons by the laser pulse with frequency $\omega_0$:

$$\langle F_{NL}\rangle = \nabla\phi_{NL}, \qquad \phi_{NL} = -\frac{m_e c^2}{2}\boldsymbol{a}^2(r,z,t), \tag{4}$$

where $\boldsymbol{a} = e\boldsymbol{A}(r,z,t)/(m_e c^2)$ is the normalized vector potential of the radiation field. The density response and the electrostatic potential in plasma oscillation are given by forced oscillator equations:

$$\frac{\partial^2 n}{\partial t^2} + \nu_{ei}\frac{\partial n}{\partial t} + \omega_p^2 n = -\frac{\omega_p^2}{4\pi e^2}\nabla^2\phi_{NL}, \tag{5}$$

and

$$\frac{\partial^2 \phi}{\partial t^2} + \nu_{ei}\frac{\partial \phi}{\partial t} + \omega_p^2 \phi = -\frac{\omega_p^2}{e}\phi_{NL}, \tag{6}$$

where $\omega_p = \sqrt{4\pi e^2 n_0/m_e}$ is the plasma frequency. The driving term arises from the ponderomotive potential. Assuming all axial and time dependencies are expressed as a function of the single variable $\zeta = z - v_p t$, with the phase velocity $v_p$ of the excited plasma wave, Eq. (6) can be written as

$$\frac{\partial^2 \phi}{\partial \zeta^2} + 2\alpha\frac{\partial \phi}{\partial \zeta} + k_p^2\phi = -k_p^2\phi_{NL} \tag{7}$$

where $k_p = \omega_p/v_p$ and $\alpha = \nu_{ei}/2v_p$. The solution for the damped harmonic oscillation can be written as

$$\phi(r,\zeta) = k_p\int_{\zeta}^{\infty} d\zeta' e^{\alpha(\zeta-\zeta')}\sin k(\zeta-\zeta')\phi_{NL}(r,\zeta'), \tag{8}$$

where $k = \sqrt{k_p^2 - \alpha^2}$. The axial and radial wakefields are given by $E_z = -\partial\phi/\partial\zeta$ and $E_r = -\partial\phi/\partial r$, respectively. Then the axial electric field is

$$E_z(r,\zeta) = \frac{m_e c^2 k_p k}{2e}\int_{\zeta}^{\infty} d\zeta' e^{\alpha(\zeta-\zeta')}\left[\cos k(\zeta-\zeta') + \frac{\alpha}{k}\sin k(\zeta-\zeta')\right]\boldsymbol{a}^2(r,\zeta'). \tag{9}$$

Considering a bi-Gaussian profile of a laser pulse given by

$$|\boldsymbol{a}(r,\zeta)| = a_0 \exp\Big(-\frac{r^2}{\sigma_r^2} - \frac{\zeta^2}{2\sigma_z^2}\Big), \tag{10}$$

where $\sigma_z$ is the rms pulse length and $\sigma_r$ is the rms spot size, the axial electric field becomes

$$\begin{aligned} E_z = &\frac{\sqrt{\pi}}{4} m_e c^2 k_p k a_0^2 \sigma_z \exp\Big(-\frac{r^2}{\sigma_r^2} - \frac{k^2\sigma_z^2}{4} + \alpha\zeta\Big) \\ &\times \Big\{ \big[C(\zeta) + \frac{\alpha}{k} S(\zeta)\big] \cos k(\zeta + \frac{\alpha\sigma_z^2}{2}) + \big[\frac{\alpha}{k} C(\zeta) + S(\zeta)\big] \sin k(\zeta + \frac{\alpha\sigma_z^2}{2}) \Big\}, \end{aligned} \tag{11}$$

where

$$C(\zeta) = 1 - \mathrm{Re}\big[\mathrm{erf}(\frac{\zeta}{\sigma_z} - i\frac{\alpha\sigma_z}{2})\big], \quad S(\zeta) = \mathrm{Im}\big[\mathrm{erf}(\frac{\zeta}{\sigma_z} - i\frac{\alpha\sigma_z}{2})\big]. \tag{12}$$

At $\zeta \to -\infty$ the accelerating gradient on the axis is expressed as

$$eE_z = \frac{\sqrt{\pi}}{2} m_e c^2 a_0^2 k_p k \sigma_z e^{-\frac{k^2\sigma_z^2}{4}} e^{\alpha\zeta} \cos[k(\zeta + \frac{\alpha\sigma_z^2}{2}) - \varphi_0] \tag{13}$$

where $\varphi_0 = \tan^{-1}(\alpha/k)$. Its maximum amplitude is $(eE_z)_{max} = m_e c^2 a_0^2/\sigma_z$ when the plasma wavelength is $\lambda_p = \pi\sigma_z$. The efficiency of plasma-wave excitation strongly depends on the pulse length and shape as well as in the PWFA where the amplitude of plasma wake-field has strong dependence on the rise and fall time of the driving bunch shape.

## RESONANT WAVE EXCITATION

The maximum amplitude of the plasma wave in the LWFA is given by

$$\Big(\frac{eE_z}{m_e c\omega_p}\Big)_{max} = \frac{a_0^2}{2}. \tag{14}$$

In the BWA the saturation amplitude due to the nonlinear oscillation is

$$\Big(\frac{eE_z}{m_e c\omega_p}\Big)_{max} = \big(\frac{16}{3} a_0^2\big)^{1/3}. \tag{15}$$

Therefore it is found that the beat wave mechanism is more efficient than the LWFA scheme for $a_0 < 2.56$. The single laser pulse with optimum length can produce a plasma wave of larger amplitude than in the BWA only with a stronger pump of the peak power density, $8\times10^{16}$W/cm$^2$ for a $CO_2$ laser and $8\times10^{18}$W/cm$^2$ for a Nd glass laser. The efficient plasma-wave excitation of the beat wave mechanism results from the resonance of the beating of electromagnetic waves with the plasma waves.

The resonant wave excitation can be achieved even by a single driving pulse on the analogy of the resonant excitation scheme of the PWFA. The resonant PWFA scheme requires a long colinear sequence of multiple bunches of which a period is equal to multiples of the plasma wavelength. An analogous scheme for the LWFA is possible through multiple round-trip propagations of a short laser pulse in a plasma. The round-trip propagation of the laser pulse can easily take place between a pair of mirrors similar to an optical resonator as shown in Fig. 1. The wake-field generated behind each traversal of the laser pulse grows to a large amplitude when plasma waves are superposed in a correct phase. A single pass wakefield behind a laser pulse propagating to the positive and negative direction can be expressed from Eq. (13).

$$E_z(z,t) = E_{z0}e^{\pm\alpha(z\mp v_p t)}\cos[k(z\mp v_p t)\pm\psi_0], \tag{16}$$

where $E_{z0}$ is the amplitude on the axis, $v_p$ is the phase velocity of the plasma wave and $\psi_0$ is an appropriate phase constant. The phase velocity of the plasma wave is equal to the group velocity of the laser pulse in a plasma.

$$v_p = \frac{\omega_p}{k_p} = v_g = c\left(1-\frac{\omega_p^2}{\omega_0^2}\right)^{1/2}. \tag{17}$$

After Mth round trip of the laser pulse between two mirrors with the spacing $L$, the amplitude of plasma waves grows up to a level given by

$$\begin{aligned} E_z(z,t+ML/v_p)/E_{z0} = &\\ & e^{\alpha(z-v_pt-ML+2mL)}\sum_{m=0}^{M/2}\cos[k(z-v_pt-ML+2mL)+\psi_0] \\ & +e^{-\alpha(z+v_pt+ML-L-2mL)}\sum_{m=0}^{M/2}\cos[k(z+v_pt+ML-L-2mL)-\psi_0] \end{aligned}$$

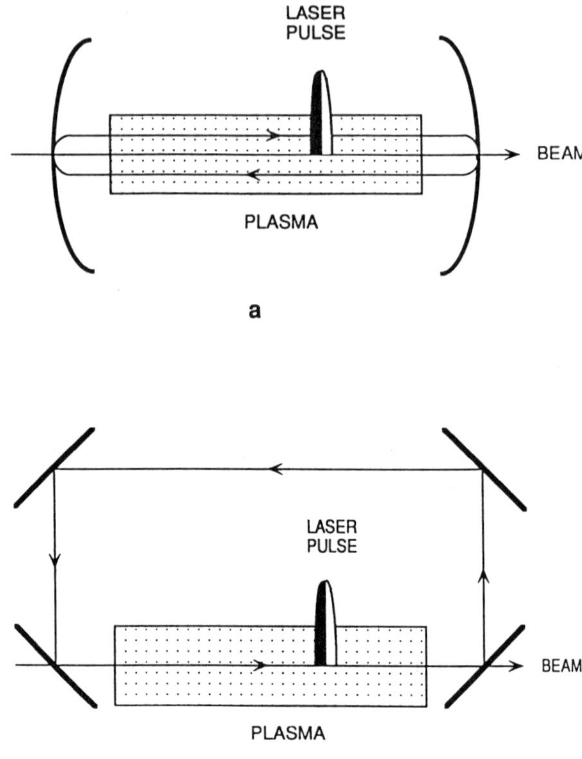


Fig. 1. Schematic of the resonant plasma-wave excitation: (a) Open resonator configuration. (b) Ring resonator configuration.

$$\begin{aligned}&\rightarrow \frac{e^{\alpha L}\sinh \alpha L}{\cosh 2\alpha L - \cos 2kL}\left(1-e^{-\alpha v_p t}\right)\left\{e^{\alpha z}\cos[k(z-v_p t)+\psi_0]\right.\\&\quad\left.+\,e^{-\alpha(z-L)}\cos[k(z+v_p t-L)-\psi_0]\right\}\qquad \text{as } M\rightarrow\infty \qquad (18)\\&\cong \frac{e^{\alpha L}}{\sinh \alpha L}\left(1-e^{-\alpha v_p t}\right)\cos kz\cos(\omega_p t+\psi_0),\quad \text{if } kL=2n\pi \text{ for } n=1,2,\ldots.\end{aligned}$$

As a result of multiple passings of the laser pulse in a plasma, the wake-field amplitude grows to a standing wave if the resonator length is taken to be a multiple of the plasma wavelength. For detuning of the resonator, the wave amplitude follows as $1/[\alpha L(1+(\delta k)^2/\alpha^2)]$, where $\delta k = k - k_p$. We note that the $Q$ value of resonator is $Q = k_p/2\alpha$.

## SATURATION AMPLITUDE OF PLASMA WAVES

For $\alpha L \ll 1$, the attainable amplitude of the axial electric field leads to

$$| E_z |_{max} \simeq \frac{E_{z0}}{\alpha L}. \tag{19}$$

The factor $1/\alpha L$ stands for an amplification gain of the resonator. This factor can be determined by taking into account the damping term of plasma oscillation. The damping of plasma waves results from collisional damping and Landau damping. The pump depletion, the energy loss of the electromagnetic pulse due to creating the plasma wave, also leads to satutration of the wave amplitude. In general the damping constant for Landau damping in the Langmuir wave is given by

$$\alpha = \frac{k_p}{2}\left(\frac{\pi}{8}\right)^{1/2} e^{-3/2}(k_p\lambda_D)^{-3}\exp\left[-1/(2k_p^2\lambda_D^2)\right], \tag{20}$$

where $\lambda_D$ is the Debye length for electrons. If we have a plasma with the electron temperature $1 \sim 10$ eV, $k_p\lambda_D \ll 1$ and therefore no Landau damping is expected.

As far as the collisional damping is concerned, a proper estimate of the collision frequency is necessary. For a large amplitude oscillation excited by a strong electromagnetic pump the collision frequency is determined not only by the thermal velocity $v_{th}$ of plasma electrons but also by the quiver velocity $v_{osc}$ of electrons in the laser field and the velocity of the plasma oscillation $v$. Thus the effective collision frequency[5] becomes

$$\nu_{eff} = \nu_{ei}\left(1 + v_{osc}^2/v_{th}^2 + v^2/v_{th}^2\right)^{-3/2}, \tag{21}$$

where $\nu_{ei}$ denotes the thermal collision frequency, given by

$$\nu_{ei} = \omega_p \frac{\ln(n_0 \lambda_D^3)}{2\pi n_0 \lambda_D^3}. \tag{22}$$

If the collisional damping is a dominant damping mechanism of the plasma wave, the amplification factor can be obtained as

$$\frac{1}{\alpha L} \simeq \frac{2n_0 \lambda_D^3}{k_p L}\left[1 + \left(\frac{v_{osc}}{v_{th}}\right)^2 + \left(\frac{v}{v_{th}}\right)^2\right]^{3/2} \sim 1.8 \times 10^{23} \frac{a_0^3}{n_0 L}. \tag{23}$$

Therefore we find that the collisional damping saturation of the plasma wave would take place only for $n_0 L > 1.8 \times 10^{23} a_0^3 \text{ cm}^{-2}$.

The pump-depletion effect can be approximately incorporated into the damping constant, using the definition of the pump-depletion[6] distance $l_{PD} = l_0 E_0^2/E_p^2$, where $l_0$ is the laser pulse length, $E_0$ and $E_p$ are the maximum electric field of laser and plasma wake, respectively. If the saturation amplitude is determined by the pump depletion, then the amplification factor is given by

$$\frac{1}{\alpha L} \cong \frac{2.3}{a_0^2}\left(\frac{\lambda_p}{L}\right)\left(\frac{\lambda_p}{\lambda_0}\right)^2 \tag{24}$$

Thereby the resonator length has to be taken so as to be $L < 2.3(\lambda_p/\lambda_0)^2 \lambda_p/a_0^2$. In addition, the resonator length must not exceed the acceleration distance over which the particle runs out of phase with the accelerating field, i.e. $L < (\lambda_p/\pi)(\lambda_p/\lambda_0)^2$. More rigorous constraint of optical guiding is imposed to the resonator length to prevent the laser beam from spreading out transversely. The distance over which the laser beam remains well collimated is given by the Rayleigh length, $l_d = \pi r_0^2/\lambda_0$, where $r_0$ is the laser spot size. Assuming $r_0 \sim \sigma_z = \lambda_p/\pi$, the resonator length is estimated to be $L \sim \lambda_p(\lambda_p/\lambda_0)$. These considerations result in the maximum attainable accelerating field excited in the resonator.

$$\left(\frac{eE_z}{m_e c\omega_p}\right)_{max} \simeq \frac{\lambda_p}{L}\left(\frac{\lambda_p}{\lambda_0}\right)^2 \sim \frac{\lambda_p}{\lambda_0} \tag{25}$$

We note that the saturation amplitude is determined by the relativistic factor of the wave phase velocity defined as $\gamma_p = \omega_0/\omega_p = \lambda_p/\lambda_0$. It means that the transverse electromagnetic field of laser has been totally converted into the longitudinal electrostatic field of plasma wave to be amplified by a factor of $E_z/E_0 \sim 1/a_0$. Then we find the maximum accelerating gradient is given by

$$(eE_z)_{max} \sim m_e c^2 k_0, \tag{26}$$

where $k_0 = \omega_0/c$.

## NONLINEAR PLASMA WAVE EXCITATION

The resonant excitation builds up the plasma wave finally to a much larger amplitude than $eE_z/m_e c\omega_p = 1$. This implies that the electron motion in plasma oscillation becomes highly relativistic even if the laser field is not sufficiently strong. The coupled equations[7] describing the nonlinear plasma-wave excitation and the wave equation are given by

$$\frac{d^2\varphi}{d\zeta^2} = \frac{k_p^2}{2}\left[\frac{1+\boldsymbol{a}^2}{(1+\varphi)^2} - 1\right], \tag{27}$$

and

$$2\frac{\partial}{\partial t}\left(i\omega_0 \boldsymbol{a} + c\frac{\partial \boldsymbol{a}}{\partial \zeta}\right) + c^2\frac{\omega_p^2}{\omega_0^2}\frac{\partial^2 \boldsymbol{a}}{\partial \zeta^2} = -\omega_p^2 \frac{\varphi}{1+\varphi}\boldsymbol{a}, \tag{28}$$

where $\varphi \equiv |e|\phi/m_e c^2$ is the normalized scalar potential. The wave equation (28) describing the distortion of laser pulse shape and its energy loss can be neglected if the condition, $a_0 \ll 2.5(\omega_0/\omega_p)$, is satisfied. Then the nonlinear wave excitation can be obtained from the solution of the equation

$$\frac{d^2\varphi}{d\xi^2} = \frac{1}{2}\Big[\frac{1+\boldsymbol{a}^2(\xi)}{(1+\varphi)^2} - 1\Big], \tag{29}$$

where $\xi \equiv k_p\zeta$. This value gives the the plasma-electron density perturbation, i.e. $n(\xi)/n_0 = d^2\varphi/d\xi^2$. The electric field is found from integration of Eq. (29):

$$\Big(\frac{eE_z}{m_e c\omega_p}\Big)^2 = \Big(\frac{d\varphi}{d\xi}\Big)^2 = \int_\infty^\xi \frac{\boldsymbol{a}^2(\xi)}{(1+\varphi)^2}\Big(\frac{d\varphi}{d\xi}\Big)d\xi - \frac{\varphi^2}{1+\varphi}, \tag{30}$$

with the boundary conditions, $\varphi = 0$ and $d\varphi/d\xi = 0$ at $\xi = \infty$. The multi-traversal of a short laser pulse can be modelled on a train of multiple pulses equally spaced with period of the resonator length. As a simple example, we consider the plasma wave excitation by an infinitesimally short $\delta$-function pulse with the power density, $I_0 = \sqrt{2\pi}a_0^2$. Then the solution is given by

$$\xi = 2\sqrt{b_N+1}\ \ \mathrm{E}(\Psi_N, K_N), \tag{31}$$

where $\mathrm{E}(\Psi_N, K_N)$ is the incomplete elliptic integral of the second kind and

$$\Psi_N = \sin^{-1}\sqrt{\frac{(b_N-\varphi)(b_N+1)}{b_N(b_N+2)}}, \qquad K_N = \frac{\sqrt{b_N(b_N+2)}}{b_N+1}, \tag{32}$$

$$b_N = \epsilon_N\big[\epsilon_N + \sqrt{\epsilon_N^2+4}\big]/2, \quad \epsilon_N \equiv \Big|\frac{d\varphi}{d\xi}\Big|_{max} = \Big[\frac{I_0^2}{(1+\varphi_N)^4} + \frac{\varphi_N^2}{1+\varphi_N}\Big]^{1/2}, \tag{33}$$

with $\varphi_N = \varphi(\xi_N)$ at $\xi_N = Nk_pL$ for $N = 0, 1, 2, \ldots$. The potential $\varphi$ oscillates in the range $b_N \geq \varphi \geq -b_N/(b_N+1)$. Fig. 2 shows the electrostatic potential, the electron density perturbation and the axial electric field for the nonlinear excitation by a single $\delta$-function pulse. In the plasma-wave excitation by the

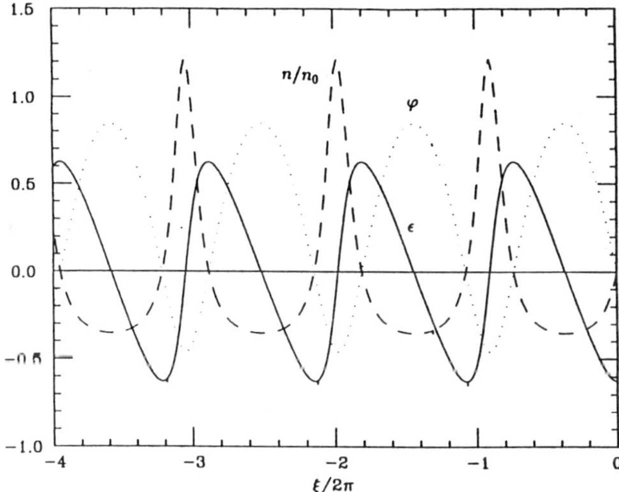


Fig. 2. The electrostatic potential $\varphi \equiv |e|\phi/m_e c^2$ (dotted line), the electron density perturbation $n/n_0$ (dashed line) and the axial electric field $\epsilon \equiv eE_z/m_e c\omega_p$ (solid line) excited by a single $\delta$-function pulse with $a_0 = 0.5$.

pulse train, the maximum wakefield is obtained when the trailing pulse follows in the phase at which the wake potential excited by the leading pulse becomes the minimum value. In this condition the wakefield rapidly grows after successive passing of pulses to a large-amplitude nonlinear wave, showing a steepening of the electric field and an lengthening of the wave period as illustrated in Fig. 3. From Eq. (33), it is noted that the condition for the amplitude buildup gives the recursive relation of the potential value:

$$\varphi_{N+1} = -b_N/(b_N + 1) = \epsilon_N\left[\epsilon_N - \sqrt{\epsilon_N^2 + 4}\right]/2, \tag{34}$$

with $\varphi_0 = 0$.

The phase matching for resonant excitation requires $n\lambda_N = M$, when $n$ is

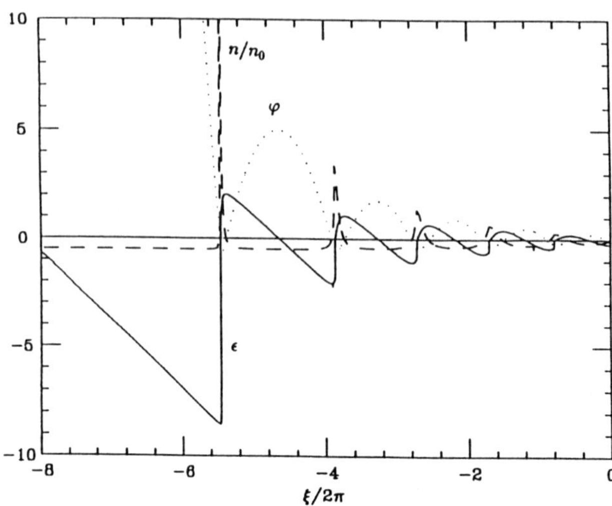


Fig. 3. The electrostatic potential (dotted line), the density perturbation (dashed line) and the electric field (solid line) in the nonlinear plasma wave resonantly excited by a pulse train.

integer, $\lambda_N$ is the plasma wavelength for the Nth pass and $M = L/\lambda_p$. In the nonlinear oscillation the period becomes

$$k_p\lambda_N = 4\sqrt{b_N + 1}\ \ \mathrm{E}(K_N), \tag{35}$$

where $\mathrm{E}(K_N)$ is the complete elliptic integral of the second kind. Then the tuning condition of the resonator is

$$\frac{\lambda_{N+1}}{\lambda_N} = \sqrt{\frac{b_{N+1}+1}{b_N+1}}\ \frac{\mathrm{E}(K_{N+1})}{\mathrm{E}(K_N)} = \frac{M-N}{M-N-1}, \tag{36}$$

taking into account increase of the period as the wave amplitude increases. This condition is satisfied if the wavelength varies as $\lambda_N/\lambda_p = M/(M-N) \simeq 1 +$

$N/M$. It is noted that if the detuning rate is appropriately controlled by means of changing the pulse intensity pass by pass, the resonance condition is automatically satisfied in the nonlinear wave excitation. As an illustration, Fig. 4 (a) shows the nonlinear growth of the electric field and the period increase for the resonator with $M = 200$. The resonance condition gives the power density variation of a laser pulse as shown in Fig. 4 (b).

## CONCLUSION

We have presented a scheme for the resonant excitation of large amplitude plasma wave by propagation of a short laser pulse in a resonator filled with plasma. This scheme has advantages the BWA, PWFA and LWFA schemes cannot overcome. It is not necessary to make fine tuning between the laser frequency and the plasma density. The wave-excitation mechanism has nothing to do with technological difficulties arising from manipulating a high-current, high-energy particle beam the PWFA scheme requires. The resonator system becomes rather simple and compact than the PWFA system. The technique on an ultra-short laser pulse[8] has recently achieved great advance in the laser technology. In the linear regime the excitation efficiency is remarkably improved, compared to a simple LWFA scheme, and even in the nonlinear regime wave saturation due to relativistic detuning would be able to be avoided.

If the electromagnetic energy of laser pulse is totally depleted in the resonator to excite plasma waves, an achievable accelerating gradient is $m_e c^2 k_0$; 3 GeV/cm for $CO_2$ laser and 30 GeV/cm for Nd glass laser. The attainable wave amplitude would be actually limited by other energy-dissipation mechanisms in both laser and plasma, in particular the plasma wave breaking arising from trapping of plasma electrons and possible instabilities of the plasma wave caused by ion motion. Extensive consideration will be pursued.

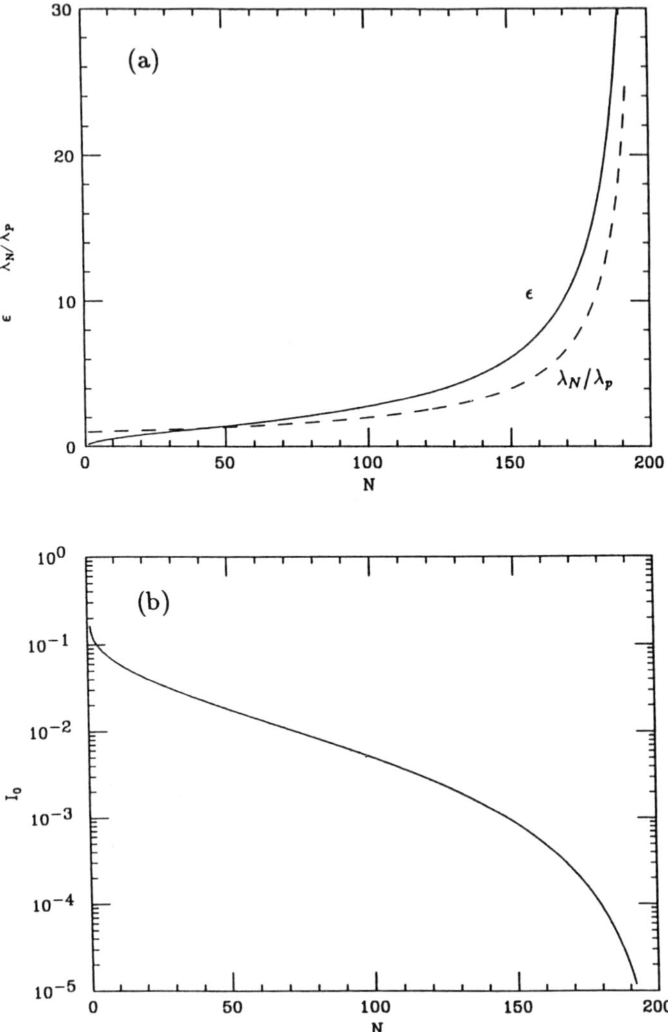


Fig. 4. (a) Growth of the electric field (solid line) and increase of the period (dashed line) for the nonlinear plasma wave in the resonator with length $200\lambda_p$. (b) Variation of the power density $I_0$ of a laser pulse resulting in the resonant excitation of the plasma wave.

## REFERENCES

1. T. Tajima and J. M. Dawson, Phys. Rev. Lett. **43**, 267 (1979).

2. P. Chen, J. M. Dawson, R. W. Huff and T. Katsouleas, Phys. Rev. Lett. **54**, 693 (1985).

3. E. Esarey, A. Ting, P. Sprangle and G. Joyce, Comments Plasma Phys. Controlled Fusion, **12**, 191 (1989).

4. K. Nakajima et al., Nucl. Instrum. Methods, **A292**, 12 (1990).

5. J. P. Matte and F. Martin, Plasma Phys. Controlled Fusion, **30**, 395 (1988).

6. W. Horton and T. Tajima, Phys. Rev. A, **34**, 4110 (1986).

7. U. de Angelis, Physica Scripta, **T30**, 210 (1990).

8. P. B. Corkum, Opt. Lett., **8**, 514 (1983); P. Maine et al., IEEE J. Quantum Electron., **24**, 398 (1988).

**Research Trends in Physics: Coherent Radiation Generation and Particle Acceleration**
Editorial Board: J.M. Buzzi, A. Prokhorov (Editor-in-Chief), P. Sprangle, and K. Wille
*La Jolla International School of Physics*, The Institute for Advanced Physics Studies, La Jolla, California

# Plasma Wakefield Acceleration and Plasma Lens Experiments at KEK

**A. Ogata**

National Laboratory for High Energy Physics
Oho, Tsukuba-shi, Ibaraki-ken, 305, Japan

## ABSTRACT

This paper describes two experiments using two linac facilities: plasma wakefield acceleration experiments and plasma lens experiments, which use the KEK (National Laboratory for High Energy Physics) PF electron linac for positron generation, and one of the twin linacs of the University of Tokyo, respectively.

The plasma wakefield acceleration experiments (PWFA) use a train of several 250 or 500MeV electron bunches with a total charge of 5-10nC of the KEK PF positron generator. They proved that the field of the plasma wave excited by preceding bunches accelerates or decelerates trailing bunches. The resonant condition between the buncher radio frequency and the plasma frequency causes a 12MeV barycenter shift in the energy distribution of a trailing bunch at the maximum. This energy shift is much larger than that predicted by the linear theory.

The plasma lens experiments are based on self-focusing, due to shielding of the space charge of a particle beam by a quiescent plasma. 18MeV, single-bunched round-shaped electron beams are used. A differential pumping technique is applied in order to separate the plasma chamber from the linac acceleration duct. Though it was shown that overdense plasmas can successfully focus electron beams, some phenomena which are difficult to be explained have also been observed.

 

## 1. Introduction

Two of the major roles expected of a plasma in accelerators are acceleration and focusing. This paper describes experiments which have been studying both of these.

A plasma wakefield accelerator(PWFA) is one of the plasma-based-type accelerators that show promise to produce ultra-high accelerating gradients. In a PWFA, a high-intensity relativistic driving beam excites a large amplitude plasma wave which, in turn, accelerates a low-intensity trailing beam.[2] Our experiments on PWFA were conceived while using a high-intensity electron beam for positron production at the KEK PF linac.[3] The linac provides a sequence of multiple bunches which generate wakefields in a plasma, resulting in acceleration or deceleration of trailing bunches. By analyzing the energy of each bunch, the energy transfer between the bunches can be observed without a test beam. Theory tells us that plasma wakefields are enhanced at certain plasma frequencies which are resonant with the frequency of spacing of the linac bunches. Because the plasma frequency is determined by the plasma density, we can probe the resonance by controlling the plasma density. Such experiments were started in 1989.[4–6] The maximum energy shift hitherto observed was approximately 12MeV.[7]

The plasma lens has been proposed as a final focus device of the next generation of linear colliders. Its focusing force can exceed those of a superconducting magnet by several orders of magnitude, since a plasma can support very large electromagnetic fields. At least three distinct particle focusing schemes exists, although the physical mechanisms and capabilities of each are quite different.[8] These are:(1) the focusing of particles by the radial field of a large-amplitude plasma wave moving with the beam; (2)focusing by the azimuthal magnetic field of a z-pinch plasma carrying a large axial current; (3)self-focusing due to the shielding of space charge of a particle beam by a quiescent plasma.[9,10] The third concept is experimentally examined in this paper. Experiments have shown that the plasma certainly has a lens effect, although certain aspects of the results remain unexplained.[11]

## 2. Plasma Wakefield Acceleration

### 2.1. THEORY

One of the features of our experiments is the use of a train of bunches. Fig. 1 shows how their use can increase the acceleration field. We first make the approximation that a bunch is impulsive. Fig.1(a) shows the space dependence of the wakefield caused by a single impulse: $G(s) = \cos k_p s$, $k_p = \omega_p/c$. The wakefield caused by a train of impulses is just the superposition of the wakefield of each impulse. If the plasma frequency is half the pulse frequency, or the frequency difference between the two has a fraction equal to half the plasma frequency, the wakefield cannot grow (Fig. 1(b)). On the other hand, if the plasma frequency is equal to the pulse frequency, or to a multiple of the pulse frequency, the wakefield increases with time(Fig. 1(c)). The rapid growth of the wakefield is shown in (d), where the injection timing of the bunches is the same as in the case (b), although the bunch intensities make a 1, 3, 5, .... sequence.[12,13] The idea of envelope modulation arose through a calculation showing that a properly shaped bunch density can result in a high transformer ratio.[14] Note that the envelope shown in Fig. 1(d) is quite similar to the doorstep-type function suggested in Ref. 14.

The real bunch trains of the KEK linac have gaussian envelopes in most cases. The bunch itself is not an impulse but rather has finite distribution. Fig. 2 shows the evolution of wakefield caused by a train of six pulses with a gaussian envelope; the plasma frequency is twice the pulse frequency. Fig. 1 shows that the relation

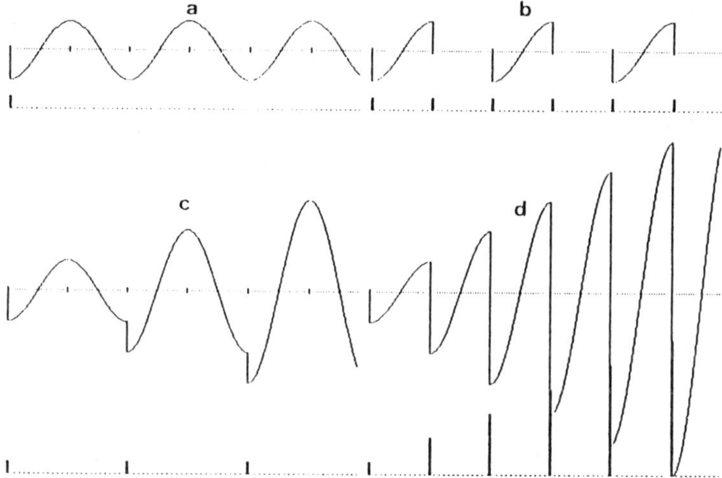


Fig. 1. (a)Wakefield caused by a single bunch; (b)a train of bunches with constant intensities whose frequency is half the plasma frequency; (c) the frequency is equal to the plasma frequency and (d) with increasing intensity, but half the plasma frequency.

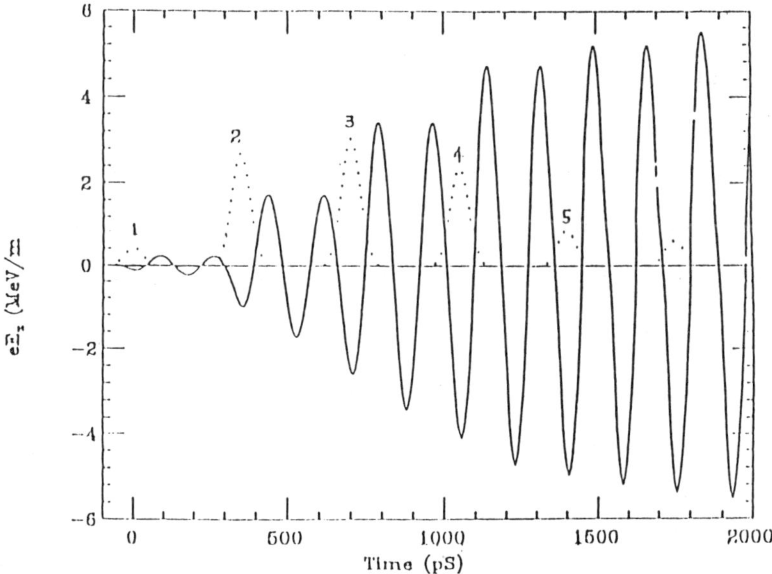


Fig. 2. Evolution of a wakefield by a train of six pulses with a gaussian envelope. The plasma frequency is twice the pulse frequency.

between two frequencies, the plasma frequency and the bunch frequency, plays an important role. Because the bunch frequency is determined by the rf buncher, we must control the plasma frequency in the experiments. Fig. 3 shows the wakefields expected at each bunch as a function of the plasma density in the range of our experiments, where the numerals denote the order of the bunches.[12] Beam profiles are taken into account in calculating the figure using a two-dimensional linear theory;[15] it is assumed that the total charge of 7nC is distributed over six bunches in a gaussian envelope, each of which has a transverse parabolic distribution with 1.5mm length, and a longitudinal gaussian distribution with 3mm rms length. At resonances, all bunches are decelerated while producing a maximum amplitude wakefield behind the bunch train. As a matter of course, the latter bunches exibit higher energy gains. We have studied the way that this figure is reproduced by experiments.

### 2.2. EXPERIMENTAL SETUP

The KEK PF electron linac for positron generation is used owing to its large current. A 4ns beam pulse emitted by a triode gun is compressed to less than 2ns in a subharmonic buncher. The 2856MHz rf buncher then separates the 2ns pulse into a train of more than 6 bunches with a 350ps spacing. They are then accelerated up to either 250 or 500MeV. Bunches with a total charge of 5-10nC are focused on a plasma by a quadrupole triplet at the end of the linac. The radius and length of the maximum bunch are 1-1.5 and 3mm, respectively. It was found that the preceding bunches cause a deceleration of the following, even in the absence of a plasma, exciting wakefields in the linac structure. The difference in the barycenter energy between the first and last bunches amounts to 15MeV without a plasma.

Another feature of our experiments is the use of a 1m long plasma, so that it is possible to directly observe the acceleration gradient in unit of MeV/m. The plasma is produced in a chamber with .3m in diameter and 1m in length by pulse discharges between the cathodes and the chamber. The cathodes are made of tungusten filaments but have been changed to comprise four $LaB_6$ blocks so as to increase the plasma intensity. The cathodes are heated by a 10V-80A direct current source. The discharge pulse has a voltage of 80-100V, a current of 10-20A, a duration of 2ms and a rate of .5Hz, equal to the linac beam repetition rate.

The multidipole field of the permanent magnets, 1kG at the inner surface of the water-cooled chamber, is applied in order to confine the plasma. Both beam windows of the plasma chamber comprise 50$\mu$m thich titanium foils. Identical foils are used at the windows of the linac duct and the spectrometer magnets to separate the vacuum from the plasma chamber. The base pressure of the chamber is typically made to be $10^{-8}$Pa by using a 300$l$/s turbomolecular pump. Argon gas is fed through a gas-flow controller to maintain a neutral gas pressure of $4 \times 10^{-6}$Pa for a plasma density of $10^{12}$cm$^{-3}$. The plasma density can be controlled by both the gas pressure and the discharge current.

The plasma electron density and temperature are measured by a Langmuir probe. The plasma density was measured to be radially homogeneous, and extending over 20cm. Though we made no measurement of the longitudinal distribution, measurements of a similar confinement device used in plasma lens experiments showed that it is also fairly homogeneous.

A combination of a bending magnet and a streak camera makes it possible to measure the energy spectrum of each bunch. Since the energy aperture of the magnet is only 15MeV, it is necessary to sweep the analyzing field to obtain a spectra for all bunches. Bunches analyzed in the bending magnet travel in air over a length of about .5m to a mirror, while radiating Cherenkov radiation. The mirror reflects only the radiation, transmitting the electron beam. The reflected radiation is finally focused on the slit of the streak camera. A Dove prism in front of the slit rotates the image so that the energy dispersion axis is perpendicular to the time axis of the camera.

The resultant streak pictures contain three-dimensional information; the horizontal dimension gives the convolution of the energy spread and the horizontal beam size; the vertical dimension shows the time structure of the bunches; the picture intensity indicates the number of electrons. By digitizing the streak picture, it is possible to calculate the mean and standard deviation of the energy of each bunch.

### 2.3. EXPERIMENTAL RESULTS

Two kinds of beams are used; low-density ($\sigma_r$ = 1.5mm), low-energy (250MeV) beams and high-density ($\sigma_r$ = 1mm), high-energy (500MeV) beams. The beam density is more meaningful than is the beam energy in these experiments. The high beam density is, however, a by-product of the beam energy, because the higher energy beams are less scattered at the beam windows. The higher energy also increased the Cerenkov radiation intensity while improving the signal-to-noise ratio of streak camera measurements. We now describe the results obtained for these two beams separately, since they illustrate some different aspects.

In low-density beams, the experimental results agree quite well with calculations, as shown in Fig. 4. Theory suggests that the resonant excitaion of plasma wakefields occurs when the plasma frequency coincides with a multiple of the radio frequency of the buncher; *i.e.*, $\omega_p/2\pi = kf_{\rm rf}$, where $k$ is an integer. The energy spectra of the 4th, 5th and 6th bunches were measured as a function of the plasma

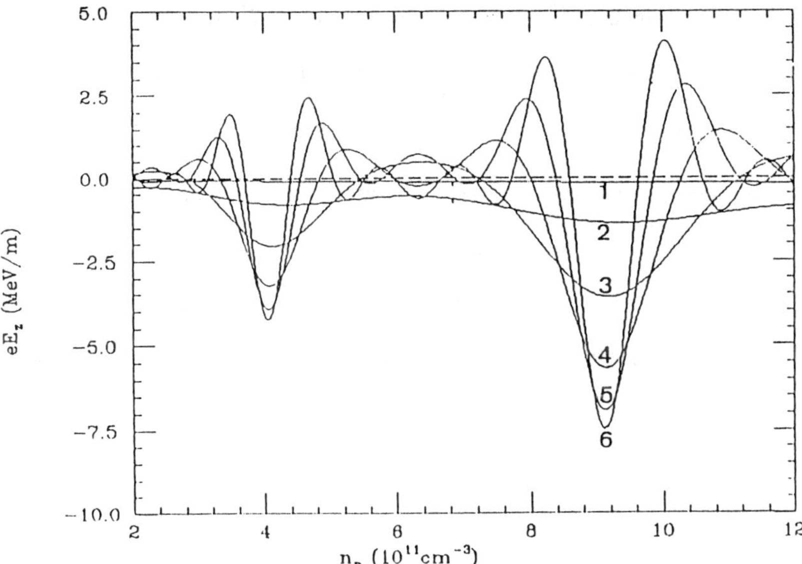


Fig. 3. The expected longitudinal wakefield at the barycenter of each bunch. The numerals denote the order of the bunches.

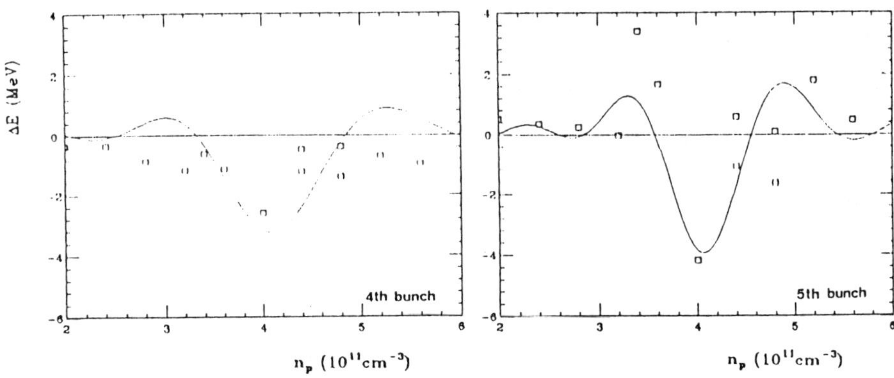


Fig. 4. The barycenter energy shifts of the 4th and 5th bunches observed around a plasma density of $4 \times 10^{11}\text{cm}^{-3}$ caused by low-density beams. The solid curves show calculations based on a two-dimensional linear theory.

density around $4 \times 10^{11}\text{cm}^{-3}$; *i.e.*, $k = 2$ resonance. We found decelerations of approximately 3 and 4MeV in the 4th and 6th bunches (Fig. 4), where solid lines give calculation. The 6th bunch was too weak to be energy-analysed.

Using high-density beams, we tried a wider plasma density region (up to $10^{12}\text{cm}^{-3}$) to meet with a new aspect; bunches with maximum intensity are strongly decelerated at the resonant plasma densities under this condition. Little deceleration, less than 2MeV, is found in a bunch following just after a maximum one. Instead, the energy spectrum of this bunch becomes broad. The energy changes of the following weak bunches are contrarily negligible.

Fig. 5(a) shows the energy spectra of a bunch with maximum intensity. The difference in the mean energy between those given by the solid and dotted lines is approximately 12MeV, where the solid line gives the energy spectrum in the presence of a plasma with the density of $9 \times 10^{11}\text{cm}^{-3}$, (the $k = 3$ resonant density given in Fig. 3), while the dotted line is the spectrum in the absence of a plasma. The observed energy shift is much greater than that predicted by the linear theory given in Fig. 3; this figure shows that the energy shift of the 3rd bunch with maximum intensity should be about 5MeV when the beam density increase in the present experiments is taken into account. The observed resonance given in Fig. 6 is, moreover, sharper than the theoretical prediction given in the curves 3 or 4 of Fig. 3. In Fig. 6, curve 6 from Fig. 3 is re-scaled and depicted, which strangely fits the experiments, in spite of the fact that it should have no physical meaning.

Fig. 5(b) shows the energy spectra of a bunch following just after the maximum. The spectrum at the resonant plasma density is 40% broader than that obtained off the resonance. The spectrum for the case without a plasma has a similar distribution to that given by the dotted line in Fig. 5(b). Fig. 7 shows

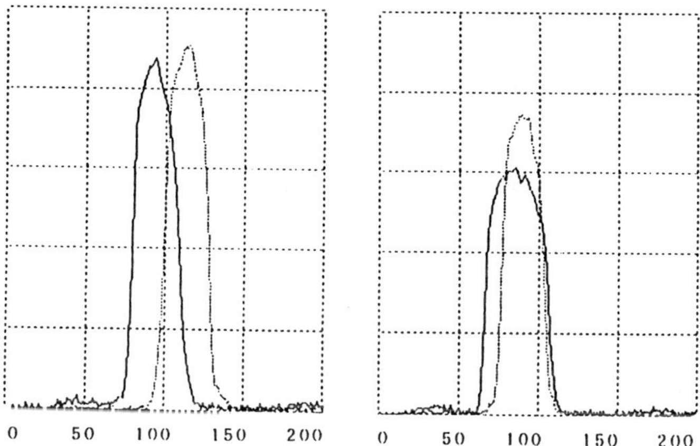


Fig. 5. Typical energy spectra, .547MeV/channel. (a) Those of a bunch with maximum intensity. The solid line is in the presence of a plasma with a resonant density. The dotted line is for the absence of a plasma. (b) Those of a bunch just after that with maximum intensity. The solid line is in the presence of a plasma with resonant density; dotted line is for a plasma with density off the resonance.

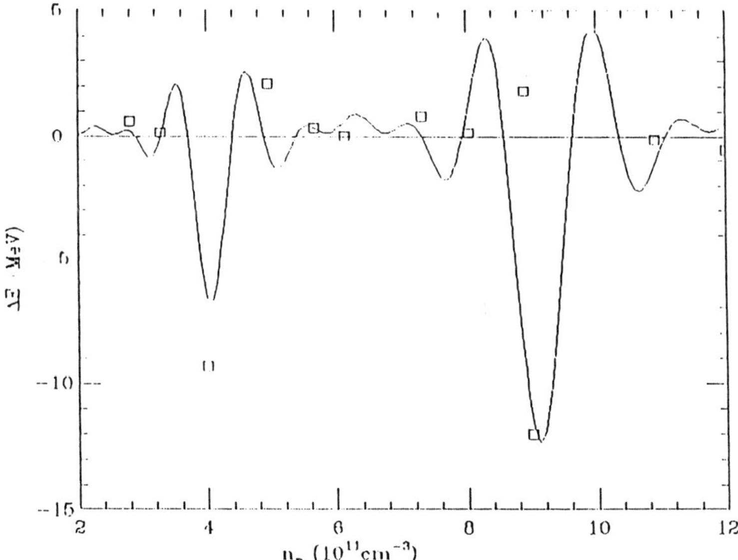


Fig. 6. Barycenter energy shifts of the maximum bunch as a function of the plasma density caused by high-density beams.

the rms energy width of this bunch as a function of the plasma density, where resonances exist but are broader than in Fig. 6. We cannot tell whether this is due to energy broadening or to defocusing of the wakefield. As shown in Fig. 5(a), no significant change was found in the width of the spectrum of the maximum bunch at the resonant plasma density.

One explanation for the anomalous deceleration of the maximum bunch is the introduction of a nonlinear mechanism. The perturbed electron density is given by[16] $n_1 = k_p N exp[-(k_p \sigma_z)^2/2]/(2\pi\sigma_r^2)$, where $k_p = c/\omega_p$ and $N$ is the number of particles in the bunch. Inserting our values and assuming that the maximum bunch has a charge of 2nC, we have $n_1 = 3.04 \times 10^{11}\text{cm}^{-3}$, or $n_1/n_o = .3$, at the second resonance. This is certainly a region where the linear theory is not applicable.

The nonlinear effect also provides an explanation for the fact that the tail bunches do not exhibit any energy change at the resonances; a phase difference is probable between the plasma wave and the bunch spacing if the plasma wave does not remain sinusoidal. It seems to correspond to the phenomenon of spectrum broadening, as shown in Fig. 7. Another plausible explanation is that the tail bunches are off-centered before they enter the plasma. Vertical offsets of tail bunches more than 1mm are often observed in this linac,[17] which is brought about by a transverse wake field inherent in the linac structure. If the offset is greater than the driving bunch beam size, its effect on the trailing bunches cannot be sufficiently strong to cause an energy shift.

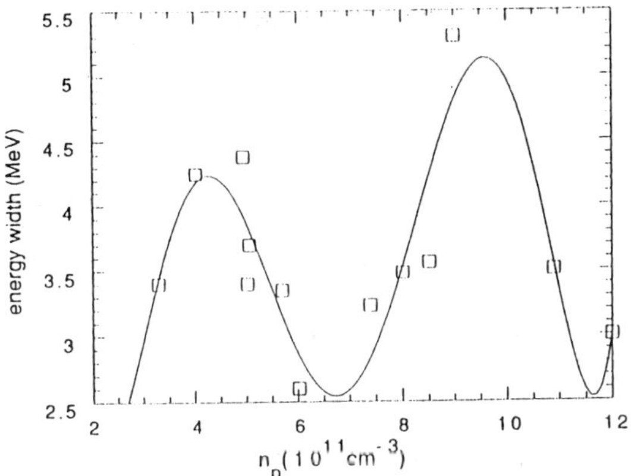


Fig. 7. The rms energy width of the bunch following just after the maximum as a function of the plasma density. The solid line shows the fitting to 6th order polynominal.

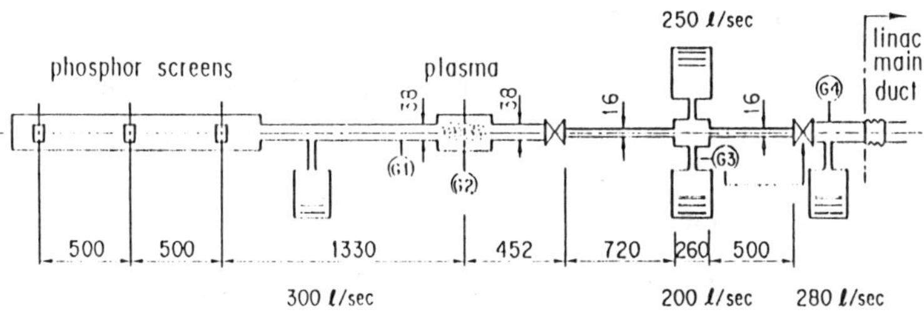


Fig. 8. Vacuum line setups. Typical pressures measured at gauge G1-G4 without gas feed are .744, 1.43, .00957 and .0186mPa, respectively, which changes to 38.5, 146, .212 and .0146mPa with gas feed, respectively.

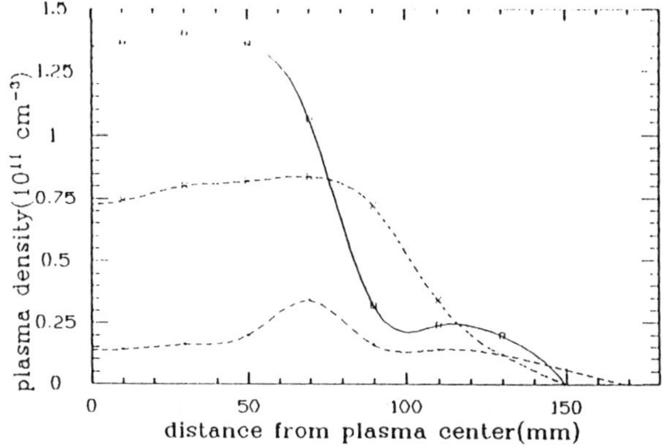


Fig. 9. Longitudinal plasma distribution.

## 3. Plasma Lens

### 3.1. THEORY

There are two regions in a self-focusing plasma lens: overdense and underdense regions; our experiments spanned these two regions. The physical mechanism of an overdense plasma lens, in which the plasma density is much larger than the beam density, is as follows. In a relativistic electron beam traveling through a vacuum, a repulsive force due to the space charge of all the electrons in the bunch is canceled by an attractive force due to a self-magnetic field of the bunch; thus, the beam almost maintains a constant radius. However, if the same beam now enters a plasma, the plasma electrons respond to the excess charge by shifting away from the beam particles. The remaining plasma ions neutralize the space-charge force within the beam; although the plasma is very effective at shielding the space charge of the beam, it is less effective at shielding its current. The beam thus experiences almost the full effect of its self-generated azimuthal magnetic field.

In an underdense plasma lens, in which the beam is denser than the plasma, the space charge of the electron beam essentially blows out all of the plasma electrons, leaving a uniform column of positive ion charge.

### 3.2. EXPERIMENTAL SETUP

The experiments were conducted at the Tokyo University 18MeV linac.[18] It produces a single pulse whose rms length (measured by a streak camera) is less than 3mm, and whose repetition rate is a few Hz. The charge of a pulse in the present experiments was about 0.5nC. The averaged electron density inside the bunch was about $1.2 \times 10^{10}\text{cm}^{-3}$. Single-bunch beams were introduced into a plasma chamber, where the plasma density could be varied from underdense to overdense regions. Transverse profiles of the beams were measured by three phosphor screens placed downstream of the chamber.

We usually separate the plasma chamber from the linac duct using metal foils in order to avoid any vacuum problems. An obstacle to the measurements in this case is multiple scattering of the beam caused by the foils and gas along the beam transport. Differential pumping solves this problem by enabling separation without having to use any hard boundaries. Fig. 8 shows the vacuum line setup. Four turbomolecular pumps are used, three of which are placed between the linac main duct and the plasma chamber. Ducts with low conductance, 16mm in diameter and 1233mm in total length, connect the linac and the plasma. An automatic gate valve closes the line whenever the pressure of this section exceeds a prescribed value.

An argon plasma was produced in the chamber, 147mm in inner diameter and 360mm in length, by a discharge between the $LaB_6$ cathodes and the plasma chamber in synchronism with the linac pulse. The plasma pulse width was about 2msec. It was then confined by the multidipole field of permanent magnets placed around the chamber periphery. The magnetic field had its maximum value 700G at the chamber wall. One of the features of this confinement is that there is no magnetic field along the beam transport. The argon plasma density ranged from $.5 - 15 \times 10^{10}\text{cm}^{-3}$ and the temperature from $2.5 - 4$eV, as measured by a Langmuir probe. Fig. 9 shows the longitudinal plasma density distribution in different conditions. It shows that the plasma length along the beam transport is about 15cm in the case $n_e = 1.3 \times 10^{11}\text{cm}^{-3}$, which increases to 20cm at $n_e = .75 \times 10^{11}\text{cm}^{-3}$.

Transverse beam profiles were observed on three phosphor screens (Desmarquest AF995R), which were located 1330, 1830 and 2330mm from the center of the

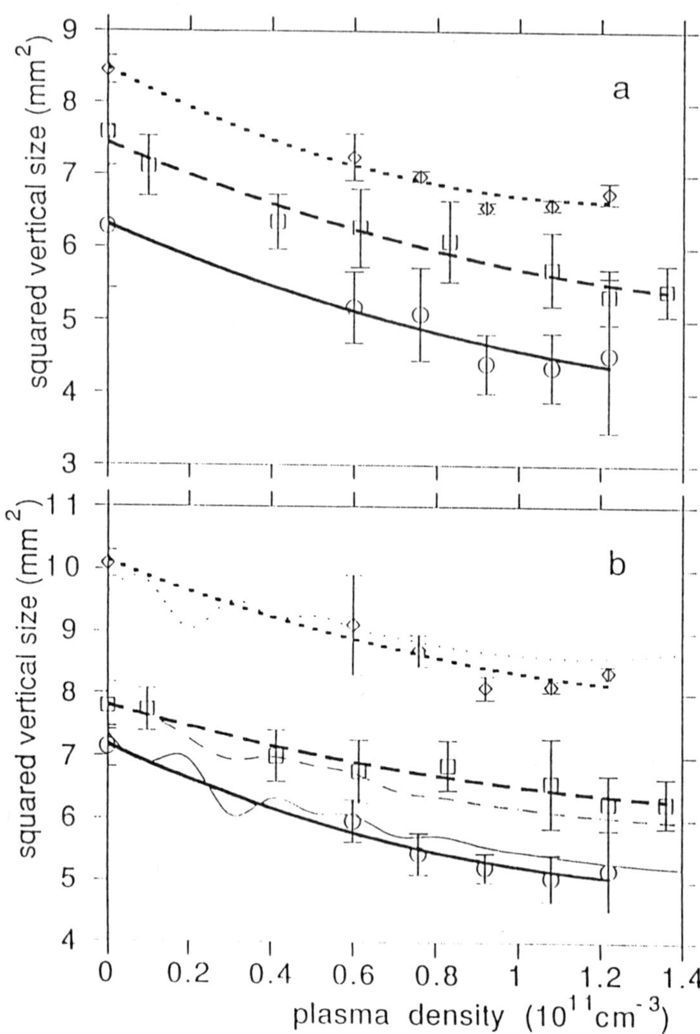


Fig. 10. Variances of the horizontal(a) and vertical(b) distributions observed on three phosphor screens. The solid lines indicate the first screen; the dashed lines the second screen; and the dotted lines the third screen. Error bars were obtained from 18 pulses the linac. The bold lines are quadratic approximations, while the fine lines in (b) indicate the density dependence of sizes derived from experiments under the constraint of constant emittance. See text.

plasma chamber. We call them the first, second and third screens in this paper, counting from the nearest one to the plasma. Simultaneous measurements using these three screens are impossible because of multiple scattering at the screens. The reproducibility of the linac beams was very good, an important factor in our measurements. The images were observed by a CCD camera with good linearity, and stored on video tapes and computer-processed. The CCD camera is triggered in synchronism with the linac pulse. Special care was taken not to saturate the CCD camera.

### 3.3. EXPERIMENTAL RESULTS

Systematic measurements were made both in the overdense and underdense regions using different beam optics. We report here only on the overdense case. It was found that, though the vertical profiles are gaussian, the horizontal profiles often deviates from a gaussian. Fig. 10 shows statistical variances or squares of the standard deviation of the horizontal and vertical distributions as a function of the plasma density. The figure shows that, firstly, the beam size is always smallest at the first screen, and that it increases with distance from the plasma; secondly, the beam sizes decrease as the plasma density increase. Three bold lines in Fig. 10 represent quadratic approximations to the density dependences.

Because there is free space between the plasma and the phosphor screens, it is possible to derive three parameters (two Twiss parameters and the emittance) at the plasma as a function of the plasma density from a set of three data at three screens. We have found, however, that real solutions are always obtainable only in the vertical direction. In the horizontal direction, some of the Twiss parameters become imaginary. The solid lines in Fig. 11 show the Twiss parameters and vertical emittance, thus calculated. As the figure shows, the plasma decreases the

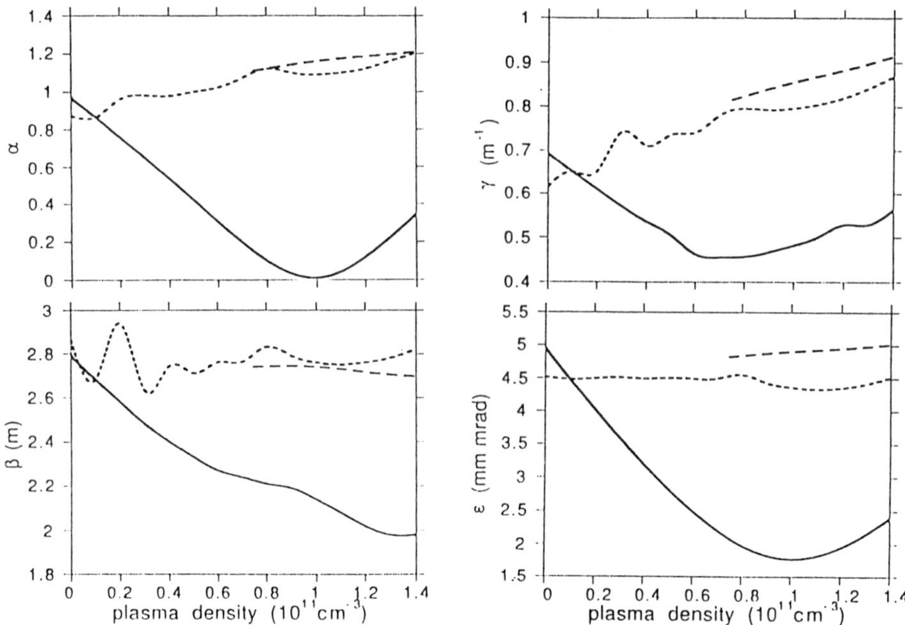


Fig. 11. Dependence of vertical Twiss parameters and emittance on plasma density. The solid lines were calculated from the quadratic approximation of Fig.10(b). The dotted lines were calculated from the fine lines in Fig. 10(b), assuming a nearly constant emittance. The dashed lines are theoretical based on the round-beam model.

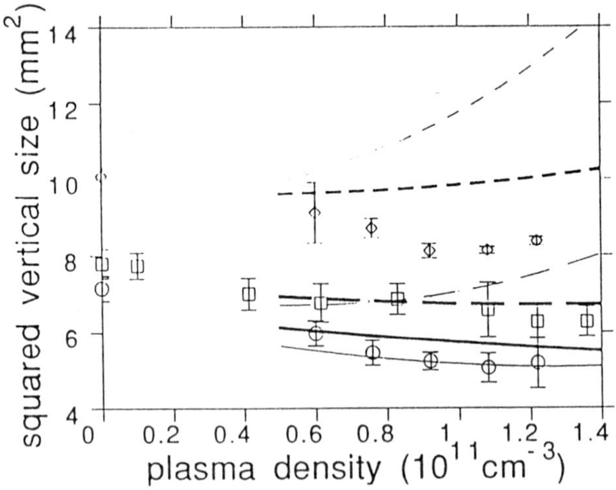


Fig. 12. Theoretical dependences of the squared beam sizes at the three screens on the plasma density (thick lines). The thin lines show the case when the focusing strength is 1.5 times that the case of thick lines. Experimental data are also given.

emittance. The dependence of $\alpha$ shows that it makes the phase ellipse erect; $\beta$ and $\gamma$ show that it reduces the size of the ellipse in phase space.

Because nonconservation of the emittance is not plausible, another method of analysis was tried which assumed a nearly constant emittance. It assumes that the true squared size exists somewhere in the range within ±5% of the value given by the qudratic approximations. A quadratic penalty function was introduced in order to: 1) make the vertical emittance be as close as possible to $4.5\pi$mm mrad (an approximate value in the absence of a plasma), and 2) make the curves showing the density dependence of the squared sizes be as close as possible to the quadratic approximation. The results depend on how we weigh these two requirements. The fine lines in Fig. 10(b) show an example of the result. Here, it is required that the size at the front screen should be larger, while the size at the middle screen should be smaller than the average of the experimental data. The dotted lines in Fig.11 show the vertical Twiss parameters and emittance calculated from this density dependence of sizes. The change in the emittance is roughly within ±5%. The increase of $\alpha$ and $\gamma$ suggests that the lens effect increases as the density increases. Though Fig. 11 gives only information about the vertical parameters, the horizontal parameters have a similar plasma density dependence.

We now try to explain the density dependence theoretically. In the range $n_e >$ $.5 \times 10^{11}\text{cm}^{-3}$, Chen's conditions of the round-beam-limit, $1/(4\pi r_e a^2) \gg n_e \gg 4Nk_p^2 b/(\pi a^2)$, are satisfied[9], where the parabolic profiles in both the transverse and longitudinal directions are assumed as approximations to a gaussian, with $a$ and $b$ denoting the bunch radius and half the bunch length, respectively. Under these conditions, the focusing force is linear in $r$, but proportional to $\zeta^3$: $F = e^2 k_p^2 N \zeta^3 r/(a^2 b^2)$, where $\zeta$ denotes the longitudinal position inside a bunch. Using the transfer matrices of free space and the thick lens, and averaging the dependence of the focusing strength on $\zeta$ with the weight of a parabolic distribution, we can calculate the beam size at any plasma density.

The bold lines in Fig. 12 give the theoretical density dependence of the squared sizes at the three screens, where we set the lens length at 15cm and assumed $a = b = (\beta_{y0}\epsilon_{y0})^{1/2}$, with $\beta_{y0}$ and $\epsilon_{y0}$ denoting values in the absence of a plasma (Fig. 11). As it shows, the experimental beam sizes are always smaller than those predicted theoretically. Although it is possible to adjust the parameters to fit the theoretical to experimental at the first or second screen, it is hard to do so at the third screen. It should be noted that the theoretical curves are also different from those given in Fig.10, which were derived by assuming a constant emittance. The dashed lines in Fig.11 show also theoretical dependence of the Twiss parameters on the plasma density, based on the round-beam model, which has a similar trend to the dotted lines experimentally obtained under the assumption of a constant emittance. It should be noted that the apparant emittance increases with the plasma density theoretically.

### 3.4. DISCUSSION

One explanation for the apparent emittance reduction of Fig.11 would be beam scraping. However, the density dependence of the total intensity of any screen image eliminates this possibility. If the emittance reduction at the high-density region is due to scraping, the total count should decrease. Experimental evidence is contrary: the total count is minimum in the absence of a plasma, and increases with the plasma density, becoming constant above $n_e > .4 \times 10^{11}\text{cm}^{-3}$. The maximum difference is only 10%. This is probably because the beam spread is so wide without the plasma that the visual field in the computer cannot cover the entire beam region on the screen.

Another plausible explanation is horizontal-vertical coupling; one may wonder if a decrease in vertical emittance is compensated by the horizontal emittance growth. Fig. 10 eliminates this possibility. We have calculated that a vertical focal point exists both near and behind the first screen in the absence of a plasma. As the focusing strength increases, the size on the first screen should become smaller, while the size on the third screen should become larger. However, the experimentally obtained vertical size at the third screen becomes smaller and smaller as the plasma density increases. This is the reason for the apparent emittance reduction. As Fig. 10(a) shows, the situation is the same in the horizontal direction; in other words, the emittance decrease takes place also in this direction. We must therefore discard the idea of a horizontal emittance increase.

## 4. Future Experiments at KEK

Although the plasma lens experiments have shown that a plasma has a lens effect, beam size reduction far from the expected focal point is puzzling. We have not as yet conceived of a systematic experimental effect which can explain this observation; neither can we explain it theoretically. Resolving this will be a principle objective when we undertake the next set of experiments in the near future. A streak camera will reveal the lens effect on the longitudinal profiles.

Two efforts are proceeding in the KEK PWFA experiments to increase the wakefield. One is to increase the plasma density, and the other involves pulse envelope modulation. In order to increase the plasma density, we must replace the present plasma chamber with a new one. A high-density plasma, up to $10^{14}\mathrm{cm}^{-3}$, has already been produced in a 5cm diameter, 1m long glass tube under a few kG solenoidal field by excitaion with a helicon wave.[19]

The modulation of pulse train envelope provides a practical method to improve the transformer ratio, as shown in Fig. 1(d). The cathode grid pulser of the KEK linac will be modified in order to produce a proper envelope of the bunch train. Although only deceleration is remarkable at the plasma resonant density at present, where bunch spacing is fixed, and the envelope of the bunch train is gaussian, an envelope modulation must make acceleration possible.

PWFA experiments using the Tokyo University linacs are also planned. The linac used in the plasma lens experiments is one of twins; there are two parallel linacs, with energies of 18 and 28MeV. The time difference between the bunches of the two linacs are adjustable. In the PWFA experiments, one linac will provide the driving bunches; the other will provide test bunches. We expect that more subtle experiments are possible in these linacs than in the KEK PF linac. It will take a while, however, to modify the beam lines so that the two beams can be joined at the plasma chamber.

## REFERENCES


1. The following people have participated: A. Ogata, K. Nakajima, H. Nakanishi, A. Enomoto, T. Kurihara, H. Kobayashi, Y. Suetsugu, T. Oogoe (KEK); Y. Yoshida, T. Ueda, T. Kobayashi, H. Shibata, S. Tagawa, T. Kozawa, K. Miya (The University of Tokyo); Y. Nishida, N. Yugami (Utsunomiya University); T. Shoji (Nagoya University) and B. S. Newberger (University of Texas-Austin).
2. P. Chen *et al.*, *Phys. Rev. Lett.* **54**(1985), 693.
3. A. Enomoto *et al.*, *Nucl. Instr. and Meth.* **A281**(1989), 1.
4. A. Ogata *et al.*, Proc. 1989 IEEE Part. Accel. Conf., Chicago (IEEE 1989) p618.
5. H. Nakanishi *et al.*, *Part. Accel.* **32**(1990), 203.
6. K. Nakajima *et al.*, *Nucl. Instr. and Meth.* **A292** (1990), 12.
7. A. Enomoto *et al.*, Proc. 2nd European Part. Accl. Conf., Nice (Editions Frontieres 1990) p634.
8. J. J. Su *et al.*, *Phys.Rev.* **A41** (1990), 3321.
9. P. Chen, *Part. Accel.* **20** (1987), 171.
10. P. Chen *et al.*, *Phys. Rev.* **D40** (1989), 923.
11. H. Nakanishi *et al.*, KEK Preprint 90-146 (1990).
12. K. Nakajima, *Part. Accel.* **32**(1990), 209.
13. B. Zotter, CERN CLIC Note 18 (1986).
14. K. L. F. Bane *et al.*, *IEEE Trans. Nucl. Sci.* **NS-32** (1985), 3254.
15. R. Keinigs and M. E. Jones, *Phys. Fluids* **30**(1987), 252.
16. J. B. Rosenzweig, FERMILAB-Conf-90/40 (1990).
17. Y. Ogawa *et al.*, *Phys. Rev.* **D43** (1991), 258.
18. H. Kobayashi and Y. Tabata, *Nucl. Instr. and Meth.* **B10/11**(1985), 1004.
19. R. W. Boswell, *Plasma Phys. Controled Fusion* **26**(1984), 1147.

**Research Trends in Physics: Coherent Radiation Generation and Particle Acceleration**
Editorial Board: J.M. Buzzi, A. Prokhorov (Editor-in-Chief), P. Sprangle, and K. Wille
*La Jolla International School of Physics*, The Institute for Advanced Physics Studies, La Jolla, California

# On the Theory of Relativistic Self-Focusing of Short Radiation Pulses and the Problem of Wake Field

**L.N. Tsintsadze,**

Institute of Physics, Georgian Academy of Sciences
Tbilisi, Georgia

ABSTRACT

A generalized set of equations for nonstationary three dimensional wake field and selffocusing of relativistic intense electromagnetic wave was derived. Modulational and filamentational instabilities at arbitrary amplitude of electromagnetic wave was studied. It was shown that the growth rate is mainly caused by relativistic effect.

Currently, one of the most important problems of plasma physics is the study of interaction of relativistic intense electromagnetic waves with plasma. Such interest to this phenomenon is caused by the possibility of plasma heating and realization of laser CRR /1-3/ as well as excitation of wake fields of the charge density. A special interest to the latter is connected with the design of new plasma methods of acceleration of charged particles /3-6/

Recent rapid development of laser UNF machinery allows to obtain such fields in which electrons will oscillate with relativistic velocities. As it has been shown, the increase of radiation intensity (above $10^{16}$ v/cm$^2$) changes essentially oscillating properties of plasma /7/, the structure of soliton /7/, leads to parametric resonance in the electron plasma /8/ and generation of quasistationary magnetic field /9/, affects the propagation of shock waves,/10/ relativistic modulational /11/ and filamentational /12/ instabilities. Recently it was shown in /13/ that relativitic effects in the problem of selffocusing of radiation lead to plasma compression in the range of moving focus, that is a new phenomenon in isotropic plasma.

Due to complexity of the set of equations describing the processes of selffocusing or excitation of the three dimensional wave fields, these processes in the above papers were investigated under numerous assumptions. Obviously the most important is the problem of derivation of more relativistic system.

A generalized set equations was obtained in the present paper for nonstationary wake field and selffocusing of relativistic intense electromagnetic waves. Modulational and filamentational instabilities are studied at arbitrary relativism.

1. Let us consiedr propagation of relativistically strong circularly polarized high frequency electromagnetic wave along axis $z$ in the collisionless plasma. We use the set of Maxwellian equations and relativistic hydrodynamics equations of electrons neglecting thermal motion. The contribution of the latter in the equation of motion is small in the case of strong fields as $T_e/m_0c^2 \cdot \delta n_e/n_0 \cdot 1/I \ll 1$ where $T_e$, $m_0$, $n_0$ the temperature, the mass at the rest and equilibrium density of electrons, respectively, and $I^2 = \left|\frac{eA_\perp}{m_0c^2}\right|^2$ where $|A_\perp|$ the amplitude of electromagnetic wave.

Plasma ions are assumed to be immobable.

$$\Delta A_\perp - \frac{1}{c^2}\frac{\partial^2}{\partial t^2}A_\perp = \frac{4\pi e^2}{m_0c^2}\frac{nA_\perp}{\gamma} \qquad (1.1)$$

$$\frac{d\vec{P}_e}{dt} = -e\left[\vec{E} + \frac{1}{c}\left[\vec{V}_e\vec{B}\right]\right] \qquad (1.2)$$

$$\frac{\partial n_e}{\partial t} + \operatorname{div} n_e\vec{V}_e = 0 \qquad (1.3)$$

$$\operatorname{div}\vec{E} = 4\pi e(n_0 - n_e) \qquad (1.4)$$

where

$$A_\perp = A_x + iA_y, \qquad \vec{E} = -\frac{1}{c}\frac{\partial\vec{A}}{\partial t} - \nabla\varphi$$

$$\vec{P}_e = \frac{m_0 \vec{V}_e}{\sqrt{1 - V_e^2/c^2}}, \qquad \gamma = \sqrt{1 + \frac{P^2}{m_0^2 c^2}}$$

Using the traditional approach, the vector potential is represented as

$$A_\perp = (\hat{e}_x + i\hat{e}_y) A_0(t, z, r_\perp) \exp i(k_0 z - \omega_0 t) \tag{1.5}$$

then equation (1.1) for slowly changing in time and coordinate of the complex amplitude are rewritten as well-known form

$$\left(\Delta_\perp + \frac{\partial^2}{\partial z^2} - \frac{1}{c^2}\frac{\partial^2}{\partial t^2}\right) A_0 + \frac{2i\omega_0}{c^2}\left(\frac{\partial}{\partial t} + \frac{k_0 c^2}{\omega_0}\frac{\partial}{\partial z}\right) A_0$$
$$+ \frac{\omega_0^2}{c^2}\left(1 - \frac{k_0^2 c^2}{\omega_0^2}\right) A_0 = \frac{\omega_p^2}{c^2}\,\frac{1 + \delta n/n_0}{\gamma} A_0 \tag{1.6}$$

where

$$\delta n_e = n_e - n_0, \qquad \omega_p^2 = \frac{4\pi e^2 n_0}{m_0}$$

Now let us discuss the problem of the group velocity for the case of avoilability of relativistic strong wave in plasma. The usual expression before the first derivative along $z$ in the second bracket of the left part of equation (1.6) is taken as group velocity ( $V_g = k_0 c^2/\omega_0$ ). Actually this equation agrees with the group velocity in the nonrelativistic case. As concerns relativistically intense waves, when oscillating velocities of the electrons become of the order of velocity of light, in this case it is impossible to introduce the motion of the group velocity. There is also no equation of dispersion. Only in the approximation of the geometric optics one can obtain equation the form of which agrees with the dispersion one

$$\frac{k_0^2 c^2}{\omega_0^2} = 1 - \frac{\omega_p^2(n)}{\omega_0^2 \gamma} \tag{1.7}$$

Since in (1.7) the density and relativistic factor change in the space and time, the latter is the equation of the state relativistically strong electromagnetic wave. In this connection definition of $K_0$ and $\omega_0$ in plasma becomes problematic. Indeed, we could have introduced the vacuum values of $K_0$ and $\omega_0$, then $V_g = C$ and any change of the value of the wave vector has to be included into the change of amplitude $A_0$. Otherwise, one may assume that the amplitude in certain point is constant, then expression (1.7) becomes dispersion equation with constant values $n_0$ and $\gamma_0 = \sqrt{1 + \left(\frac{eA_{0L}}{m_0c^2}\right)^2}$

It follows from (1.7) that if $\omega_0 \gg \omega_p/\gamma^{1/2}$ then $V_g \simeq C$ (almost vacuum situation) and $\frac{\partial^2 A_0}{\partial \xi^2} \simeq 1/c^2 \frac{\partial^2}{\partial t^2} A_0$ is in equation (1.6). Further, the structure of equation (1.6) indicates at the character of dependence $A_0$ upon coordinates: 1. If nonstationary process of propagation of a strong wave is considered,

$$A_0 = A_0(z - vt, t, \vec{r}_\perp) \tag{1.8}$$

with account of $V = \frac{K_0c^2}{\omega_0} \sim C$, equation (1.6) will be of the form

$$\Delta_\perp A_0 + \frac{2i\omega_0}{c^2}\frac{\partial A_0}{\partial t} + \frac{2V}{c^2}\frac{\partial^2}{\partial \xi \partial t}A_0 + \frac{\omega_0^2}{c^2}\left(1 - \frac{K_0^2c^2}{\omega_0^2}\right)A_0 = \frac{\omega_p^2}{c^2}\frac{1 + \delta n/n_0}{\gamma}A_0 \tag{1.9}$$

In the case of nonstationary selffocusing

$$A_0 = A_0(z - vt, z, \vec{r}_\perp) \tag{1.10}$$

for $A_0$ we have

$$\Delta A_0 + 2iK_0\frac{\partial}{\partial z}A_0 + 2\frac{\partial^2 A_0}{\partial z \partial \xi} + \frac{\omega_0^2}{c^2}\left(1 - \frac{K_0^2c^2}{\omega_0^2}\right)A_0 = \frac{\omega_p^2}{c^2}\frac{1 + \delta n/n_0}{\gamma}A_0 \tag{1.11}$$

where $\xi = z - vt$

If $\omega_0 \simeq \omega_p/\gamma^{1/2}$, equation (1.6) becomes the three-dimensional Schrödinger equation without term $\sim 1/c^2 \frac{\partial^2}{\partial t^2}$, since in this case

$$\frac{\partial^2}{\partial t^2} \ll c^2\frac{\partial^2}{\partial z^2}, \qquad V = \frac{K_0c^2}{\omega_0} \ll C$$

Obviously, in equations (1.9) and (1.11) it is assumed that $\partial A_0/\partial t \ll v \frac{\partial A_0}{\partial \xi}$ and $\frac{\partial A_0}{\partial z} \ll \frac{\partial}{\partial \xi} A_0$ respectively. Note that in some papers relationship $\omega_0^2 = \omega_p^2 + k_0^2 c^2$ was used incorrectly, which in relativistic plasma is of no physical sence that leads to the erroneous Schrődinger equation.

Also, third terms in (1.9) and (1.11) are small in comparison, say, with second ones, but just account of these terms leads to the essential fast variable phase

$$\omega = \omega_0 + v \frac{\partial \varphi}{\partial \xi} \qquad \text{or} \qquad k = k_0 + \frac{\partial \varphi}{\partial \xi}$$

If one represents $A_0$ as $A_0 = a \exp(i\varphi)$ (where $a$ and $\varphi$ are real functions), it follows from (1.9) and (1.11)

$$\frac{\partial}{\partial t} \int d\xi \, a^2 \left(\omega_0 + v \frac{\partial \varphi}{\partial \xi}\right) + c^2 \nabla_\perp \int d\xi \, a^2 \nabla_\perp \varphi = 0 \tag{1.12}$$

$$\frac{\partial}{\partial z} \int d\xi \, a^2 \left(k_0 + \frac{\partial \varphi}{\partial \xi}\right) + c^2 \nabla_\perp \int d\xi \, a^2 \nabla_\perp \varphi = 0 \tag{1.13}$$

Obviously, if we integrate (1.12) and (1.13) over we find first integrals

$$\int d\xi \, d\vec{z}_\perp \, a^2 \left(\omega_0 + v \frac{\partial \varphi}{\partial \xi}\right) = \text{const} \tag{1.14}$$

$$\int d\xi \, d\vec{z}_\perp \, a^2 \left(k_0 + \frac{\partial \varphi}{\partial \xi}\right) = \text{const} \tag{1.15}$$

Up to-date there is no precise equations for three dimensional relativistic intensive electromagnetic wave, describing nonlinear dynamics. As a matter of fact, in previous papers the expression contained only transverse momentum, caused by the field of pumping and $\gamma$ was represented as

$$\gamma = \sqrt{1 + \frac{\tilde{p}_\perp^2}{m_0^2 c^2}}$$

where

$$\tilde{P}_\perp = \frac{1}{\sqrt{2}}(\hat{x} + i\hat{y})\, \tilde{P}_\perp(t, z, \vec{r}_\perp)\, \exp i(k_0 z - \omega_0 t) + c.c.$$

$\omega_0$ , $k_0$ - frequency, wave number of electromagnetic wave of pumping, respectively. The transverse coordinate

$$r_\perp = \sqrt{x^2 + y^2}$$

As it was shown in /4-6/ concerning the laser accelerator on wake field, the longitudinal momentum $\langle P_\parallel \rangle$ (caused by ponderomotive force) becomes of the same order that $|\tilde{P}_\perp|$ , but in ultrarelativistic case is much more, meaning that $\gamma$ should be represented as $(1 + (\tilde{P}_\perp^2 + \langle P_\parallel \rangle^2)/m_0^2 c^2)^{1/2}$ . All this concerned the one dimensional plasma motion in the wave field. The problems becomes complicable when the wave amplitude depends upon three coordinates. As a matter of fact, in this case a ponderomotive force appears, acting in transverse direction relative to propagation of the pumping wave. It results in formation of the mean (over the period of pumping field $\tau = 2\pi/\omega_0$ ) transverse momentum, i.e. $\vec{P}_\perp = \langle \vec{P}_\perp \rangle + \tilde{\vec{P}}_\perp$ Presence of $\langle \vec{P}_\perp \rangle$ in equations of electron motion and Maxwellian makes the problem practically unsolvable.

The present paper gives relationship which allows to simplify the study of propagation of relativistic intensive wave in three dimensional case, not loosing the generality.

In order to show the above said, let us write the equation of electron motion without account of gas dynamic pressure in the form

$$\frac{\partial \vec{P}}{\partial t} = -e\vec{E} - m_0 c^2 \nabla \gamma \tag{1.16}$$

where $\vec{E} = -\frac{1}{c}\frac{\partial \vec{A}}{\partial t} - \nabla\varphi$ , $\gamma = \sqrt{1 + \vec{P}^2/m_0^2 c^2}$

$\vec{P}$ - total momentum,

Let us represent $\vec{P}$ as

$$\vec{P}_\perp = \langle \vec{P}_\perp \rangle + \vec{\tilde{P}}_\perp$$

$$P_{\parallel} = \langle P_{\parallel} \rangle \tag{1.17}$$

Substituting expression for momentum $\vec{P}$ (1.17) into (1.16) and averageing over period $\tau = 2\pi/\omega_0$, we obtain

$$\frac{\partial}{\partial t} \langle \vec{P}_\perp \rangle = \nabla_\perp \left( e\varphi - m_0 c^2 \langle \gamma \rangle \right) \tag{1.18}$$

$$\frac{\partial}{\partial t} \langle P_{\parallel} \rangle = \nabla_{\parallel} \left( e\varphi - m_0 c^2 \langle \gamma \rangle \right) \tag{1.19}$$

Here it is accounted for that $\langle \vec{E} \rangle = -\nabla \varphi$

$$\langle \gamma \rangle = \frac{\omega_0}{2\pi} \int_0^{\frac{\omega_0}{2\pi}} \sqrt{1 + \frac{\left( \langle \vec{P}_\perp \rangle + \vec{\tilde{P}}_\perp \right)^2 + P_{\parallel}^2}{m_0^2 c^2}} \tag{1.20}$$

From (1.18) and (1.19) we can easily get

$$\nabla_{\parallel} \langle \vec{P}_\perp \rangle = \nabla_\perp \langle P_{\parallel} \rangle \tag{1.21}$$

Relationship (1.21) allows to compare $\langle P_\perp \rangle$ with $\langle P_{\parallel} \rangle$. If the averaged physical values along propagation of the pumping change faster than in the transverse direction, it follows from (1.21) that $\langle P_\perp \rangle \ll \langle P_{\parallel} \rangle$ since $\langle P_\perp \rangle / \langle P_{\parallel} \rangle \approx \ell_{\parallel} / \ell_\perp \ll 1$ (1.22). In this case $\langle \vec{P}_\perp \rangle^2$ can be neglected in (1.20) as compared with $\langle P_{\parallel} \rangle^2$ and $\langle \gamma \rangle$ is written as

$$\langle \gamma \rangle = \gamma \left( 1 - \frac{\langle P_\perp \rangle^2 \, |\tilde{P}_\perp|^2}{2 m_0^4 c^4 \gamma^4} \right)$$

$$\gamma = \sqrt{1 + \left( \tilde{P}_\perp^2 + \langle P_{\parallel} \rangle^2 \right) / m_0^2 c^2} \tag{1.23}$$

Taking into account the smallness of the term in (1.23), $\langle \gamma \rangle$ will be further replaced by $\gamma$ . Condition ($\ell_{\parallel} < \ell_\perp$) for such problems as selffocusing of selfchannelling or generation of wake field is fulfilled well. Just inequality led the authors of paper /13/, when investigating selffocusing of weakly relativistic circularly, polarized Gaussian pulse in a trasparent plasma in aberratoionaless paraxial approximation, that at nonstationary self-focusing of short pulses with sufficient wide transverse dimensions in the range of running focuses concentration of plasma increases. This can be of a great importance at heating and contraction of plasma in the laser thermonuclear synthesis devices and for design of new type accelerator. In order to close the set of equations it is necessary to obtain the equation for the potential $\varphi$ . One should find nonlinear perturbation of the electron density $\delta n = n_e - n_0$ . Due to strong relativism it is impossible to express in reasonable functions or to write the equation, which could be solved in three dimensional case. But such possibility exists if one uses the above condition and $\frac{\partial f}{\partial \xi} \gg \nabla_\perp f \gg \frac{\partial f}{\partial z}$ . Then the Poisson equation (1.4) and continuity equations will be of the form

$$\frac{\partial^2 \varphi}{\partial \xi^2} = -4\pi e \delta n \tag{1.24}$$

$$\frac{\delta n}{n_0} = \frac{v_\parallel}{V - v_\parallel} \tag{1.25}$$

where $v_\parallel = \langle P_\parallel \rangle / m_0 \gamma$ ,

It follows from (1.19) that

$$-V \langle P_\parallel \rangle = e\varphi - m_0 c^2 \gamma + C \tag{1.26}$$

where $C$ - the integration constant. To determine $C$ let us consider two cases. Let on the infinity minus $\langle P_\parallel \rangle = \varphi = A_\perp = 0$ , then $C = m_0 c^2$ and in dimensionless limits (1.26) is written as

$$\beta u_\parallel = -(1 + \Phi) + \gamma \tag{1.27}$$

where $\beta = \frac{V}{c}$ , $u_\parallel = \frac{\langle P_\parallel \rangle}{m_0 c}$ , $\Phi = \frac{e\varphi}{m_0 c^2}$

If on the infinity $\langle P_\parallel \rangle = \varphi = 0$, $A_\perp = A_{0\perp}$

$$\beta u_\parallel = \gamma - \left(\Phi + \sqrt{1 + u_{0\perp}^2}\right) \tag{1.28}$$

where $u_{0\perp} = \dfrac{eA_{0\perp}}{m_0 c^2}$

In the refracted plasma or in the case of strong relativism, when $\omega_0^2 >> \frac{\omega_p^2}{\gamma}$, one can assume $\beta \simeq 1$ in (1.9) and (1.11). Then form (1.4) and (1.28),(1.25), (1.9) and (1.11) we obtain the set of two equations in dimensionless units

$$\frac{\partial^2 \Phi}{\partial \rho_z^2} = \frac{1}{2}\left[\frac{1+|u_\perp|^2}{(1+\Phi)^2} - 1\right] \tag{1.29}$$

$$2i\frac{\omega_0}{\omega_p}\frac{\partial}{\partial \tau}u_\perp + \Delta_{\rho_\perp} u_\perp + 2\frac{\partial^2}{\partial \rho_z \partial \tau}u_\perp + \frac{\omega_0^2}{\omega_p^2}\left(1 - \frac{k_0^2 c^2}{\omega_0^2}\right)u_\perp - \frac{u_\perp}{1+\Phi} = 0 \tag{1.30}$$

$\rho_z = \omega_p (z-ct)/c$, $\rho_\perp = \omega_p z_\perp / c$, $\tau = \omega_p t$.

In the case when $\Phi(\rho_z, \rho_\perp, z)$, equation (1.31) is as follows

$$2i\frac{k_0 c}{\omega_p}\frac{\partial}{\partial \rho_z}u_\perp + \Delta_{\rho_\perp} u_\perp + 2\frac{\partial^2 u_\perp}{\partial \rho_z \partial \rho_z} + \frac{\omega_0^2}{\omega_p^2}\left(1 - \frac{k_0^2 c^2}{\omega_0^2}\right)u_\perp - \frac{u_\perp}{1+\Phi} = 0 \tag{1.31}$$

If on the infinity $u_\perp$ differs from zero, in (1.25), (1.30) and (1.31) $1+\Phi$ should be replaced by

$$\Phi + \sqrt{1+u_{0\perp}^2}$$

In the general case when $\beta$ is arbitrary value, equation for $\Phi$ becomes

$$\frac{\partial^2 \Phi}{\partial \rho_z^2} = \gamma_p^2 \left\{ \left(1 - \frac{1}{\gamma_p^2}\right)^{1/2} \frac{y}{\left[y^2 - \gamma_\perp^2/\gamma_p^2\right]^{1/2}} - 1 \right\} \tag{1.32}$$

where

$$\gamma_p = \frac{1}{\sqrt{1-\beta^2}}, \quad y = \Phi + 1, \text{ or } y = \Phi + \sqrt{1+u_{0\perp}^2}$$

Obviously, at $\gamma_p \to \infty$, equation (1.32) becomes (1.29).

Note one important thing. The authors of /4-6/ in

order to obtain maximum electron energy, assumed that $\beta \sim 1$. But in this case increases hard the length on which electrons should accelerate, since $\ell = \frac{\omega_0^2}{\omega_p^2}\frac{c}{\omega_p}$ . This would be insufficient for the new type of laser accelerators. Besides, at $\beta \sim 1$ (that on the other hand means that the group velocity of the field of pumping is close to $c$ , i.e. $(\omega_0/\omega_p)^2 \gg 1$ ), the connection of the passed transvers wave of electromagnetic field with plasma is very weak. Therefore more interesting is the study of the case $\beta < 1$ .

2. Recently, many papers were published on modulational and filamentational instabilities of strong electromagnetic wave in plasma. Such an interest is caused that in plasma strongly nonlinear structures are formed, for instance, fine filaments at the decay of intense light beams. Mechanisms of generation of such formations are not yet clear. Modulational or filamentational instabilities may be one of the mechanisms. Investigation of these phenomena are also important from the point of view of plasma heating by means of lasers. Filamentation of electromagnetic waves in the flame of laser plasma can noteable affect the efficiency, of wave damping and homogeneity of implosion.

Using conventional procedure of equation linearization (1.6),(1.2),(1.3),(1.4) and considering perturbations as plane circularly polarized electromagnetic wave ( $\delta A_\perp \sim exp\,i(kz-\omega t)$ ), we obtain relationship of dispersion for arbitrary amplitude $A_{0\perp}$ of the pumping field

$$\left\{(\omega - k_\parallel V_g)^2 - \frac{k_\perp^2 c^2}{4\omega_0^2}\left(k_\perp^2 c^2 - \frac{\omega_p^2\alpha^2}{\gamma_0^2}\right)\right\}(\omega^2-\omega_p^2) = \frac{\omega_p^2}{4\omega_0^2} k_\perp^2 c^2 k^2 c^2 \frac{\alpha^2}{\gamma_0^2} \tag{2.1}$$

where $V_g = \frac{\partial \omega_0}{\partial k_0}$, $\alpha^2 = \frac{e^2|A_{0\perp}|^2}{m_0^2 c^4}$, $\gamma_0^2 = 1+\alpha^2$

2.1.1. Modulational instability

Modulational instability was considere in /11/ in assumption of $\omega \ll \omega_p$ . It will be shown below that the most interesting case is $\omega \sim \omega_p$ . Just in this case, when electrons are oscillating and ions are assumed to be immovable, there is a strong association of laser radiation with plasma. Calculations show that in this case the increment is much higher than that of /11/.

In the dispersion equation (2.1) in the left part in brackes there is a pure relativistic term, which promotes the modulational instability, in particular, compensates diffracting terms, and in a strong relativism it is the only mechanism of instability.

Let us consider different interesting cases:

a) Assuming that diffractive term $K_\perp^2 c^2$ is compensated by relativistic term $\omega_p^2 \alpha^2/\gamma_0^2$ and $\omega = K_\parallel V_g + \delta$ and $\omega = \omega_p + \delta$ , we obtain expression for the increment from (2.1)

$$\mathrm{Im}\,\delta = \frac{\sqrt{3}}{4}\left(\frac{K^2 c^2}{\omega_0^2}\frac{\alpha}{\gamma_0}\right)^{1/3}\frac{\alpha}{\gamma_0}\,\omega_p \tag{2.2}$$

2.2. In second case we assume that diffracting term prevails relativistic one, i.e. $K_\perp^2 c^2 > \omega_p^2 \alpha^2/\gamma_0^2$ then equation of dispersion (2.1) is solved as

$$\omega = \omega_p + \delta, \quad \omega = K_\parallel V_g - \frac{K_\perp c}{2\omega_0}\left[K_\perp^2 c^2 - \omega_p^2\frac{\alpha^2}{\gamma_0^2}\right]^{1/2} + \delta$$

In this case for imaginary part $\delta$ we obtain

$$\mathrm{Im}\,\delta = \frac{1}{2^{3/2}}\left(\frac{\omega_p}{\omega_0}\right)^{1/2}\frac{\alpha}{\gamma_0}KC\left(\frac{K_\perp^2 c^2}{K_\perp^2 c^2 - \omega_p^2\alpha^2/\gamma_0^2}\right)^{1/4} \tag{2.3}$$

In the case of small relativism $\alpha^2 \ll 1$ the increment becomes

$$\mathrm{Im}\,\delta = \frac{1}{2^{3/2}}\left(\frac{\omega_p}{\omega_0}\right)^{1/2}\alpha\cdot KC \tag{2.4}$$

2.3. At almost perpendicular propagation of perturbation, i.e. $K_\parallel \ll K_\perp$ the relativistic term prevails the strictional effect and simple calculation show that there is aperiodic instability with increment

$$\mathrm{Im}\,\omega = \frac{K_\perp c}{2\omega_0}\frac{\alpha}{\gamma_0}\left(1-\frac{K_\perp^2 c^2}{\omega_p^2\alpha^2}\right)^{1/2}\omega_p \tag{2.5}$$

This expression has maximum

$$\mathrm{Im}\,\omega_{max} = \frac{1}{4}\left(\frac{\omega_p}{\omega_0}\right)\frac{\alpha^2}{\gamma_0}\omega_p \tag{2.6}$$

at $K_{0\perp} = \frac{\omega_p}{\sqrt{2}c}\alpha$

b) Filamentational instability

For demonstration, let us consider traditional case, i.e. $\omega = 0$ , Then we obtain from (2.1) an expression for the increment of filamentational instability

$$\mathrm{Im}\,K_\parallel = \frac{K_\perp c}{2\omega_0 V_g\gamma_0}\left(\omega_p^2\alpha^2 - K_\perp^2 c^2\right)^{1/2} \tag{2.7}$$

Maximum value of the increment $\left(\frac{\partial\,\mathrm{Im}\,K_\parallel}{\partial K_\perp}\Big|_{K_\perp=K_{0\perp}} = 0\right)$ is achieved at $K_{0\perp} = \frac{1}{\sqrt{2}}\frac{\omega_p}{c}\alpha$ . It is

$$(\mathrm{Im}\,K_\parallel)_{max} = \frac{1}{4V_g}\left(\frac{\omega_p}{\omega_0}\right)\frac{\alpha^2}{\gamma_0}\omega_p \tag{2.8}$$

As it has been mentioned above, nonstationarity of the process when $\omega \neq 0$ and especially $\omega$ is close to $\omega_p$ becomes important for filamentational instability. Also, to develop filamentational instability one should not superimpose conditions on the value $\omega/\omega_p$ . No doubt, the increment value depends upon $\omega/\omega_p$ , but instability takes place for any values of this $\omega/\omega_p$ ratio.

Following from the equation of dispersion (2.1) for nonstationary filamentational instability, we obtain the expression for the increment

$$\mathrm{Im}\,K_\parallel = \frac{K_\perp c}{2\omega_0 V_g}\left(\frac{\omega_p^2\alpha^2}{\gamma_0^2} - K_\perp^2 c^2\left[1-\frac{\omega_p^2}{\omega_p^2-\omega^2}\frac{\alpha^2}{v_0^2}\right]\right)^{1/2} \tag{2.9}$$

We determine the value of threshold amplitude of the pumping field $(\mathrm{Im}\, K_{\parallel} = 0)$ from (2.9)

$$\left(\frac{\alpha}{\gamma_0}\right)_{tr} = \frac{K_\perp C}{\omega_p \left(1 + K_\perp^2 C^2/(\omega_p^2 - \omega^2)\right)^{1/2}} \qquad (2.10)$$

Expression (2.10) gives dependence of the threshold field upon $\omega/\omega_p$. It follows that at $\omega \sim \omega_p$ the threshold of filamentational instability tends to zero.

When $\omega_p^2\left(1 - \frac{\alpha^2}{\gamma_0^2}\right) > \omega^2$, the transverse wave number is determined from (2.9), at which $\left(\frac{\partial\, \mathrm{Im}\, K_{\parallel}}{\partial K_\perp}\right)\Big|_{K_\perp = K_{0\perp}} = 0$

$$K_{0\perp} = \frac{\omega_p}{\sqrt{2}C} \frac{\alpha}{\gamma_0} \frac{1}{\left(1 - \omega_p^2/(\omega_p^2 - \omega^2) \cdot \alpha^2/\gamma_0^2\right)^{1/2}}$$

For maximum parameter we have

$$(\mathrm{Im}\, K_{\parallel})_{Max} = \frac{\omega_p}{\sqrt{2}C} \frac{\alpha^2}{\gamma_0^2} \frac{\omega_p}{\omega_0} \frac{1}{\left(1 - \omega_p^2 \alpha^2/(\omega_p^2 - \omega^2)\gamma_0^2\right)^{1/2}} \qquad (2.11)$$

In the ultrarelativistic regime, i.e. when $\alpha/\gamma_0 \simeq 1$ we obtain for the increment

$$\mathrm{Im}\, K_{\parallel} = \frac{K_\perp C}{2\omega_0 v_g} \left(\omega_p^2 + K_\perp^2 C^2 \frac{\omega^2}{\omega_p^2 - \omega^2}\right)^{1/2} \qquad (2.12)$$

Expression (2.12) shows that if $\omega^2 > \omega_p^2$, the increment reaches maximum with increase of $K_\perp$, and then drops and vanishes at $K_\perp^2 > \frac{\omega_p^2}{\omega^2} \frac{\omega^2 - \omega_p^2}{C^2}$. In the inverse case the increment increases with the enhancement of $K_\perp$.

In conclusion I would like to mention that if one assumes $\alpha^2/\gamma_0^2 = 1$ in equation (2.1) then amplitude of the wave vanishes in equation of dispersion (2.1). Indeed, in ultrarelativistic case the oscillating velocity in the external field is represented as $V_\sim \simeq C \exp i\varphi$ i.e. the role of the amplitude is formally played by the light velocity C.

## REFERENCES


1. C.E.Max, In: Laser Plasma Interaction, (edited by R.Balian and J.C.Adam), North-Holland Publishing CO., Amsterdam, (1982).
2. J.L.Bobin, Phys.Rep. 122, 173(1985).
3. T.Tajima and J.M.Dawson, Phys.Rev.Lett., 43, 267(1979).
4. T.Katsouleas et al. In: New Developments in Particle Acceleration Technique, edited by S.Turner, ICEA-CAS Workshop proceedings, v.II, p.401, (1987).
5. V.I.Berezhiani, I.G.Murusidze, Phys.Lett., 148, 2146 (1990).
6. S.V.Bulanov, V.I.Kirsanov, A.C.Sakharov, Sov.plasma Physics, 16, 935(1990).
7. V.I.Berezhiani, N.L.Tsintsadze, D.D.Tskhakaya, J. Plasma Phys. 24, 15(1980).
8. N.L.Tsintsadze, Zh.Eksp.Teor.Fiz., 59, 1250(1970).
9. V.I.Berezhiani, D.D.Tskhakaya, G.Auer, J. Plasma Phys. (1987), vol.38, part 1, pp.139-153.
10. J.F.Drake, Y.C.Lee, K.Nishikawa, N.L.Tsintsadze, Phys. Rev.Lett., 36, 196(1976).
11. P.K.Shukla, R.Bharuthram, N.L.Tsintsadze, Phys.Rev.A, 35, 4889(1987).
12. P.K.Shukla, R.Bharuthram, N.L.Tsintsadze, Physica Scripta, 38, 578(1988).
13. D.P.Garuchava, N.L.Tsintsadze. D.D.Tskhakaya, Zh.Eksp: Teor.Fiz., 98, 1558(1990).

# Part III

# ABSTRACTS

**Research Trends in Physics: Coherent Radiation Generation and Particle Acceleration**
Editorial Board: J.M. Buzzi, A. Prokhorov (Editor-in-Chief), P. Sprangle, and K. Wille
*La Jolla International School of Physics*, The Institute for Advanced Physics Studies, La Jolla, California

# Nonlinear Dynamics of Radiation in a Free Electron Laser

**T.M. Antonsen, Jr.**

Laboratory for Plasma Research
Department of Electrical Engineering,
and Department of Physics and Astronomy
University of Maryland, College Park
College Park, Maryland 20742-3511

The behavior of the radiation field in a free electron laser oscillator is described by a dissipative nonlinear system of equations. As beam current is raised the time asymptotic radiation field in the resonator changes from being single frequency, to multifrequency, to broadband and chaotic. Simple models can be derived which predict the route to chaos as occurring either through a sequence of period doublings, a sequence of frequency splittings involving incommensurate frequencies, or a combination of the two. The present paper describes these models and their predictions.

**Research Trends in Physics: Coherent Radiation Generation and Particle Acceleration**
Editorial Board: J.M. Buzzi, A. Prokhorov (Editor-in-Chief), P. Sprangle, and K. Wille
*La Jolla International School of Physics*, The Institute for Advanced Physics Studies, La Jolla, California

# The Advanced Light Source at Lawrence Berkeley Laboratory

**J. Marx and the ALS Project Team***

Accelerator and Fusion Research Division
Lawrence Berkeley Laboratory
1 Cyclotron Road
Berkeley, CA 94720

The Advanced Light Source is a national user facility for the production of high brightness and partially coherent x-ray and ultraviolet synchrotron radiation, which is now under construction at Lawrence Berkeley Laboratory. The facility is based on a low emittance electron storage ring, long insertion devices, photon beamlines, and user support facilities. The lattice optics is optimized for undulator operation and can accommodate up to 11 insertion devices in the straight sections and up to 48 ports in the bending magnets. The nominal electron energy is 1.5 GeV, the horizontal emittance $10^{-8}$m rad, the circulating current 400 mA in the multibunch mode of operation. The parameters are chosen to cover the photon spectrum from about 8 eV to over 1 keV with undulators, and up to 10 keV with wigglers. The facility is now in its third year of construction and is planned to be completed in mid-1993 at a total cost of $99.5 million. The performance capability, layout, and status of the ALS will be described with an emphasis on the generation of partially coherent soft x-rays.

* This work was supported by the Director, Office of Energy Research, Office of Basic Energy Sciences, Materials Sciences Division, of the U.S. Department of Energy, under Contract No. DE-AC03-76SF00098.

 

**Research Trends in Physics: Coherent Radiation Generation and Particle Acceleration**
Editorial Board: J.M. Buzzi, A. Prokhorov (Editor-in-Chief), P. Sprangle, and K. Wille
*La Jolla International School of Physics*, The Institute for Advanced Physics Studies, La Jolla, California

# Sun as an Accelerator

**J.J. Su, J.M. Dawson and R. Bingham***

University of California, Los Angeles
Department of Physics
Los Angeles, CA 90024

*Rutherford Laboratory

We have investigated a potential electron acceleration mechanism in solar flares. Ion rings are produced by perpendicular shocks, which can be generated by coronal mass ejections (CME). Such ion ring distributions are known to be unstable to the generation of lower hybrid waves, which have high phase velocity parallel to the field and therefore resonantly accelerate electrons in that direction. 2½D fully relativistic electromagnetic PIC simulations are compared with theory and observation.

**Research Trends in Physics: Coherent Radiation Generation and Particle Acceleration**
Editorial Board: J.M. Buzzi, A. Prokhorov (Editor-in-Chief), P. Sprangle, and K. Wille
*La Jolla International School of Physics*, The Institute for Advanced Physics Studies, La Jolla, California

# A Channeling Radiation X-Ray Sources for Angiography, X-Ray Lithography, X-Ray Structure Determination and Elemental Analysis

**H. Überall[(1)*], B.J. Faraday[(1)], X.K. Maruyama[(2)], and B.L. Berman[(3)]**

[(1)]SFA Inc.
1401 McCormick Dr.
Landover, MD 20785

[*]also at
Naval Research Laboratory, Code 4694
Washington, D.C. 20375, USA and

Catholic University of America
Department of Physics
Washington, D.C. 20064, USA

[(2)]Naval Postgraduate School, Department of Physics
Monterey, CA 20943

[(3)]George Washington University, Department of Physics
Washington, DC 20052

Sychrotron radiation of 33 keV ($\lambda \cong 0.4$ Å) has been tentatively employed for coronary angiography studies [1]; the results were successful, but so far lacked clinical quality. Synchrotron radiation has also been proposed as a tool for microlithography and the efficient production of integrated circuits [2], using energies of 0.5 - 2.5 keV ($\lambda \cong$ 5-20A). This application, although not yet carried out in earnest, appears promising enough so that a number of synchrotron sources dedicated to this purpose are now being set up world-wide [3]. These applications require storage rings for ~ 1 GeV electrons, costing 20 - 60 or more millions of dollars.

Kumakhov has pointed out [4,5] that channeling radiation of electrons in monocrystals constitutes the most intense known radiation source for soft or hard x-rays. Using diamond targets, a 20-MeV electron linac (available for approximately one million dollars or less) can produce intense, monochromatic, tuneable and polarized x-rays up to the 30 keV range [6]. Less expensive lower energy linacs, e.g. the Varian Linatron, affordable by smaller institutions, which can produce ~ 1 keV channeling radiation photons [7] are suitable for microlithography. Channeling radiation intensities, especially for electrons of some tens of MeV and up will surpass synchrotron radiation intensities [8,9].

We have carried out studies confirming the above conclusions regarding comparisons of channeling radiation vs. synchrotron radiation sources for the mentioned applications in quantitative detail. These studies also extend to applications for purposes of macro-molecular and crystal structure determinations, so far carried out with conventional x-ray sources [10]; exposure times can reduce several orders of magnitude using channeling radiation [9]. Further, channeling radiation requiring 10 – 20 MeV electron linacs also constitues a good source for trace elemental analysis, so far proposed for a synchrotron source [11]. It should be pointed out that an alternate x-ray source using electron beams, namely transition radiation [12] requires much larger electron energies (hundred or more MeV) for a given photon energy (30 keV) and its spectrum, in contrast to that of channeling radiation, is rather broad-band (as is also synchrotron radiation); if monochromaticity is required here, monochromators must be used which cut down intensities by large factors.

## References

1. E. Rubenstein et al., Nucl. Inst. Meth. A291, 80 (1990).

2. A.R. Neureuther, in Synchrotron Radiation Research (H. Winick and S. Doniach, eds.) Plenum, New York. 1982, p. 223.

3. J.B. Murphy, Proc. 1989 IEEE Particle Accelerator Conference, March 20 - 23, 1989, Chicago, IL, vol. 2, p. 757.

4. M.A. Kumakhov, Phys. Letters 57, 17 (1976); Sov. Phys. Doklady 21, 581 (1976).

5. V.V. Beloshitsky and M.A. Kumakhov, in Coherent Radiation Sources (A.W. Saenz and H. Uberall, eds.), Springer, Berlin/Heidelberg 1985, p. 91; R.W. Terhune and R.H. Pantell, Appl. Phys. Lett. 30, 265 (1977).

6. R.K. Klein, J.O. Kephart, R.H. Pantell, H. Park, B.L. Berman, R.L. Swent, S. Datz, and R.W. Fearick, Phys. Rev. B31, 68 (1985).

7. J.U. Andersen, K.R. Ericksen, and E. Laegsgaard, Physica Scripta 24, 588 (1985).

8. R.H. Pantell et al., Proc. 6th Conf. on Application of Accelerators in Research and Industry, Denton, Texas, Nov. 1980.

9. R. Wedell, Phys. Stat. Sol. 99, 11 (1980).

10. See e.g., I.L. Karle and J.L. Flippen-Anderson, Acta Cryst. C46, 303 (1990).

11. H.G. Haubold et al., Rev. Sci. Instrum. 60, 1943 (1989).

12. M.A. Piestrup et al., Phys. Rev. A32, 917 (1985).

## Index Of Contributors

# Index Of Names

# Key Word Index

## H



## I



## J



## K



## L

## M

## Z